Biotechnology and Nutrition

HAMAKER LIBRARY
NORTHWESTERN COLLEGE
ORANGE CITY, IOWA 51041

PROCEEDINGS OF THE
THIRD INTERNATIONAL SYMPOSIUM

Jointly Sponsored by:
The University of Maryland,
The United States Department of Agriculture
and E. I. du Pont de Nemours & Co.

Editorial Board

Co-Editors-in-Chief: DONALD D. BILLS, USDA, ARS, Beltsville, MD
SHAIN-DOW KUNG, University of Maryland

Editors: ANTHONY KOTULA, USDA, ARS, Beltsville, MD
ALLEY WATADA, USDA, ARS, Beltsville, MD
DENNIS WESTHOFF, University of Maryland
BRUNO QUEBEDEAUX, University of Maryland
GARY FADER, E. I. du Pont de Nemours & Co.
JOHN GOSS, E. I. du Pont de Nemours & Co.

Butterworth–Heinemann
Boston London Oxford Singapore Sydney Toronto Wellington

Copyright © 1992 by Butterworth-Heinemann, a division of Reed Publishing (USA) Inc.
All rights reserved.

No part of this publication may be reproduced, stored in a retrieval system, or transmitted, in any form or by any means, electronic, mechanical, photocopying, recording, or otherwise, without the prior written permission of the publisher.

Recognizing the importance of preserving what has been written, it is the policy of Butterworth-Heinemann to have the books it publishes printed on acid-free paper, and we exert our best efforts to that end.

ISBN 0-7506-9259-6
Library of Congress Catalog Card Number 91-077736

Butterworth-Heinemann
80 Montvale Avenue
Stoneham, MA 02180

10 9 8 7 6 5 4 3 2 1

Printed in the United States of America

Contents

Molecular Approaches in the Modification and Production of Edible Oils

PREFACE

Genetic manipulation will lead to plants and animals that resist diseases, pests, and stresses, require less fertilizer or feed, and yield more food and fiber. These changes will increase the availability of agricultural products while decreasing their cost. These are important goals, but they are not new goals. Classical breeding and selection already have made great progress along these lines.

More exciting, and probably of greater ultimate consequence, is the use of biotechnology to alter the chemical composition of plants and animals to improve the nutritional value of food obtained from them. Transferring genetic material between species can provide characteristics difficult or impossible to obtain through classical breeding. Improving nutrition is a goal in developed countries, but it is even more important to the developing countries. In the mid-term, the most important contribution of biotechnology to nutrition in developing countries will be improved yields of food crops. In the long-term, however, improvement of nutritional quality may be more significant. For example, adequate protein, with a balanced composition of essential amino acids, is lacking in diets in most developing countries. Conventional plant breeding techniques have produced some new varieties with a better profile of amino acids; biotechnology can accomplish such improvements more rapidly.

One of the great challenges will be to decide what compositional changes of crop plants and meat animals are desirable from a nutritional standpoint. In many cases, the basic information needed to make sound decisions is lacking. For example, the nutritional influence of dietary fatty acids of various chain-lengths and degrees of saturation remains incompletely understood. Without adequate knowledge, changing the composition of nutrients in food could result in poorer nutrition instead of better nutrition. For example, while it might be possible to mimic the fatty acid composition of fish in many other foods, to do so on a widespread basis may not make sense—consumption of 20 g or more per day of fish oil can cause physiological harm. Furthermore, the highly unsaturated fatty acids of fish are subject to rapid autoxidation with accompanying off-flavors, loss of nutritional value, and generation of toxic products. To impart the fatty acid composition of fish to other foods also would impart the delicate, perishable nature of fish to those foods.

In considering compositional changes, there is no reason to try to make every food nutritionally complete in itself. In fact, enhancing the vitamin and mineral content of every food would be undesirable, because some vitamins and minerals are toxic when consumed at high levels. Precedents for making such decisions exist in the rationales that were used to support the fortification of a few foods, such as milk and bread, with vitamins.

In this volume, the authors explore a number of areas related to biotechnology and nutrition. They begin with the broad consideration of human nutrition and the ability of biotechnology to improve nutrition. Specific subjects appear under the headings of Carbohydrates, Proteins, Vitamins and Minerals, and Edible Oils.

Biotechnology will improve the nutritional quality of foods, but the full realization of its potential will require the participation of scientists with many backgrounds. In organizing this symposium, our intent was to provide a forum for nutritionists, molecular biologists, animal and plant biochemists, food scientists, policymakers, and others who will influence the improvement of nutrition through the application of biotechnology. We also wanted to bring together scientists from universities, government, and industry. We believe that our intentions were fulfilled, and most of the credit must be given to the excellent speakers who shared their diverse knowledge and experience with us and, now, with you the reader. The remainder of the credit must be divided between the Program Committee, which used such excellent judgment in assembling a panel of such well-qualified speakers, and the Arrangements Committee, which provided such a favorable environment for our symposium.

We are indebted to the University of Maryland, the Agricultural Research Service of the USDA, and the du Pont Company, who sponsored the symposium and provided financial support and other services that made the Third International Symposium on Biotechnology and Nutrition a success. The symposium provides a continuing example of a successful and mutually beneficial cooperation between a public university, a governmental agency, and private industry.

Donald D. Bills, *Agricultural Research Service, USDA*
Shain-dow Kung, *University of Maryland*

Acknowledgment

We gratefully acknowledge the efforts of the members of the Program Committee and the Local Arrangements Committee. Special appreciation is due to Dr. Anthony Kotula, who chaired the Program Committee and to Mrs. Helen Phillips, who chaired the Local Arrangements Committee.

Donald D. Bills
Shain-dow Kung

PROGRAM COMMITTEE

Chair Anthony Kotula, *USDA, ARS, Beltsville, MD*

Alley Watada, *USDA, ARS, Beltsville, MD*
Dennis Westhoff, *University of Maryland*
Bruno Quebedeaux, *University of Maryland*
Gary Fader, *E. I. du Pont de Nemours & Co.*
John Goss, *E. I. du Pont de Nemours & Co.*

LOCAL ARRANGEMENTS COMMITTEE

Chair Helen Phillips, *University of Maryland*

Patricia Moore, *UM*
Marilyn Stetka, *USDA, ARS, Beltsville, MD*
Claudia Volz, *USDA, ARS, Beltsville, MD*

ENHANCED HUMAN NUTRITION

An International Perspective on Biotechnology and Nutrition

Alvin L. Young
Office of Agricultural Biotechnology
U.S. Department of Agriculture
14th and Independence Ave., SW
Washington, DC 20250

The new tools of biotechnology will have a broad and diverse impact on world agriculture and human nutrition. For industrialized countries, biotechnology may provide solutions to food safety problems, as well as lead to improved nutritional qualities of some agricultural products. It may also lead to the development of new products with unique nutritional qualities. In the medium-term, the most important contribution of biotechnology for developing countries will be continued improvement of the yields of basic food crops. Improved nutritional qualities of traditional crops also hold great promise for combatting malnutrition, but these applications are likely to take longer to develop. Special efforts in research, training, and technology transfer will be necessary if the developing world is to fully benefit from the new products of agricultural biotechnology.

Introduction

The new tools of biotechnology will have a broad and diverse impact on world agriculture and human nutrition. For centuries, people have sought to improve plants, animals and microorganisms to produce food and fiber for their own needs. The process of genetic improvement is the backbone of agriculture, and the crux of our ability to feed the growing world population.

The recent discoveries of biotechnology offer a powerful yet precise set of new tools to use in this endeavor. We are only beginning to explore the diverse applications of these tools. Indeed, biotechnology may one day have a greater impact on our lives than any of the other revolutionary technological advances of the 20th century.

The world needs more and better food. This is a simple and yet easily forgotten fact. It is estimated that 75 million people are malnourished—75,000 people die each day from the effects of malnutrition and related diseases. And, because of continued population growth, this depressing picture will grow even more gloomy, if we do not increase agricultural productivity.

According to the Consultative Group on International Agricultural Research, world cereal yields must increase 35 to 40 percent in the next decade just to keep hunger and malnutrition from getting worse. If we expect to make headway, then we need to do better than 40 percent.

Biotechnology offers one of our best hopes of meeting this challenge. The diversity of topics being presented at this Third International Symposium on Biotechnology and Nutrition is evidence of the breadth of direct applications of biotechnology to food safety, food production, processing and digestion of food by man and animal. Research on molecular approaches to carbohydrate modification, amelioration of protein, enhancement of vitamins and minerals and modification and production of edible oils offers important opportunities to improve our food supply and to tailor products to meet specialized nutritional needs, and food preferences.

Also of great importance to improving the nutritional status of the world's people is what biotechnology can do to increase the quantity of food being produced, especially in the Third World. In this regard biotechnology may be a key technology for increasing yields of basic food crops and decreasing post-harvest losses, a critical problem in developing countries.

These improvements will not come about without concerted efforts to translate the potential of biotechnology into reality. If we are to fully benefit from biotechnology, the United States and other industrialized countries must set the tone by dealing effectively with two key issues—regulatory policy and consumer acceptance.

As a leader in the development of biotechnology, the United States, scientific community must actively communicate with the public and inform them about both the benefits and the risks associated with biotechnology. We also must develop systems of oversight that are as coherent and transparent as possible, so that the public gains confidence in the processes that we are using to ensure their safety and protect the environment. But equally important is developing systems of oversight based on risk, which do not overburden the research community with regulatory requirements in areas we already know are safe.

This is a difficult balance to achieve. If we are successful in achieving this balance in the United States, biotechnology will flourish, and this should help policy-makers in other countries pursue a similar path.

In addition to achieving the general prerequisites of consumer acceptance and sound regulatory policy, we must also be sure that we devote adequate resources to supporting the science base upon which biotechnology is being built. While it is true that much of the research supporting product development is being done in the private sector, continued success will not be possible if we do not strengthen public research. The needs in this area can be summed up rather succinctly—more and better trained people, new types of research institutions, and, of course, increased funding.

In addition, there is a need to make special efforts to ensure the application of biotechnology to the developing world. These efforts may involve increased funding for national and international research centers, creating international networks of researchers to address certain commodities which are not commercially important to the industrialized world, and stepping up training for developing countries' scientists and technicians.

Benefits to the Consumer

During the next three days you will be hearing many impressive examples of how biotechnology is being applied to improve the food supply and insure food safety; thus it would be redundant for me to go over these now. What I would like to emphasize is that many of the areas being addressed offer direct benefits to consumers. This is good news for agricultural biotechnology, since one problem to date in trying to communicate with the public has been that products which benefit consumers directly have been slow to emerge from the research and development pipeline.

Biotechnology products which offer nutritional advantages should appeal to consumers, both here and abroad. According to the National Research Council's report, *Designing Foods*, recent public opinion surveys show a growing trend toward consumer interest in the nutritional composition of food and the diet as a whole. Two-thirds of consumers indicated they had recently decided to alter their diets by making such changes as reducing salt, sugar or fat intake, or increasing consumption of foods which contain fiber, or calcium.

Of course, consumer attitudes studies do not always coincide with real consumer behavior. Indeed new high-calorie, high-fat foods such as gourmet ice cream bars, frozen french toast and pancakes and potato chips are selling better than ever. This paradox has led the National Research Council to hypothesize that in the United States we are seeing the emergence of a new philosophy of nutrition: that a balanced diet can be achieved from a variety of foods—high-fat as well as low-fat—consumed over several days or even a week, compared with the more traditional thinking of three square meals a day.

In addition, consumers are demanding that food be free of additives and residues, exciting, palatable, appetizing, available all year round, move from supermarket shelf to stomach in record time, and be entirely "natural."

Public education programs are to prevent misinformation about nutrition and food safety. However, in areas where real nutritional needs exist, or where there are food safety problems to be solved, biotechnology can be part of the solution.

Producing More Food

But improving food and making it safer are not enough. The world needs to increase agricultural production as well. For the past few decades we in the industrialized world have been shielded from the reality of the increasing world demand for food and fiber because of the success we have enjoyed in our domestic agricultural production. One might say our vision has been blocked by warehouses of butter and cheese and silos overflowing with wheat.

Some Western European nations are even cutting back on research which would increase the efficiency of food production, because they believe they simply don't need more food. Such policies are shortsighted. As Eastern Europe moves to a market economy demand for high quality food and fiber products, our capacities to produce them may be outstripped. World wheat reserves are at the lowest point in years, and in 1989 and 1990, the United States experienced a fluid milk deficit in some regions of the country.

Developing countries must dramatically increase their food production just to keep pace with population growth. The most promising approach may be to apply biotechnology to increase yields of basic food crops. For example, *in vitro* culture for rapid multiplication of disease free planting materials is now a well understood, yet labor-intensive technique. According to a recent FAO report, China already uses this approach to produce about 250,000 hectares of virus-free potato seedlings each year or about 10 percent of all potato production in the country. Using *in vitro* culture has allowed the Chinese to increase potato yields by as much as 150 percent in recent years.

An international network of scientists which stretches from Africa to Europe to the Americas is striving to apply the newest techniques of biotechnology to cassava. Cassava is a basic staple in much of the world which has proved difficult to improve using conventional techniques. One goal of the research network is to increase the protein content of the plant. This could be extremely helpful in fighting protein malnutrition in Africa and elsewhere. In addition, the researchers are striving to produce cultivars resistant to viral diseases and pests which often plague cassava production in Africa and Latin America. They are also trying to improve stress tolerance to make cassava more drought resistant.

The Rockefeller Institute is supporting an international effort to apply biotechnology to improve rice yields. The International Rice Research Institute (IRRI) in the Philippines is spearheading this effort by creating an interdisciplinary team of scientists to work on transforming rice. Beyond improving yields, scientists in Pakistan and the Philippines are working on introducing nitrogen fixation genes into rice. If this breakthrough occurs, it could dramatically reduce the need for fertilizer in many developing countries.

Another strategy which may assist in meeting the nutritional requirements of the third world is using biotechnology to help reduce post harvest losses. Some estimates are that as much as fifty percent of food produced in developing countries is lost during storage, transport, and preparation.

One example of how biotechnology may assist is "anti-sense" technology being developed by firms in Europe and the United States. Simply by reversing a piece of RNA, the researchers have been able to block the creation of an enzyme which causes tomatoes to spoil. Thus, the antisense tomato can stay ripe at room temperature without getting soft and mushy. This technology may be widely applicable to fruits and vegetables and a real boon to developing countries where refrigeration is largely unavailable.

These are but a few examples of the application of biotechnology to improving the nutrition of the world's poor. There are many other examples. Yet, critics of biotechnology say that such an optimistic appraisal is naive. They point to the "green revolution" which some contend was over-sold as a solution to Third World problems. Some observers even go so far as to say that most technologies which Third World countries adopt from the industrialized world only exacerbate the differences between the haves and the have-nots.

Although I agree that it would be naive to portray biotechnology alone as an answer to developing countries' food problems, it should be pointed out that biotechnology has characteristics which make it especially promising for Third World conditions.

For example, it may actually reduce the amount of inputs needed to produce food crops. Varieties with built-in resistance to insects and diseases will obviate the need for some agricultural chemicals. If breakthroughs occur in transferring nitrogen fixation genes to cereals, fertilizer requirements would also be reduced.

Biotechnology is a broad set of techniques applicable to many species with the potential to affect practically every agro-climatic zone from the Amazon to the Sahel. Although the first plant species was transformed less than ten years ago, over 50 plant species have now been modified. New techniques such as microinjection have overcome earlier barriers. The range of these new techniques is extremely broad, as is the range of genes which may prove useful.

Biotechnology is size-neutral. Because we can incorporate improvements in the seed, small farmers will benefit as well as large farmers. In addition, farmers will not have to learn to grow new crops or drastically change their practices in order to reap these benefits.

Lastly, biotechnology fits into agricultural systems which are sustainable and conserve soil and water. Current work with genetically modified rhizobia may lead to environmentally sound methods of improving soil fertility. The goals of modifying crops to increase their tolerance to drought and salinity are also seeming more and more attainable. These traits are controlled by more than one gene, but researchers have recently reported that they have successfully transformed plants with as many as five genes at once.

The Policy Context

Technologically, the future seems bright. But as I indicated earlier, the promise of biotechnology to improve our food supply will not become reality, unless the countries which are biotechnology leaders come to grips with two crucial issues—public acceptance and regulatory policy.

The U.S. Department of Agriculture (USDA) has decided to make public information about biotechnology a priority. We publish a monthly newsletter called "Biotechnology Notes" which is written in layman's language. We also publish press releases which describe new research discoveries.

Since some environmental groups have expressed concerns about the environmental consequences of testing genetically modified organisms outdoors, USDA has made a special effort to involve the public in decisions about the oversight of field testing.

For example, all meetings of the Agricultural Biotechnology Research Advisory Committee (ABRAC) which advises USDA on biosafety issues involved with field testing are open to the public. ABRAC meetings are attended by representatives of environmental groups, the press, representatives of private firms and other members of the public. These groups frequently comment on both the processes and the decisions made by ABRAC and USDA.

USDA is currently publishing research guidelines which will help USDA research agencies conduct adequate environmental assessments of field tests involving organisms with modified heritable traits outdoors. Once the guidelines are published, we will conduct a series of regional public meetings throughout the United States to gather public comment.

Thus far, these efforts have been successful in building public confidence, with the result that field testing is moving forward smoothly in the United States. As of August 1990, the USDA Animal and Plant Health Inspection Service had issued over 90 permits for over 150 test plots with transgenic plants. Tests have been conducted in more than 20 states. The plant species involved include tomato, tobacco, corn, cotton, potato, soybeans, alfalfa, rice, poplar, cucumber, walnut, cantaloupe and squash. All these tests have gone smoothly with little expression of concern by the public.

The next major public information effort needed will be to focus on the food safety issues related to biotechnology. Unfortunately, bovine somatotropin (BST) has become an example of what can go wrong if the public latches on to misinformation about a product. BST, as many of you know, is a naturally occurring hormone in cattle, which can now be produced in the laboratory using recombinant DNA technology. When BST is administered to lactating cows it improves the milk-to-feed ratio by 5 to 15 percent. It does not necessarily mean more milk. It can also mean the same amount of milk from fewer cows.

The milk produced by BST-treated cows is no different than milk from non-treated cows. Although the Food and Drug Administration is still reviewing BST, it, along with other regulatory agencies in Europe, has issued investigational findings stating that milk and meat products from BST-treated cows are safe for human consumption. Yet, some critics continue to call for a ban on BST, and have confused the public about safety issues.

Moreover, some groups in the European Community are trying to ban BST on socio-economic grounds. The United States has strongly objected to this. If BST is approved for commercial use in the United States, such a ban would create major trade problems between the United States and the European Community. Banning BST would do little to protect small dairy producers in Western Europe from the forces of economic change, while creating a terrible precedent for the fledgling agricultural biotechnology industry.

Which brings me to the question of regulatory policy. The United States has been reasonably successful in creating a system of oversight which allows for research and development of biotechnology while protecting human health and the environment. While continuing to refine this system, we must now step up our interactions with other nations to ensure that our systems of regulation are compatible.

For example, the United States has played a major role in the Organization for Economic Cooperation and Development in successfully setting international norms for contained use and for outdoor research involving genetically modified organisms. These documents are extremely useful in delineating the scientific principles which should serve as the base for national systems of oversight.

But we need to do more. If agricultural biotechnology is to reach its potential, the international scientific community must agree on the scientific principles which underpin sound decision-making. We need to work together to harmonize risk assessment processes and to define together what, if any, additional safety questions need to be addressed as we move toward commercialization of agricultural biotechnology products.

If we can agree on the scientific aspects of risk assessment, then it will be much easier to make sure our risk management procedures do not end up as impediments to development and trade of biotechnology products.

Support for Research

I would like to conclude by returning to the question of research, to address what we will need to do here in the United States, as well as what special efforts need to be made to ensure that the benefits of biotechnology reach the Third World.

One crucial issue for the United States is the need to attract more young men and women to the agricultural and food sciences. The National Academy of Sciences projects that the United States could be short by almost a million scientists and engineers by the year 2000. Human capital—creative and ingenious people—is the primary natural resource of biotechnology. Thus, we need to work harder to attract young men and women, including minority students, to science. We know we are working in an exciting field, but we need to convey our enthusiasm to primary and secondary level students through innovative curricula and hands-on science education.

We also need to continue to reorganize and revitalize our public agricultural research institutions. Progress in biotechnology demands intellectual contributions from a number of scientific specialties. Molecular biologists need to work more closely with food technologists, animal and plant breeders, physiologists and even ecologists.

Right now we have many new techniques for genetic modification at our disposal, but we are having a difficult time putting these new techniques to work because we still lack important knowledge about basic plant and animal physiology. Institutions need to be developed which encourage interdisciplinary research, through facilities which allow physical proximity, new communications systems, and through novel funding mechanisms.

Which brings me to the subject which is always difficult—money. Agricultural research in the United States needs an infusion of new funding. The United States, in recent years, has consistently underinvested in agricultural research. Clearly, new funding is needed if we expect to continue to make discoveries in biotechnology, and to apply these new discoveries to the many challenges that face agriculture. That is why we have placed before the Administration and the Congress the National Initiative for Agricultural Research. It is the intent of this new initiative to dramatically increase (annual increments of $100 million) the funding available to the research community.

We will also need to make special efforts to reach developing countries with biotechnology. I have already alluded to the important work being conducted at IRRI on improving rice. IRRI, the Centro Internacional de Mejorimiento de Maiz y Trigo (CIMMYT), and the other international centers have been crucial in improving Third World agriculture in the last few decades. These centers have the unique ability to draw on the world scientific community to work on Third World problems in the geographic areas which are most affected. Thus, products of biotechnology can be researched and tested in the environment where they will eventually be used.

It is also imperative that scientists in national research centers in the Third World have the opportunity to get involved in biotechnology. This can be achieved in several ways. The U.S. Agency for International Development awards grants to U.S. scientists who wish to work on projects of mutual scientific interest with researchers in developing countries. These grants often include funds to bring young researchers from the developing country to work in the laboratory of the U.S. counterpart scientist. The Governments of France, Germany, the Netherlands and several other industrialized countries are operating similar programs.

Likewise, some private biotechnology companies are forging direct links with institutions in the Third World. A firm in Belgium has committed its resources to support the cassava research network I mentioned earlier, while a major U.S. company is working with Mexican researchers to introduce genes for virus-resistance into the varieties of potatoes which are commonly grown in Mexico.

The needs of the Third World are great, but with energy and ingenuity they can be met. One very positive factor is that policymakers in the Third World are fully committed to applying biotechnology to the problems of their countries. Their interest is not to debate the pros and cons of this exciting new technology, but to employ it to feed their people. I am confident that biotechnology will help them meet this challenge.

Impact of Diet on Human Oncology

Peter Greenwald
Division of Cancer Prevention and Control
National Cancer Institute
National Institutes of Health
Bethesda, MD 20892

The application of rapidly occuring advances in food biotechnology is bringing revolutionary change to the food industry and may significantly impact on overall health and disease prevention. Research on diet and cancer prevention at the National Cancer Institute, including an area of research called "chemoprevention," explores the potential of these advances to reduce the risk of cancer by changing the nutrient or chemical characteristics of foods. Biotechnology has been used to enhance the nutritive quality of fruits and vegetables, develop synthetic fat substitutes, and produce leaner beef; products that may reduce cancer risk. Based on results from epidemiologic and laboratory studies, NCI has collaborated with the USDA in several clinical metabolic studies to investigate the influence of dietary fat on fecal mutagenicity, and the dose and toxicity of beta-carotene and selenium tissue levels for intervention trials. A new program at NCI centers around foods fortified with phytochemicals that are being studied to prevent cancer in humans. Close collaboration with industry is necessary to ensure a continuum of biotechnological advances.

Introduction

The scientific evidence from laboratory and epidemiological research studies strongly indicates that most cancer incidence may be related to lifestyle and environmental factors. In theory, therefore, cancer is largely preventable (Doll and Peto, 1981; Higginson, 1988). Cancer prevention research, a relatively new area of cancer research, focuses on reducing cancer incidence by avoiding or minimizing exposure to those factors thought to increase cancer occurrence and precancerous progression. Until recently, the elimination of carcinogenic agents and the early detection of premalignant abnormalities dominated the field of cancer prevention. However, insights gained from epidemiological and basic studies of biochemical mechanisms and biologic processes of carcinogenesis are presenting new directions and opportunities for cancer prevention.

Contributing to the opportunities of cancer prevention is the application of advances in biotechnology that are bringing revolutionary changes to the food industry. These changes may significantly affect overall health and disease prevention. Changing the nutrient or chemical content of fruits and vegetables, developing synthetic fat substitutes, producing leaner beef, and lowering the fat content of milk products are recent advances in food and agricultural technologies that indicate a trend that, as a whole, may reduce cancer risk.

The National Cancer Institute (NCI), using the weight of the scientific evidence for the role of nutrition in cancer prevention as a basis, explores effective interventions to prevent and control cancer mostly in medical settings. After a relatively short time, positive results from these efforts may expand the scope of the practice of medical oncology. Other prevention research is being accomplished in the public health sector: this includes research within high-risk and minority populations; research performed by the agricultural, food, and pharmaceutical industries; and research within the public health agencies and public health community.

A recent example of the potential health benefits from preventive research efforts is that of hypertension, where high blood pressure is detected for the prevention of strokes and heart disease. Medical intervention is available that can reduce that risk. It may be feasible in the future to accomplish the same reduced risk for the progression of cancer. Cancer dysplasias or other premalignant abnormalities that indicate high probabilities of risk are being identified. Studies are in progress whose concept is to titrate or lower the probability of that risk.

Diet and Cancer

Despite the incomplete knowledge available concerning the mechanisms of nutrition and disease, there are many research leads to justify an aggressive approach to prevention research related to nutrition and cancer. For this reason, the National Cancer Institute directs a program for diet, nutrition, and the prevention of cancer and has greatly increased its commitment to this program. Within the Cancer Prevention Research Program, NCI conducts extramural and intramural research directed toward the development, evaluation, and demonstration of approaches to prevent cancer. This research includes studies aimed at the identification of persons at increased risk of cancer, interventions designed to reduce cancer risk, and the dietary manipulation or administration of chemopreventive agents that may inhibit aspects of the carcinogenic process. Figures 1 and 2 present the research phases of the Diet and Cancer and Chemoprevention programs: two major research efforts at NCI for reducing cancer incidence discussed in this paper.

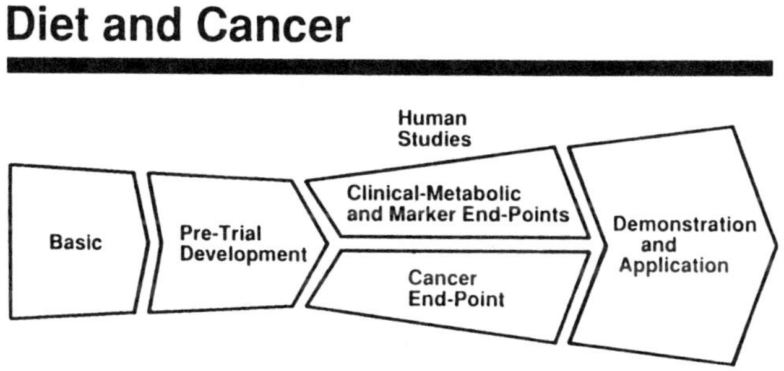

Diet and Cancer

Human Studies

Basic

Pre-Trial Development

Clinical-Metabolic and Marker End-Points

Cancer End-Point

Demonstration and Application

Figure 1.

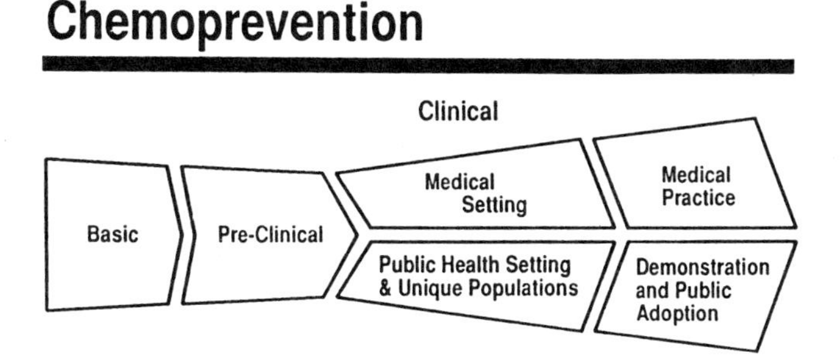

Figure 2.

A newly developed diet and cancer program will focus in part on the potential human benefits of dietary supplementation, with particular attention to fruits and vegetables. This "Designer Foods" program is concerned with phytochemicals in edible fruits and vegetables and how they may modulate neoplastic development. This basic research effort extends beyond NIH and includes U.S. Department of Agriculture (USDA)-sponsored research, dietary surveys and food composition data bases, and laboratory and food science research within the food industry and the Food and Drug Administration. This effort is involved with identifying and characterizing the elements that metabolically influence some of the biological systems. There are safety studies aimed at understanding multigenerational growth and development, bone mineral metabolism, and vitamin absorption.

Scientific Rationale for Dietary Modification

As the American diet has changed, so has the prevalence of different types of cancer. Lung, large bowel, and breast cancers now account for almost half of the annual cancer mortality in the United States, while stomach and esophageal cancers are decreasing in this country but remain prevalent in many underdeveloped countries. While there are some inconsistencies, generally compelling laboratory results and epidemiological observations provide strong support for the relationship between specific dietary factors and certain cancers—notably cancers of the breast, prostate, colon, stomach, and possibly lung (Doll, 1988). The evidence indicates a positive correlation in different populations between the incidence of cancers of the breast and colon and high fat consumption (Prentice *et al.*, 1989; Schatzkin *et al.*, 1989) and a negative correlation between the consumption of vegetable-rich diets and gastric and colorectal cancer (Howson *et al.*, 1986; Greenwald *et al.*, 1987; Trock *et al.*, 1990).

Large intakes of foods high in beta-carotene appear to have protective benefits against several cancers. Present in large amounts in fruits and vegetables, vitamin A and its synthetic analogs and beta-carotene are being subjected to intense investigation for their anticarcinogenic activity (Willet *et al.*, 1984; Ziegler, 1989).

Population studies have shown an inverse relationship between intake of foods high in vitamin A or beta-carotene and cancer incidence. More than 20 studies have evaluated this relationship, with 9 retrospective reports indicating a significant increase in cancer risk at various sites associated with diminished vitamin A intake (Greenwald, 1989). Other retrospective data confirm the inverse association between ingestion of foods containing vitamin A or beta-carotene and relative cancer risk with risk levels 2 to 2.5 times higher in the low-intake groups than in the high-intake groups (Peto *et al.*, 1981). Protective roles for foods high in selenium, alpha-tocopherol, vitamin C, and green or yellow vegetables have also been suggested by some epidemiological reports (Bertram *et al.*, 1987). Reduced tumor activity has been demonstrated in animal systems using pharmacological amounts of vitamins C, E, A and related synthetic analogs (retinoids), and beta-carotene.

Dietary fat has been studied extensively and is associated frequently with various cancers. Human population data show strong positive correlations between total dietary fat and caloric intake and cancer incidence (NAS/NRC, 1982; Prentice *et al.*, 1988). Comparisons of dietary fat intake by country indicate, for example, that populations with the highest per capita fat consumption have the highest breast cancer mortality: Denmark has a fivefold greater rate of mortality from breast cancer than Japan (Wynder *et al.*, 1976). Migration data among women from areas with low breast cancer rates to areas with higher rates show increased breast cancer risk after migration. This trend is seen among Japanese women migrating to Hawaii (Dunn, 1977) and Italian women migrating to Australia (McMichael and Giles, 1988). Time trend data, in which increasing fat intake among certain population groups over time is measured, also reveal a correlation between dietary fat and breast cancer incidence (Kurihara *et al.*, 1984). Dietary fat is consistently observed to be a tumor promoter in animal studies. Restricted caloric intake was found by Tannenbaum and Silverstone (1953) to inhibit tumor development in a series of studies using mouse mammary tumor models. Increasing levels of dietary fat increased both tumor yield and size. Animals fed high-fat diets *ad libitum* developed tumors earlier and with greater frequency than animals fed low-fat diets. Thus, data from both experimental and human correlation studies strongly support the association between dietary fat and breast cancer. There are, however, some studies that do not conclude this association (Goodwin and Boyd, 1987; Willet *et al.*, 1987). Many investigators believe that clinical trials are needed to provide a direct test of the dietary fat-breast cancer hypothesis and to document the number of years needed to achieve a benefit after reducing dietary fat.

Observations in Africa (Burkitt, 1971), where the intake of dietary fiber is very high, show a relative lack of large bowel cancer and other chronic gastrointestinal diseases common in Western countries. These observations led to later studies of a possible protective role by fiber against colorectal cancer. A review of 40 epidemiologic studies indicated an inverse association between total dietary fiber intake and colon cancer incidence in 32 of the 40 studies (Greenwald *et al.*, 1987), and meta-analyses of 37 epidemiologic studies and 16 case-control studies support the protective effect associated with fiber-rich diets (Trock *et al.*, 1990). Although results from numerous case-control, international, and within-country correlation studies generally reflect an inverse or protective association between dietary fiber and colon cancer risk, several studies show no association or a direct association (Pilch, 1987). Differences may exist because dietary fibers from different food sources are heterogeneous mixtures of components, and in addition, fiber components are difficult to accurately quantitate.

Chemoprevention

When specific nutrients or their analogs or other potential inhibitors are studied as single chemical agents for cancer prevention purposes, the National Cancer Institute refers to this research as "chemoprevention." NCI has a major chemoprevention program to identify compounds with cancer prevention activity for modifying the initiation and progression of neoplastic development. At present, hundreds of chemopreventive agents are under consideration for development; more than 123 agents are being studied *in vitro* and 95 *in vivo,* and 5 agents are undergoing animal toxicity testing. These candidate chemopreventive compounds are derived from a wide range of sources, represent varied chemical structures, and exhibit diverse physiological effects. Chemoprevention can be directed against several phases of carcinogenesis: exposure to carcinogens resulting in the biological system incorporating and metabolizing the carcinogen, interfering with key carcinogenic events within the cell (mutational), promoting the mechanisms of intracellular repair, and preventing the selective growth of a carcinogen-altered cell.

During the past year, a number of chemical and pharmaceutical agents have shown the potential to be further evaluated in clinical chemoprevention trials. As a group, the synthetic retinoids remain one of the most promising chemopreventives. Under investigation are all-trans-N-(4-hydroxyphenyl) retinamide (4-HPR); a prostaglandin synthesis inhibitor, difluoromethylornithine (DFMO); and several other pharmaceuticals and naturally occurring constituents and trace minerals (Boone *et al.*, 1990).

The NCI has developed an orderly system for the identification and testing of promising chemopreventive compounds, including research involving a broad range of modalities from tissue culture to clinical trials in humans. Although a large number of agents may be evaluated initially, only a limited number meet the rigorous criteria required for clinical testing in humans.

Clinical Intervention

Clinical Metabolic Studies. Although the metabolic contribution of any nutrient, alone or in combination, may not be fully determined, the NCI recognizes the importance of and supports human metabolic studies to clarify clinical metabolic issues important to intervention research. Some of these studies are collaborative projects with agencies such as the U.S. Department of Agriculture. Other studies are conducted by individual investigators.

A recently reported study indicates that indole-3-carbinol influences estradiol metabolism in humans, suggesting that estrogen metabolism may be modulated by dietary factors. Chemopreventive strategies directed at reducing the risk of estrogen-dependent diseases such as breast cancer may be developed as a result of dietary manipulations that affect human metabolism. Results from this particular study appear to suggest that estradiol metabolism in humans may be somewhat modulated (Michnovicz and Bradlow, 1990). Confirmation is needed as to whether diet, or in this case indoles from cruciferous vegetables, actually does modify human estrogen metabolism (Longcope, 1990).

Metabolic studies are under way to investigate the form, dose, and human toxicity of selenium. Results of several epidemiological studies indicate that blood selenium levels of individuals who later developed cancer are lower than those who remained disease free. Numerous animal studies demonstrate that pharmacologic doses of selenium can inhibit the development of a variety of tumors. However, toxicity of selenium is a concern. To resolve issues of efficacy and safety, the Cancer Prevention Research Studies Branch, NCI, coinvestigators at Harvard University, and the Human Nutrition Research Center, USDA, conducted a study of healthy adults residing in a seleniferous area of South Dakota. The goal of these studies was to determine a range of acceptable dose levels and chemical forms of selenium to be given in human cancer prevention trials. Over a 2-year period, 142 subjects were recruited. Parameters examined were blood, physical examinations, and dietary intake data. The kinetics of two chemical selenium forms were compared by modeling the appearance and disappearance of labeled selenite and selenomethionine. Data analysis will focus on the use of urinary selenium forms as an index of dietary intake. Selenium rate constants will be studied for subjects in the fasting and nonfasting state for inferences on their effects on metabolism.

Because of the possible association of carotenoids and lower incidence rates of several epithelial cell cancers, two controlled feeding studies were designed to better understand the individual response to varying doses and forms of carotenoids, to assess potential carotenoid toxicity, and to determine the usefulness of epidemiologic dietary assessment methods in estimating intake of individual carotenoid components in food. The primary objective of these studies was the determination of the blood carotenoid pattern after single and daily ingestion of selected vegetables or capsules containing beta-carotene. Secondary objectives included assessment of potential toxicity associated with daily ingestion of high levels of carotenoids and determination of the association between the plasma concentration of various carotenoids and dietary intake. Results from these carotenoid-feeding studies indicated wide variation in carotenoid absorption in normal subjects. Beta-carotene and total carotenoid plasma levels increased only at the higher intake; however, carotenemia developed in the group fed the highest level of beta-carotene supplementation. Intrasubject variation in plasma carotenoid levels has not been previously studied and is an important consideration in monitoring subject response and compliance in intervention trials.

Several collaborative clinical metabolic studies were conducted with NCI and USDA to examine the influence of low dietary fat intake (20 percent fat calories) compared with high dietary intake (40 percent fat calories) on hormone status, bile acid metabolism, fecal mutagenicity, and serum lipids in pre- and postmenopausal women. Results of premenopausal women randomized from a high-fat to a low-fat diet indicated that the concentration and total excretion of primary and secondary bile acids decreased on the low-fat diet. Reduction of dietary fat was also accompanied by a significant decrease in fecal mutagenicity. Two low-fat feeding studies with differing P:S ratios were conducted with postmenopausal women. The analyses of lipid values for low-fat diets compared with high-fat intakes were not significant. Overall, serum lipids were not strong indicators of dietary fat intake. However, significant changes were observed in fatty acid profiles, suggesting that fatty acids may be useful biochemical markers of adherence to a low-fat diet. Further analyses and inferences will be made of the steroid hormone responses for both pre- and postmenopausal women.

Clinical Intervention Trials

The feasibility of clinically assessing the efficacy and safety of chemopreventive agents has been established. More than 20 human intervention chemoprevention trials sponsored by NCI to test using randomized, controlled clinical trial designs for determining whether cancer risk can be reduced, are currently in progress in medical and community settings. Agents under study include vitamin and mineral dietary supplements and several synthetic retinoids. For example, NCI is conducting a multicenter, randomized, double-blind basal carcinoma study designed to evaluate the effectiveness of low-dose isotretinoin for reducing the incidence of basal cell carcinoma in a high-risk population. The test population includes 1,000 subjects, ages 40 to 75, with two or more biopsy-proven basal cell carcinomas diagnosed during the 5 years preceding recruitment. Any toxicity associated with the 3 years of isotretinoin intervention will also be examined. If isotretinoin is found to be effective and relatively nontoxic, it will likely have an effect on further work related to the role of synthetic retinoids in the prevention of other epithelial cell carcinomas.

In a recently conducted clinical trial, Kraemer *et al.* (1988) found that isotretinoin is markedly effective in preventing or delaying the onset of new skin cancers in patients with xeroderma pigmentosum, an inherited condition in which the defective repair of ultraviolet-damaged DNA results in a thousandfold increase in the frequency of skin cancer compared with the general population. During the period of chemopreventive therapy, there was a marked decrease in the number of new skin tumors. Although the high dosage used produced serious toxicities in some patients, protocol refinement and the development of new synthetic retinoids may lead to further reduction of cancer occurrence in persons at high risk for this skin cancer.

The NCI and the National Public Health Institute of Finland are conducting a large-scale lung cancer prevention trial testing the oral administration of beta-carotene and alpha-tocopherol in a population of heavy smokers (Albanes *et al.*, 1986). With a lung cancer incidence among the highest in the world coupled with marginal per capita intake of several micronutrients, Finland offers a unique environment for the study of lung cancer prevention. Four separate treatment groups are being evaluated in a population of 29,000, ages 50 to 69, using a 2 by 2 factorial design. The use of factorial designs, which evaluate two or more hypotheses in a single trial with a minimal increase in cost, is particularly suited to prevention trials. Clinical follow-up procedures for the Finland study will be used to identify cancers, and a reduction in cancer incidence as demonstrated by the trial will be compared with national trends by monitoring Finland's unique government-operated health registers.

Of the many other cancer prevention trials in progress, 11 are examining the effects of beta-carotene, alone or in combination with other agents on cancer incidence. For example, Hennekens and co-workers are conducting a study of 22,000 healthy physicians, ages 40 to 84, to evaluate the influence of beta-carotene and aspirin on both total cancer incidence and cardiovascular disease, using a factorial study design. Because of the benefits of aspirin against cardiovascular disease, study subjects have been informed about whether they were on aspirin, ending that arm of the randomization.

In Linxian, China, an area with a particularly high incidence of esophageal cancer, two ongoing intervention studies are testing the efficacy of multiple vitamins and minerals in the prevention of esophageal cancer incidence and mortality (Li *et al.*, 1986). The General Population Trial, which is a fractional factorial design, randomized 30,253 individuals, ages 40 to 69, into 8 intervention groups receiving placebos or 1 to 4 times the RDA of specified combinations of vitamin A, beta-carotene, zinc, riboflavin, niacin, vitamin C, molybdenum, selenium, and vitamin E. Periodic examinations are conducted and include the assessment of esophageal cytology, histology, cell proliferation, and DNA content as well as measures of immune function and other studies. A second study, the Dysplasia Trial, will examine the effect of multiple vitamins/minerals in a high-risk population with severe esophageal dysplasia. Using a dichotomized clinical trial design, 3,400 men and women, ages 40 to 69, will receive a multiple vitamin/mineral supplement at levels 1 to 4 times the RDA or a placebo. Disease progression or regression will be evaluated during treatment. In both studies, cancer incidence and mortality will be determined through hospital records, pathology slides, and radiographic reviews by a multinational panel of experts.

A highlight of this past year has been the recently published study by Dr. Jerome Decosse (1989) and his group from the Cornell Medical Center investigating the effects of dietary supplements on colon cancer risk. This three-armed study of placebo control, vitamins C and E alone, and vitamins C and E and a wheat fiber supplement showed a statistically significant reduction (adjusting for compliance) in colon polyps in the fiber/vitamin group. Using a polyp endpoint, not cancer incidence, this study, on patients with familial polyposis, is the first demonstration of an intervention trial that is randomized and blinded and provides prospective evidence that adding a fiber supplement reduces a risk marker important in the progression to colon cancer. A more extensive NCI study aims to prevent the recurrence of previously diagnosed adenomatous polyps through a low-fat, high-fiber, fruit- and vegetable-enriched dietary intervention.

Biotechnology—Health Implications of the Changing Food Supply

New technologies may reduce cancer risk because of their potential to change the nutrient or chemical constituents of foods. For example, of all the dietary components studied, fat intake offers the strongest evidence for a causal relationship with cancers of the breast, colon, and prostate. The food industry has already taken steps to incorporate existing knowledge concerning diet and disease in the development of new lower fat products. Low-fat choices for commonly consumed high-fat foods are increasingly available.

Recently developed fat substitutes, such as Simplesse and Olestra, may replace fat or reduce the percentage of fat in numerous fat- and oil-based products (Boggs, 1988). Low calorie foods are also being developed. If used as part of an overall change to a lower calorie, lower fat diet, these may be helpful as part of a varied, balanced diet in reducing cancer risk.

In addition to the development of fat substitutes, reductions have been made in the total fat content of beef and pork through breeding and genetic improvement. Further improvement is possible with the use of growth hormones such as somatotropin (Breidenstein, 1988). Somatotropin can be digested like any other protein, yet it can have a major effect on the protein to fat ratio in meat and meat products. Although it is still experimental, industry has possible ways to produce chickens with more low-fat meat and lower cholesterol eggs.

Advances in product development, preservation, and marketing have also been used to increase the availability and variety of vegetables, fruits, and other fiber-rich foods. A number of studies have reported that daily consumption of vegetables and fruits is associated with decreased incidence of colon, lung, bladder, esophageal, and stomach cancers. Enhancing the anticancer effect of fruit and vegetable constituents such as carotenoids; vitamins A, C, and E; minerals; fiber; and nonnutritive phytochemicals such as isothiocyanates, flavones, and indoles is a possible area of research.

Development of new or improved food ingredients through genetic engineering has already led to the enhancement of characteristics such as taste, shelf life, and nutritive quality (Gasser and Fraley, 1989). For example, the new germination techniques can produce large increases in the vitamin C content of peas and beans, and enzymatic treatment of carrot juice increased beta-carotene content from 7 to 60 percent (Teutonico and Knorr, 1985). Recently, researchers have developed a way of increasing the growth and proliferation of citrus fruit vesicles as well as using bioregulators to alter the constituency content of the juice. Also, the knowledge gained by investigators studying the regulation of ripening enzymes will lead to the availability of tomatoes that taste vine-ripened yet do not soften before reaching the market (Wasserman *et al.*, 1988). This trend toward improving vegetables and fruits can be favorable because, in addition to their beneficial nutritive and nonnutritive constituents, these foods have value in displacing high-fat foods in the diet.

These developments in biotechnology in the food and agricultural industries are examples of where collaboration with biomedical research may produce broad public benefits. Dietary guidance is useful for today, while collaborative research can provide more healthful food choices and more specific dietary guidance for the future. A favorable trend is the availability and variety of fiber-rich foods and low-fat and low-calorie foods offered in the marketplace. However, there is not yet sufficient evidence available to specify the type and amount of fat, fiber, or micronutrients these products should include to reduce cancer risk. An intensive research program is needed to provide us with these answers and to minimize the chance of unexpected adverse effects. As industry rapidly progresses with new technical developments, we need to understand the implications of these products for health promotion and disease prevention.

References

Albanes, D., Virtamo, J., Rautalahti, M., *et al.* 1986. Pilot study: the U.S.-Finland lung cancer prevention trial. J. Nutr. Growth Cancer 3: 207.

Bertram, J.S., Kolonel, L.N. and Meyskens, F.L. 1987. Rationale and strategies for chemoprevention of cancer in humans. Cancer Res. 47(6): 3012.

Boggs, R.W. 1988. Sucrose polyester (SPE)—A non-caloric fat. Fette Seifen Anstrichmittel 4: 154.

Boone, C.W., Kelloff, G.J. and Malone, W.E. 1990. Identification of candidate cancer chemopreventive agents and their evaluation in animal models and human clinical trials: A review. Cancer Res. 50: 2.

Breidenstein, B.C. 1988. Changes in consumer attitudes toward red meat and their effect on marketing strategy. Food Tech. 42(1): 112.

Burkitt, D.P. 1971. Epidemiology of cancer of the colon and rectum. Cancer 28: 3.

Decosse, J.J., Miller, H.H. and Lesser, M.L. 1989. Effect of wheat fiber and vitamins C and E on rectal polyps in patients with familial adenomatous polyposis. JNCI 81: 1290.

Doll, R. 1988. Epidemiology and the prevention of cancer: some recent developments. J Cancer Res. Clin. Oncol. 114: 447.

Doll, R. and Peto, R. 1981. The causes of cancer: quantitative estimates of avoidable risks of cancer in the United States today. JNCI 66: 1192.

Dunn, J.E. 1977. Breast cancer among American Japanese in the San Francisco Bay area. NCI Monogr. 47: 157.

Gasser, C.S. and Fraley, R.T. 1989. Genetically engineering plants for crop improvement. Science 244(4910): 1293.

Goodwin, P.J. and Boyd, N.F. 1987. Critical appraisal of the evidence that dietary fat intake is related to breast cancer risk in humans. JNCI 79: 473.

Greenwald, P. 1989. Principles of carcinogenesis: dietary factors. In "Cancer, Principles and Practice of Oncology," Ed. 3, p. 167. J.B. Lippincott, Philadelphia, PA.

Greenwald, P., Lanza, E. and Eddy, G.A. 1987. Dietary fiber in the reduction of colon cancer risk. J. Am. Diet. Assoc. 87(9): 1178.

Higginson, J. 1988. Changing concepts in cancer prevention: limitations and implications for future research in environmental carcinogenesis. Cancer Res. 48: 1381.

Howson, C.P., Hiyama, T. and Wynder, E.L. 1986. The decline in gastric cancer: epidemiology of an unplanned triumph. Epidemiol. Rev. 8: 1.

Kraemer, K.H., DiGiovanna, J.J., Moshell, A.N., *et al.* 1988. Prevention of skin cancer in xeroderma pigmentosum. N. Engl. J. Med. 318: 1633.

Kurihara, M., Aoki, K. and Tominaga, S. 1984. "Cancer Mortality Statistics in the World." University of Nagoya Press, Nagoya, Japan.

Li, J.-Y., Taylor, P.R., Li, G.-Y., *et al.* 1986. Intervention studies in Linxian, China: an update. J. Nutr. Growth Cancer 3: 199.

Longcope, C. 1990. Relationships of estrogen to breast cancer, of diet to breast cancer, and of diet to estradiol metabolism. JNCI 82(11): 896.

McMichael, A.J. and Giles, G.G. 1988. Cancer in migrants to Australia: extending the descriptive epidemiologic data. Cancer Res. 48: 751.

Michnovicz, J.J. and Bradlow, H.L. 1990. Induction of estradiol metabolism by dietary indole-3-carbinol in humans. JNCI 82(11): 947.

National Academy of Sciences, National Research Council, Committee on Diet, Nutrition, and Cancer. 1982. "Diet, Nutrition, and Cancer." Assembly of Life Sciences. National Academy Press, Washington, D.C.

Peto, R., Doll, R., Buckley, J.D., *et al.* 1981. Can dietary beta-carotene materially reduce human cancer rates? Nature 290: 201.

Pilch, S.M. (ed.) 1987. "Physiological Effects and Health Consequences of Dietary Fiber." Life Sciences Research Office, FASEB, p. 118. Bethesda, Md.

Prentice, R.L., Kakar, F., Hursting, S., *et al.* 1988. Aspects of the rationale for the Women's Health Trial. JNCI 80: 802.

Prentice, R.L., Pepe, M. and Self, S.G. 1989. Dietary fat and breast cancer: a quantitative assessment of the epidemiological literature and a discussion of methodological issues. Cancer Res. 49: 3147.

Schatzkin, A., Greenwald, P., Byar, D.P., *et al.* 1989. The dietary fat-breast cancer hypothesis is alive. JAMA 261(22): 3284.

Tannenbaum, A. and Silverstone, H. 1953. Nutrition in relation to cancer. Adv. Cancer Res. 1: 451.

Teutonico, R.A. and Knorr, D. 1985. Impact of biotechnology on nutritional quality of food plants. Food Tech. 39(10): 127.

Trock, B., Lanza, E. and Greenwald, P. 1990. Dietary fiber, vegetables, and colon cancer: critical review and meta-analyses of the epidemiologic evidence. JNCI 82(8): 650.

Wasserman, B.P., Montville, T.J. and Korwek, E.L. 1988. Food biotechnology: a scientific status summary by the Institute of Food Technologists' expert panel on food safety and nutrition. Food Tech. 42(1): 133.

Willett, W., Polk, R., Underwood, B.A., *et al.* 1984. Relation of serum vitamin A and E and carotenoids to the risk of cancer. N. Engl. J. Med. 310: 430.

Willett, W.C., Stampfer, M.J., Colditz, G.A., *et al.* 1987. Dietary fat and the risk of breast cancer. N. Engl. J. Med. 316: 22.

Wynder, E.L., MacCormick, F., Hill, P., *et al.* 1976. Nutrition and the etiology and prevention of breast cancer. Cancer Detect. Prev. 1: 293.

Ziegler, A.G. 1989. A review of epidemiologic evidence that carotenoids reduce the risk of cancer. J. Nutr. 119(1): 116.

Low Calorie Diets and Obesity

Richard L. Atkinson
Department of Internal Medicine
Eastern Virginia Medical School
Veterans Administration Medical Center
Hampton, Virginia 23667

Obesity is a multifactorial disease that is difficult to treat. Genetic factors play a major role in the etiology, as do alterations of both energy intake and energy expenditure. The fat content of the diet influences complications and response to treatment. Diets containing 400 to 1400 kcal per day may rapidly improve complications of obesity such as diabetes, hypertension, hyperlipidemia, and sleep apnea; although complications of dieting include gout, gallstones, cardiac arrhythmias, and sudden death. The use of very low calorie diets should be restricted to very obese patients and requires close follow up. It is clear that successful treatment of obesity requires a long term change in dietary and exercise habits and continued psychological support. More research is needed into the mechanisms of obesity to facilitate development of better treatment modalities.

Introduction

Obesity is a major public health problem in the United States, and is becoming a problem in the rest of the world. Unfortunately, many people, including physicians and scientists, have a general perception of obesity as a condition due to a failure of willpower and an inability to control eating. Investigators in the field have increasing evidence that obesity is a multifactorial disease of diverse etiologies characterized by an imbalance in the energy balance equation. Obesity may be caused by an increased intake of energy with normal expenditures, or decreased expenditures with a normal intake. The mechanisms of the imbalance that lead to excess energy storage in adipose tissue are unclear, but there is clear evidence that obesity has a strong genetic component. It is also quite clear that obesity is a serious disease associated with numerous complications, including heart disease, hypertension, diabetes mellitus, strokes, cancer, gall bladder disease, and sleep apnea. Finally, it is unfortunately clear that current methods of treatment of obesity are inadequate and most significantly obese people will not be able to achieve long term weight loss. This review will discuss the low and very low calorie diets that have been used extensively in recent years, and the use of behavioral modification of lifestyle to preserve this weight loss.

Obesity as a Disease

Debates have raged for many years about whether obese people eat more calories than do lean people. With the availability of modern respiratory chambers, studies in several laboratories have shown that, on average, obese people have a somewhat higher resting metabolic rate and total 24 hour energy expenditure (Jequier and Schutz, 1983; Zed and James, 1986; Bogardus *et al.*, 1986; Ravussin *et al.*, 1982). Since weights of these subjects were stable, the differences in energy expenditure must reflect differences in energy intake. These authors suggest that older studies showing that obese people ate less than lean are inaccurate. However, analysis of these studies suggests that the range of body weights for the same energy intake may vary widely (Jequier and Schutz, 1983; Zed and James, 1986; Bogardus *et al.*, 1986; Ravussin *et al.*, 1982; Ravussin *et al.*, 1988). Differences in body weight may be almost twofold for the same energy intake. This implies that there are differences in whole body energy expenditure, perhaps due to lower levels of activity in obese people. Adipose tissue has a relatively low metabolic rate, so if energy expenditure is expressed in terms of the lean body mass, obese and lean people are approximately the same (Segal *et al.*, 1987; Segal *et al.*, 1985). Ravussin *et al.* (1988) and Roberts *et al.* (1988) have shown that people with low levels of energy expenditure are more likely to gain weight than those with high energy expenditure. Stunkard *et al.* (1986, 1990) and Bouchard *et al.* (1990) have shown that there is a strong genetic component in the etiology of obesity. Children adopted as infants have body weights closer to those of their natural parents, rather than those of the adoptive parents (Stunkard *et al.*, 1986). With overfeeding, body weight and energy expenditure of identical twins correlate closer within pairs than between pairs (Bouchard *et al.*, 1990). These authors conclude that genetic factors may account for about 33 percent to about 70 percent of the occurrence of obesity (Stunkard *et al.*, 1986; Stunkard *et al.*, 1990; Bouchard *et al.*, 1990).

Complications of Obesity

Obesity is associated with numerous physiological and pathophysiological changes. Many of these pathophysiological changes lead to complications that are responsible for the morbidity and mortality of obesity (Van Itallie, 1979). The first three complications listed in Table 1 appear to share a common mechanism of insulin resistance for a significant part of their etiology.

Table 1. Complications of obesity

1. Diabetes mellitus
2. Hypertension
3. Hyperlipidemia
4. Cardiac disease: angina pectoris, congestive heart failure
5. Sleep apnea
6. Cerebrovascular accidents
7. Cancer (breast, colon, prostate)
8. Gall bladder disease
9. Pregnancy risks (toxemia)
10. Surgery risks (pneumonia, wound infection, thrombophlebitis)
11. Gout
12. Decreased fertility
13. Arthritis, early
14. Early mortality

Diabetes mellitus is one of the most common complications of obesity (Van Itallie, 1979). Diabetes appears when the capacity of the pancreas to secrete insulin is insufficient to overcome the increasing insulin resistance that occurs with weight gain. The mechanisms of the alterations in insulin and glucose metabolism are not well understood, but involve both receptor and postreceptor defects in many or all of the cells of the body sensitive to insulin (Amatruda *et al.*, 1975; Caro *et al.*, 1987). Insulin resistance correlates with increasing size of the fat cells (Salans *et al.*, 1968). Hypertension also occurs very commonly in obesity, and recent work suggests that insulin resistance plays a significant role (Ferrannini *et al.*, 1987). Insulin causes sodium retention by the kidney and may be in part responsible for the increased fluid volume in obesity (Ferrannini *et al.*, 1987; DeHaven *et al.*, 1982; Boulter *et al.*, 1973; Saudek *et al.*, 1973). Insulin resistance also may contribute to an increased sympathetic tone which contributes to the hypertension (Ferrannini *et al.*, 1987). Obesity and insulin resistance may cause an increase in plasma triglycerides, cholesterol, LDL, VLDL, and apolipoprotein B, while decreasing HDL and apolipoprotein A (Albrink *et al.*, 1980; Avogaro *et al.*, 1979; Zimmerman *et al.*, 1984). Diabetes, hypertension, and hyperlipidemia are major risk factors for the development of coronary artery disease (CAD) and myocardial infarctions. Obesity alone increases the risk for CAD, and the presence of one or more of these complications of obesity greatly increases the risk of CAD (Van Itallie, 1979). Fortunately, weight loss reduces or even eliminates all three of these major risk factors (Van Itallie, 1979; Avogaro *et al.*, 1979; Zimmerman *et al.*, 1984; Atkinson and Kaiser, 1985).

Sleep apnea is a little known but serious complication of obesity. The terminal stage of sleep apnea in obesity is the Pickwickian syndrome. Miller and Granada (1974) identified 10 patients hospitalized with Pickwickian syndrome and followed them for a period of 5 years. They found that 7 of the 10 died during the follow up period, indicating the gravity of this problem. Suratt *et al.* (1987) and Sugerman *et al.* (1986) noted rapid improvement in sleep apnea or obesity hypoventilation syndrome with weight loss following very low calorie dieting or gastric surgery for obesity.

The other complications of obesity listed in Table 1 may respond to weight loss, but the response is not as acute or dramatic as with the diseases discussed above.

Treatment of Obesity: Low Calorie Diets

The author arbitrarily defines a "low calorie diet" as one that provides between 10 and 20 Kcal per kg of "desirable" weight according to the 1959 Metropolitan Life Insurance Tables (1959). The substrates humans can utilize for fuel are limited to protein, carbohydrate, fat and ethyl alcohol. Alcohol is rarely prescribed during weight losing diets, thus all diets may consist of a reduction of one or more of the three main components of food. Hundreds of combinations of the three main nutrients have been devised as diets with catchy names designed to sell books, magazines, or more recently, videotapes. Diets may be divided into a few main categories listed below (Table 2).

Table 2. Classification of low calorie diets

1. Balanced diets
2. Reduction or elimination of one or more nutrients
3. Excessive amounts of one or more nutrients or food components
4. Formula diets
5. Miscellaneous and magic diets

Balanced Diets. These diets consist of a mixture of protein, carbohydrate (CHO), and fat in approximately the same ratios as consumed by the nondieting population. This type of diet probably is the most frequently used of all diets because it is self prescribed by the patient or prescribed by health professionals who are unfamiliar or uninterested in the nature of the disease, obesity. The standard advice is to "eat smaller portions." The rationale is that simple reduction of the amount of food eaten is the treatment of obesity. The success rate of balanced diets in the treatment of obesity is poor. Over 30 years ago, Stunkard and McLaren-Hume (1959) noted that few obese patients lose weight and of those who do, few keep it off. The success rate has improved somewhat due to better techniques of lifestyle modification but is still very poor with this type of diet. Three factors limit success in maintenance of long term weight loss. The first is difficulty in adherence to the diet. Americans are surrounded by food; and the stimulus to eat produced by environmental cues, social pressures, and habits is difficult to overcome. Development of behavioral modification techniques (Stuart, 1967; Stunkard, 1978; Penick *et al.*, 1971) has improved the success rate somewhat, but a high percentage of patients still fail (Stunkard, 1978). A second factor that limits success is the tendency of an obese individual to defend body weight at a level above desirable. Under equilibrium conditions body weight is maintained at a stable level. Keesey (1980, 1986) proposed a "set point" theory for the regulation of body weight. He postulates that restrictions in food intake of humans or animals under otherwise equilibrium conditions results in the activation of compensatory mechanisms that tend to minimize weight loss such as increased hunger, reduction of activity levels, and reduction of energy expenditure (Keesey, 1980; Keesey and Powley, 1986; Keys *et al.*, 1950). Increased hunger and reduced energy expenditure promote rapid weight gain when normal intake resumes, since a greater proportion of excess ingested calories can be stored. The third factor that limits usefulness of balanced low calorie diets is the fact that a high percentage of the calories comes from fat. The typical American diet consists of about 40-45 percent of calories from fat (Danforth, 1985). About one-third of these fat calories are "visible" fats such as butter, cooking oil, etc. The majority of fat calories are contained in the food and unless major changes are made in dietary patterns, the percent of fat calories in the diet remains high. There is evidence that dietary fat is processed more efficiently by the body than carbohydrate (Jequier *et al.*, 1987). Studies by Sims *et al.* (1973) and Danforth (1985) showed that overfeeding with fat results in a more rapid weight gain and greater efficiency of gain than overeating a higher carbohydrate diet. Although there is disagreement in the literature, a number of studies show dietary induced thermogenesis (the increase in energy expenditure after a meal) is less on a high fat diet than on a high carbohydrate diet (Jequier *et al.*, 1987; Flatt *et al.*, 1985; Welle *et al.*, 1981). We have preliminary data that suggest that pair feeding an equal number of calories of a high fat diet to the calorie intake of rats on a high carbohydrate diet still results in increased weight gain and increased body fat (Boozer *et al.*, 1990). The end result of these observations, coupled with the data on defense of body weight, suggests that obese people who try to maintain long term weight loss on a typical American balanced low calorie diet will need to eat significantly fewer calories than their thinner counterparts.

Reduction or Elimination of One or More Nutrients. Diets which eliminate nutrient categories generally are not useful for the long term, since a large number of food types are forbidden and patients will not adhere to these regimens. However, the novelty value of such diets may produce an early weight loss. The most drastic of these diets eliminates all but one type of food. The "one food" diets require great discipline on the part of the dieter. Usually the dieter may have an unlimited quantity of the one food, but boredom and sensory specific satiety (Rolls, 1986) act to limit total calorie intake, even if the one food was a preferred food of the patient (e.g., the "Ice Cream Diet"). One-food diets also may be effective, at least in the short term, because they require no decisions on the part of the patient. Perhaps the most famous of the one-food diets was the "Rice Diet" popularized by Kempner (1949). Originally intended as a low sodium diet for the treatment of hypertension, it later was used as a weight reduction diet. A major problem with the use of a single food is that some essential nutrients may be deficient. These diets are rarely prescribed by health professionals, with the exception of formula diets as noted below. Because one-food diets are not suitable for long term use, teach the patient little about changes in lifestyle, and have the potential for danger, their use should be discouraged.

Rarely, a fad diet recommends the elimination of protein. Since protein is an essential nutrient, such diets are dangerous and should never be used. Low CHO diets are by far the most popular "elimination" type diet, and each year new versions of a low CHO diet are popularized in the lay press. The original description of a low CHO diet was published by Banting in 1863. The repeated popularity of low CHO diets is due to the rapid weight loss early in the diet due to the diuresis of sodium and water (DeHaven *et al.*, 1982; Boulter *et al.*, 1973; Saudek *et al.*, 1973). As noted above, insulin is a sodium retaining hormone. When CHO in the diet is eliminated or reduced below about 50 gm per day, insulin levels fall to very low levels. This reduces sodium retention by the renal tubules, enhances sodium loss, and is accompanied by water loss. In addition, glucagon secretion is increased in an attempt by the body to preserve the levels of blood glucose. Glucagon is a natriuretic hormone (Saudek *et al.*, 1973). Glucagon also favors the production of ketones, which are excreted in the urine. The excretion of negatively charged ketone ions in the urine is accompanied by the excretion of cations, which tends to promote a diuresis (Boulter *et al.*, 1973; Saudek *et al.*, 1973). This combination of low sodium, high glucagon, and high ketones causes a large natriuresis and diuresis which results in large losses in weight in the initial stages of the ketogenic diet. This rapid weight loss is quite impressive to the patient and raises false hopes of quick achievement of desirable weight. Once a new steady state fluid balance is achieved, weight loss is dependent on the calorie deficit achieved by the diet, with only occasional fluctuations due to fluid shifts.

Ketogenic diets are also said to reduce the sensation of hunger (McLean *et al.*, 1974). Two controlled studies compared low calorie ketogenic vs. nonketogenic diets and found opposite results, so the question remains controversial (Rosen *et al.*, 1982; Waden *et al.*, 1985).

The use of low CHO diets for long term weight loss has not been well studied. In addition, the artificial nature of such diets and severe restriction of food choices may reduce long term compliance.

Perhaps the most sensible of the diets that reduce one type of nutrient are those that reduce the consumption of fat. For the reasons discussed above, high fat diets promote obesity, and reduction of fat in the diet leads to weight loss, usually because of the reduced total caloric load. The Surgeon General recommends reducing fat in the diet to 35 percent of total calories to help decrease heart disease. This is still a fairly high level of fat and is significantly higher than the diet humans have eaten over the millennia (Danforth, 1985). A reduction to about 20 percent of total calories as fat is achievable and compatible with reasonable palatability, but requires changes in eating habits which most Americans are not willing to make at this time. Levels of fat below 20 percent of total calories are generally not very palatable and rarely are followed for the long term.

Excessive Amounts of One or More Nutrients or Food Components. Diets which mandate eating large amounts of one nutrient or food component have the advantage that subjects do not have to feel deprived, but the boredom factor enters again; and compliance may not be optimal. The most common diets of this type are the high fiber diets. There are many studies in the literature showing that the addition of fiber to the diet results in weight loss. However, a careful examination of these studies reveals that most were poorly controlled or uncontrolled (Life Sciences Research Office, 1987). The few that were carefully controlled presented a mixed picture of success. We found that obese patients did not comply to a diet high in natural fiber, and there was no difference in weight loss from the low fiber control diet (Russ and Atkinson, 1985). We then conducted a trial with a fiber supplement and again found no difference in weight loss (Russ and Atkinson, 1986). On the other hand, numerous studies of vegetarians suggest that people who eat a diet high in fiber from foods rather than from supplements weigh less than do nonvegetarians (Life Sciences Research Office, 1987). The increased time required to eat a higher fiber meal and the caloric dilution from the fiber may result in fewer calories eaten at a given meal (Weinsier *et al.*, 1982). High fiber diets may also produce equal or greater meal induced satiety than low fiber diets (Weinsier *et al.*, 1982). The benefits to the gastrointestinal tract in reducing constipation and perhaps decreasing the incidence of diverticula of the large intestine (Life Sciences Research Office, 1987), coupled with the limitation of energy intake and high micro-nutrient level of most high fiber foods, are good reasons to recommend high fiber diets to most obese patients.

Formula Diets. Most currently used formula diets have a balanced composition of protein, carbohydrate, and fat. Formula diets may have excellent early success due to the removal of the necessity to make choices about foods and the boredom factor. However, non-compliance and dropout rates may be high (Miller *et al.*, 1978; Anderson *et al.*, 1983). Substitution of a liquid formula meal for one or two regular meals may be aceptable strategy for the longer term (Atkinson *et al.*, 1984). Many commercial formulas are available with a variety of compositions. Formulas such as Sustacal^{R} or Ensure^{R}, which were originally designed as food supplements or for tube feeding are reasonably palatable, but do not contain adequate protein, vitamins or minerals if used in a low calorie diet. To obtain the RDA for these nutrients requires a calorie intake of 1200-1500 Kcal per day. Formulas originally designed to be mixed with water for a very low calorie diet (Cambridge Diet^{R}, Columbia Diet^{R}, Optifast^{R}, etc.) contain the RDA of vitamins and minerals. They may be mixed with milk instead of water to provide a low calorie diet. Newer formulations such as Optifast 800^{R} and Metabase II^{R} are designed specifically for meal replacements and contain 50-70 gm per day of protein, 800 to 1000 Kcal, and the RDA of vitamins and minerals. Long term follow up with frequent visits for psychological support are necessary to obtain compliance with regimens that use formula diet as a replacement for one or more meals.

Miscellaneous and Magic Diets. Most of the diets in this category are not based on a scientific rationale. These diets promise easy weight loss with minimal effort and are usually promoted by unqualified or self-proclaimed "nutritionists." Many have scientifically based speculations for which there are no controlled experiments (e.g., the "Fit for Life Diet"). Other diets, such as the "Beverly Hills Diet," seem completely fanciful and have little or no accepted scientific basis. Many promise magical properties purported to "burn off fat." Mirkin and Shore (1981) warn of the hazards of such diets and Berland (1983) has published a summary of fad diets.

Such diets may be harmless, but prevent the patient from obtaining care from a trained professional. Diets found to be dangerous should be strongly condemned by health professionals.

Treatment of Obesity: Very Low Calorie Diets. The use of very low calorie diets (VLCD) is based on the principle that intake of predominantly protein foods containing very few calories diminishes the negative nitrogen balance associated with complete starvation, but results in weight loss nearly as great. VLCD were popularized by a series of articles in the early 1970's (Apfelbaum *et al.*, 1970; Blackburn *et al.*, 1973; McLean *et al.*, 1974; Genuth *et al.*, 1974). Protein intakes of as little as 30 gm per day were said to result in attainment of nitrogen balance (Howard, 1985). Publication of "The Last Chance Diet" by Linn (1976) in the lay press induced thousands of Americans to go on a diet based on hydrolysed collagen as a protein source. For reasons that are unclear, but may have to do with toxic properties of the protein, 60 people on this diet died of cardiac arrhythmias (Sours *et al.*, 1981; Isner *et al.*, 1979). Most of these diets did not contain adequate vitamins, electrolytes, or minerals, and many of the dead were not followed by a physician, so the exact etiologies of the deaths remain unknown. Subsequent investigations of VLCD have used high quality protein (casein or soy based), and adequate vitamins, minerals and 5Zr. Drenick (1987) reviewed the literature and concluded that both protein intake and energy intake are important in preserving body protein, but that most subjects will come into nitrogen balance after an initial period of nitrogen loss. The various studies disagree somewhat, but it appears that greater than 44 gm per day of high quality protein are needed for nitrogen balance. Intakes of 66 gm per day of good quality protein gave a less negative nitrogen balance than the lower levels of intake that were tested (Fisler *et al.*, 1987).

As noted above, rapid weight reduction results in rapid correction of complications of obesity such as diabetes mellitus, hypertension, hyperlipidemia, and sleep apnea. VLCD is an option for initial therapy in very obese individuals with these problems. Our studies found that in diabetics, blood sugar falls within 1 to 2 weeks in most patients (Atkinson and Kaiser, 1985). Hypertension also may respond in 1 week or less (DeHaven *et al.*, 1982; Atkinson and Kaiser, 1981). Sleep apnea may disappear within 2 weeks (Suratt *et al.*, 1987). However, VLCD are not without hazards (Sours *et al.*, 1981; Isner *et al.*, 1979; Atkinson and Kaiser, 1981; Wadden *et al.*, 1983; Fisler and Drenick, 1987). Table 3 lists a number of complications of these diets, the most serious of which is sudden death due to cardiac arrhythmia.

Table 3. Complications of very low calorie diets

1. Central - headache, difficulty concentrating
2. Cardiovascular - postural hypotension, cardiac arrhythmias, myocardial atrophy
3. Gastrointestinal - nausea, vomiting, constipation, diarrhea, abdominal discomfort, exacerbation of gall bladder disease
4. Genitourinary - menstrual irregularity, loss of libido, uric acid renal stones
5. General - lethargy, fatigue, loss of stamina, cold intolerance, halitosis, hunger, dry skin, hair loss, acute gout, negative nitrogen balance, mineral and electrolyte abnormalities
6. Sudden death

The risk of sudden death with modern formulas appears to be slight. We found no increase in arrhythmias in 24 patients undergoing a variety of stress tests while on VLCD for 6 weeks (Moyer *et al.* 8, 1989). Other complications of VLCD are generally benign, with the exception of precipitation of acute gout in patients with a prior history of gout. Such patients must be placed on drug treatment before starting a VLCD. Also, VLCD, as with any weight loss regimen, can precipitate gall stones and cholecystitis (Bennion and Grundy, 1975; Broomfield *et al.*, 1988). The mechanisms of gall bladder disease with weight loss appear to be increased cholesterol turnover causing sludging of the bile and gall stone formation, and appearance in the bile of prostaglandin E2 and a glycoprotein, both of which favor precipitation of cholesterol (Broomfield *et al.*, 1988).

The potential for complications mandates that a thorough history and physical examination be performed on all patients before starting VLCD. VLCD are contraindicated in patients with malignant arrhythmias, unstable angina, major systemic disease, treatment with potassium-losing diuretics, and body weight less than 20 percent over desirable. Patients with diabetes and hypertension should be followed particularly closely. Patients on potassium-wasting diuretics should be taken off before starting VLCD, and insulin or oral hypoglycemic dosages should be reduced by half or discontinued in diabetic patients going on a VLCD. Blood glucose levels should be followed closely in patients previously on insulin or oral agents, preferably with outpatient self-glucose monitoring, and drug therapy restarted if glucose levels do not come under control. Hyperglycemia with ketosis is a contraindication to initiation of a VLCD. The ketosis should be corrected before starting the VLCD as such patients may be prone to develop diabetic ketoacidosis on VLCD. All patients should be seen at least every 2 weeks. There is disagreement about the need for routine blood analyses for electrolytes, renal and liver function tests, and for electrocardiograms. Symptoms suggestive of an arrhythmia should alert the physician to examine the patient carefully, consider a Holter monitor, or consider terminating the VLCD. The author recommends that patients remain on VLCD for no longer than 12-16 weeks, but other investigators allow patients to remain on VLCD for much longer periods of time.

In summary, health professionals treating obese patients should prescribe lower fat diets (20-30 percent of total calories as fat) and provide long term access to a comprehensive program consisting of psychological support, nutrition counseling, behavioral modification techniques, and exercise therapy (Atkinson *et al.*, 1984). For rapid weight loss in very obese patients, a VLCD may be used initially.

However, without a commitment of both patient and therapist to long term treatment, success is not likely.

References

Albrink, M.J., Krauss, R.M., Lindgren, F.T., von der Graeben, J., Pan, S. and Wood, P.D. 1980. Intercorrelations among plasma high density lipoprotein, obesity and triglycerides in a normal population. Lipids 14: JH668-676.

Amatruda, J.M., Livingston, J.N. and Lockwood, D.H. 1975. Insulin receptor: role in the resistance of human obesity to insulin. Science 188: 264-266.

Andersen, T., Hyldstrup, L. and Quaade, F. 1983. Formula diet in the treatment of moderate obesity: Int. J. Obesity 7: 423-430.

Apfelbaum, M., Boudon, P., Lacatis, D. and Nillus, P. 1970. Effects metaboliques de la diete proteique chez 41 subjets obeses. La Presse Medicale 78: 1917-1920.

Atkinson, R.L. and Kaiser, D.L. 1985. Effects of calorie restriction and weight loss on glucose and insulin levels in obese humans. J. AM. Coll. Nutr. 4: 411-419.

Atkinson, R.L., Russ, C.S., Ciavarella, P.A., Owsley, E.S. and Bibbs, M.L. 1984. A comprehensive approach to outpatient obesity management. J. Am. Diet Assoc. 84: 439-444.

Atkinson, R.L. and Kaiser, D.L. 1981. Nonphysician supervision of a very low-calorie diet: results in over 200 cases. Int. J. Obesity W: 237-241.

Avogaro, P., Cazzolato, G., Bittolo Bon G., Quinci, B.G. 1979. Variations of plasma lipoproteins and apolipoproteins B and AI in obese subjects fed with hypocaloric diet. Obesity/Bariatric Med. 8: 158-161.

Banting, W. 1863. Letter on corpulence. Addressed to the public. John Ebers, London, Harrison.

Bennion, L.J. and Grundy, S.M. 1975. Effects of obesity and caloric intake on biliary lipid metabolism in man. J. Clin. Invest. 56: 996-1011.

Berland, T. 1983. "Consumer Guide: Rating the Diets." Signet, New York.

Blackburn, G.L., Flatt, J.P., Cloves, G.H.A. *et. al.* 1973. Protein sparing therapy during period of starvation with sepsis or trauma. Ann. Surg. 177: 588-594.

Bogardus, C., Lillioja, S., Ravussin E. *et al.* 1986. Familial dependence of the resting metabolic rate. N. Eng. J. Med. 315(2): 96-100.

Boozer, C.N., Schoenbach, G. and Atkinson, R.L. 1990. High fat diets fed isocalorically promote increases in rat fad pad weights, despite similar body weights. FASEB J. 4: A511.

Bouchard, C., Tremblay, A., Depres, J.P., Nadeau, A., Lupien, P.J., Theriault, G., Dussault, J., Moorjani, S., Pinault S. and Fournier, G. 1990. The response to long-term overfeeding in identical twins. N. Engl. J. Med. 322: 1477-1482.

Boulter, P.R., Hoffman, R.S. and Arky, R.A. 1973. Pattern of sodium excretion accompanying starvation. Metabolism 22: 675-683.

Broomfield, P.H., Chopra, R., Sheinbaum, R.C., Bonorris, G.G., Silverman, A., Schoenfield, L.J. and Marks. J.W. 1988. Effects of ursodeoxycholic acid and aspirin on the formation of lithogenic bile and gallstones during loss of weight. N. Eng. J. Med. 319: 1567 1572.

Caro, J.F., Sinha, M.K., Raju, S.N., Ittoop, O., Pories, W.J., Flickinger, E.G., Meelheim, D. and Dohm, G.L. 1987. Insulin receptor kinase in human skeletal muscle from obese subjects with and without noninsulin dependent diabetes. J. Clin. Invest. 79: 1330-1337.

Danforth, E. Jr. 1985. Diet and obesity. Am. J. Clin. Nutr. 41: 1132-1145.

DeHaven, I., Sherwin, R., Hendler, R. and Felig, P. 1982. Nitrogen and sodium balance and sympathetic-nervous-system activity in obese subjects treated with a low-calorie protein or mixed diet. New Engl. J. Med. IG302: 477-482.

Ferrannini, E., Buzzigoli, G., Bonadonna, R., Giorico, M.A., Olegini, M., Graziadei, L., Pedrinelli, R., Brandi, L. and Bevilacqua, S. 1987. Insulin resistance in essential hypertension. N. Engl. J. Med. 317: 350-357.

Fisler, J.S. and Drenick, E.J. 1987. Starvation and semistarvation diets in the management of obesity. Ann. Rev. Nutr. 7: 465-484.

Flatt, J.P., Ravussin, E., Acheson, K.J. and Jequier, E. 1985. Effects of dietary fat on postprandial substrate oxidation and on carbohydrate and fat balances. J. Clin. Invest. 76: 1019-1024.

Genuth, S.M., Castro, H.J. and Vertes, V. 1974. Weight reduction in obesity by outpatient semistarvation. JAMA 230: 987-991.

Howard, A.N. 1985. The development of a very low calorie diet: a historical perspective. In "Management of Obesity by Severe Caloric Restriction," Blackburn, G.L. and Gray, G.A., Eds. p. 3-20. PSG Publishing, Littleton, MA.

Isner, J.M., Sours, H.E., Paris, A.L. *et al.* 1979. Sudden, unexpected death in avid dieters using the liquid-protein-modified fast diet: observations in 17 patients and the role of the prolonged QT interval. Circulation 60: 1401-1412.

Jequier, E., Acheson, K. and Schutz, Y. 1987. Assessment of energy expenditure and fuel utilization in man. Ann. Rev. Nutr. 7: 187-208.

Jequier, E. and Schutz, Y. 1983. Long term measurements of energy expenditure in humans using a respiration chamber. Am. J. Clin. Nutr. 38: 989-998.

Keesey, R.E. and Powley, T.L. 1986. The regulation of body weight. Ann. Rev. Psychol. 37: 109-133.

Keesey, R.E. 1980. A set point analysis of the regulation of body weight. In "Obesity," Stunkard, A.J., Ed. p. 144-165. W.B. Saunders Co., Philadelphia, Pa.

Kempner, W. 1949. Treatment of heart and kidney disease and of hypertensive and arteriosclerotic vascular disease with rice diet. Ann. Intern. Med. 31: 821-856.

Keys, A., Brozek, J., Henschel, A. *et. al.* 1950. "The Biology of Human Starvation," Vols. 1 and 2. University of Minnesota Press, Minneapolis, Mn.

Life Sciences Research Office. 1987. Physiological effects and health consequences of dietary fiber. Federation of American Societies for Experimental Biology, Contract Number FDA 223-84-2059, p. 160. Washington, D.C.

Linn, R. and Stuart, S.L. 1976. "The Last Chance Diet." Lyle Stuart, Secaucus, NJ.

McLean, B.I., Parsons, R.L. and Howard, A.N. 1974. Clinical and metabolic studies of chemically defined diets in the management of obesity. Metabolism 23: 645-657.

Metropolitan Life Insurance Co. Statistical Bulletin. 1959. 40: 1-4.

Millar, J.W.U.S., Innes, J.A. and Munro, J.F. 1978. An evaluation of the efficacy and acceptability of "Slender" in refractory obesity. Int. J. Obesity 2: 53-58.

Miller, A. and Granada, M. 1974. In-hospital mortality in the Pickwickian syndrome. Am. J. Med. 56: 144-150.

Mirkin, G.B. and Shore, R.N. 1981. The Beverly Hills Diet, Managers of the newest weight loss fad. JAMA 246: 2235-2237.

Moyer, C.L., Holly, R.G., Amsterdam, E.A. and Atkinson, R.L. 1989. The effects of cardiac stress during a very low calorie diet and exercise program in obese women. Am. J. Clin. Nutr. 50: 1324-1327.

Penick, S.B., Filion, R., Fox, S. and Stunkard, A.J. 1971. Behavior modification in the treatment of obesity. Psychom. Med. 33: 49-55.

Ravussin, E., Lillioja, S., Knowler, W.C., Christin, L., Freymond, D., Abbott, W.G.H., Boyce, V., Howard, B.V. and Bogardus, C. 1988. Reduced rate of energy expenditure as a risk factor for body-weight gain. N. Engl. J. Med. 318: 467-472.

Ravussin, E., Acheson, K.J., Vernet, O., Danforth, E. and Jequier, E. 1985. Evidence that insulin resistance is responsible for the decreased thermic effect of glucose in human obesity. J. Clin. Invest. 76: 1268-1273.

Ravussin, E., Burnand, B., Schutz, Y. and Jequier, E. 1982. Twenty-four-hour energy expenditure and resting metabolic rate in obese, moderately obese, and control subjects. Am. J. Clin. Nutr. 35: 566-573.

Roberts, S.B., Savage, J., Coward, W.A., Chew, B. and Lucas, A. 1988. Energy expenditure and intake in infants born to lean and overweight mothers. N. Engl. J. Med. 318: 461-466.

Rolls, B.J. 1986. Sensory-specific satiety. Nutr. Rev. 44: 93-101.

Rosen, J.C., Hunt, D.A., Sims, E.A.H. and Bogardus, C. 1982. Comparison of carbohydrate-containing and carbohydrate-restricted hypocaloric diets in the treatment of obesity: effects on appetite and mood. Am. J. Clin. Nutr. 36: 463-469.

Russ, C.S. and Atkinson, R.L. 1986. No effect of dietary fiber on weight loss in obesity. Am. J. Clin. Nutr. 43: 136A.

Russ, TRC. S. and Atkinson, R.L. 1985. Use of high fiber diets for the outpatient treatment of obesity. Nutr. Rep. Inst. 32: 193-198.

Salans, L.B., Knittle, J.L. and Hirsch, J. 1968. The role of adipose cell size and adipose tissue insulin sensitivity in the carbohydrate intolerane of human obesity. J. Clin. Invest. 47: 153-165.

Saudek, C.D., Boulter, P.R. and Arky, A.R. 1973. The natriuretic effect of glucagon and its role in starvation. J. Clin. Endocrinol. Metab. 36: 761-765.

Segal, K.R., Gutin, B., Albu, J. *et. al.* 1987. Thermic effects of food and exercise in lean and obese men of similar lean body mass. Am. J. Physiol. 252: (Endocrinol Metab.) 15: E110-E117.

Segal, K.R., Gutin, B., Nyman, A.M. *et. al.* 1985. Thermic effects of food at rest, during exercise, and after exercise in lean and obese men of similar body weight. J. Clin. Invest. 76: 1107-1112.

Sims, E.A.H., Danforth, E. Jr., Horton, E.S. *et. al.* 1973. Endocrine and metabolic effects of experimental obesity in man. Recent Prog. Horm. Res. 29: 457-496.

Sours, H.E., Frattali, VXV.P., Brand, C.D. *et al.* 1981. Sudden death associated with very low calorie weight reduction regimens. Am. J. Clin. Nutr. 34: 453-461.

Stuart, R.B. 1967. Behavioral control of overeating. Behav. Res. Ther. 5: 357-365.

Stunkard, A.J., Harris, J.R., Pedersen, N.L. and McLearn, G.E. 1990. The body-mass index of twins who have been reared apart. N. Engl. J. Med. 322: 1483-1487.

Stunkard, A., Sorensen, T.I.A., Hanis, C., Teasdale, T.W., Chakraborty, R., Schull, W.K. and Schulsinger, F. 1986. An adoption study of human obesity. N. Engl. J. Med. 314: 193-198.

Stunkard, A.J. 1978. Behavioral treatment of obesity: the current status. Int. J. Obesity 2: 237-249.

Stunkard, A.J. and McClaren-Hume, M. 1959. The results of treatment for obesity. Arch. Int. Med. 103: 79-85.
Sugerman, H.J., Fairman, R.P., Baron, P.L. and Qwentus, J.A. 1986. Gastric surgery for respiratory insufficiency of obesity. Chest 90: 81-86.
Suratt, P.M., McTier, R.F., Findley, L.J., Pohl, S.L. and Wilhoit, S.C. 1987. Changes in breathing and the pharynx after weight loss in obstructive sleep apnea. Chest 92: 631-637.
Van Itallie, T.B. 1979. Obesity: adverse effects on health and longevity. Am. J. Clin. Nutr. 32: 2723-2733.
Wadden, T.A., Stunkard, A.J., Brownell, K.D. and Day, S.C. 1985. A comparison of two very-low-calorie diets: protein-sparing-modified fast versus protein-formula-liquid diet. Am. J. Clin. Nutr. 41: 533-539.
Wadden, T.A., Stunkard, A.J. and Brownell, K.D. 1983. Very low calorie diets: their efficacy, safety, and future. Ann. Intern. Med. 99: 675-684.
Weinsier, R.L., Johnston, M.H., Doleys, D.M. and Bacon, J.A. 1982. Dietary management of obesity: evaluation of the time-energy displacement diet in terms of its efficacy and nutritional adequacy for long-term weight control. Br. J. Nutr. 47: 367-379.
Welle, S. Lilavivat, U. and Campbell, R.G. 1981. Thermic effect of feeding in man: increased plasma norepinephrine levels following glucose but not protein or fat consumption. Metabolism 30: 953-958.
Zed, C. and James, W.P.T. 1986. Dietary thermogenesis in obesity: fat feeding at different energy intakes. Int. J. Obesity 10: 375-390.
Zimmerman, J., Kaufman, N.A., Fainaru, M., Eisenberg, S., Oschry, Y., Friedlander, Y. and Stein Y. 1984. Effect of weight loss in moderate obesity on plasma lipoprotein and apolipoprotein levels and on high density lipoprotein composition. Arteriosclerosis 4: 115-123.

Bioengineering of Meat

Morse B. Solomon
U.S. Department of Agriculture
Agricultural Research Service
Beltsville Agricultural Research Center
Meat Science Research Laboratory
Beltsville, MD 20705

Americans are inundated with warnings about the health risks of consuming certain types/classes of foods. A recent report from the National Research Council, funded by the U.S. Department of Agriculture, cites the recommendations of both medical and health professionals who urge reduced consumption of dietary fat; particularly that of animal origin. Furthermore, consumers are becoming more health and weight conscious with a growing preference for leaner meat. In the search to offer consumers foods which are lower in fat content, new technologies and production methods appear to hold promise for improving the nutritional attributes of animal products (meat).

Potential for manipulation of growth of farm animals has never been greater than at present. The confirmation of the growth-promoting and repartitioning effects of somatotropin, somatomedins, select β-adrenergic agonists, immunization of animals against target circulating hormones or releasing factors and gene manipulation techniques offer a wider range of strategies than ever before available. Eating quality and safety of meats must not be sacrificed as leaner animals are developed. Animals must go to market at their "optimal slaughter potential" with just enough fat for optimal carcass and eating quality. It is clear that meeting or approaching recommendations by medical and health professionals, without sacrificing palatability/quality, will increase consumer acceptance of red meat, and research towards this end is of paramount importance.

Introduction

Americans are inundated with warnings about the health risks of consuming certain types/classes of foods. Approximately 36 percent of the total food energy and between 36 and 100 percent of each of the major nutrients in the food supply come from animal products. In a recent publication of a 3-year study by the National Research Council (1988), funded by the U.S. Department of Agriculture, animal products were stated to contribute more than half the total fat, more than one-third of the calories, nearly three-fourths of the saturated fatty acids, and all the cholesterol that is suggested to adversely affect human health (e.g., increase risk of heart disease and cancer). In 1988, the Surgeon General's Report on Diet and Health was released, which stated that "the primary priority for dietary change is the recommendation to reduce intake of total fats, especially saturated fat, because of their relationship to the development of several chronic disease conditions" (USDHHS, 1988).

Americans consume about 37 percent of their calories from fat - far more than the 30 percent or less recommended by most health experts. Specific target levels for caloric intake and nutrients in the diet have been suggested by major national health organizations (NRC, 1988). Recommendations include: (1) thirty percent or less of total calories from fat; (2) saturated fat intake should be less than 10 percent of calories; (3) ten percent or less of these calories from polyunsaturated fat; (4) fifteen percent or less of these calories from monounsaturated fats; (5) total cholesterol intake should be less than 100 mg/1000 cal, not to exceed 300 mg/day.

U.S. consumers are becoming more health and weight conscious and desire fewer calories in their diet. In fact, the '80s were considered the decade of "nutrition." Therefore, they prefer products which contain less fat as a means of limiting their caloric intake. In the search to offer consumers foods which are lower in fat content, new technologies and production methods appear to hold promise for improving the nutritional attributes of animal products.

A wide range of strategies for altering the balance between lean and adipose tissue growth in meat-producing animals is available. These include genetic selection, management (production) strategies (e.g., intact males, limit feeding, lighter slaughter weights). More recently the confirmation of the growth-promoting and repartitioning effects of somatotropin, somatomedins, select β-adrenergic agonists, immunization of animals against target circulating hormones or releasing factors and gene manipulation techniques have given rise to a technological revolution for altering growth and development. This paper will discuss research at USDA-ARS-Beltsville where objectives were to lower and modify (alter) the fat content of red-meat-producing animals.

Genetic Selection and Management Strategies for Ruminants

Research in progress has shown that the lean yield (fat/lean ratio) of lambs can be improved by the use of young rams fed diets with a high alfalfa (roughage = 77 percent) content (Solomon *et al.*, 1986;

Solomon *et al.*, 1988; Lynch and Solomon, 1989). Evaluation of lipid deposition and metabolism of these young growing ruminants may help reduce the quantity of specific hyperlipidemic and hypercholesteremic fatty acids in the edible tissues of the carcass. With this in mind, we found it necessary to obtain data on the comparison of young ram (intact) lambs with wether (castrates) and cryptorchid (short scrotum) lambs (Solomon *et al.*, 1990a) of the same age (6 months) and fed a similar diet (high roughage (alfalfa)); low concentrate). Results in Table 1 show that ram lambs were the leanest with the least subcutaneous fat cover (55 percent less than the castrates) and also had 22 percent less intramuscular fat than wethers. A very interesting and important difference was that the intramuscular fat from ram lambs contained 36 percent more polyunsaturated fatty acids (PUFA) than wethers. When based on 100g serving of meat, the consumer would consume 22 percent less saturated fat (SFA) from these lean ram lambs than from the traditional castrates. Meat from wethers is the primary type of lamb marketed in this country. However, results indicate that meat from ram lambs is less tender (28 percent) than meat from wethers.

Table 1. Lipid composition of longissimus lean and subcutaneous fat by sex type.

Component	Ram	Cryptorchid	Wether	SEM
Subcutaneous fat thickness, cm	.26^{b}	.36^{B}	.56^{a}	.03
W-B shear force, kg	6.66^{a}	5.72^{b}	5.19^{c}	.31
Longissimus muscle				
Total lipid, g/100g	3.82^{b}	4.70^{ab}	4.91^{c}	.32
Total SFA^{a}, %	45.37	45.77	45.09	.60
Total MUFA, %	41.75^{b}	42.18^{ab}	44.92^{a}	.59
Total PUFA, %	7.06^{a}	6.24^{ab}	5.21^{b}	.40
Saturated fat^{a}/100g	1.73^{b}	2.15^{ab}	2.21^{a}	.19
Cholesterol, mg/100g	64.18^{b}	69.94^{a}	65.66^{b}	2.10
Subcutaneous fat				
Total lipid, g/100g	62.46^{b}	64.91^{b}	76.29^{a}	3.95
Total SFA^{a}, %	43.30^{b}	46.80^{b}	51.38^{a}	1.63
Total MUFA, %	36.69	36.32	36.6	.67
Total PUFA, %	4.07^{a}	4.35^{ab}	3.60^{b}	.27
Saturated fat^{a}/100g	26.97^{c}	30.58^{b}	39.21^{a}	.71
Cholesterol, mg/100g	79.51^{a}	74.70^{ab}	70.84^{b}	4.73

^{a,b,c}Means with different superscripts within a row are different ($P<.01$).

In recent studies in our laboratory, we attempted to alter, through diet modifications, the lipid composition of carcass tissues. Because ruminant animals convert unsaturated fatty acids in their diet into SFA by the action of rumen microorganisms, it is difficult to modify carcass fat composition. In the first experiment (Solomon *et al.*, 1989) we determined the effect of alternate sources of dietary protein supplementation for lambs fed a high roughage diet. Three different dietary sources for protein supplementation were used: rapeseed meal, whole rapeseed and soybean meal. Diets contained 75 percent alfalfa, 14 percent crude protein and 2.0 Mcal ME/kg and were fed for 35 days. The ram lambs averaged 46 kg body and 5.5 months of age at the onset of the experiment. Lean tissue and corresponding subcutaneous fat from three different locations on the carcass were evaluated. These locations represented: leg (semimembranosus muscle), loin (longissimus muscle) and shoulder (triceps brachii muscle). Results (Tables 2, 3 and 4) indicated that lean tissue cholesterol content was highest for lambs fed soybean meal and lowest for lambs fed rapeseed meal, regardless of muscle location. Palmitic acid content was highest in the lean tissue of lambs fed rapeseed meal compared to those fed either whole rapeseed or soybean meal. Palmitic acid has been reported to be hyperlipidemic and thus may increase serum cholesterol levels (Keys *et al.*, 1965). There were no significant differences for total intramuscular fat content or total SFA, monounsaturated fatty acids (MUFA) or PUFA of the lean as a result of the different dietary protein supplements. In the subcutaneous fat, lambs fed rapeseed meal had the least cholesterol content. No difference was observed for palmitic acid content; however, a substantial effect on subcutaneous stearic acid was observed. Whole rapeseed supplementation resulted in more stearic acid compared to the two sources of meal used. On the contrary, rapeseed meal tended to yield the least stearic acid. Lipid deposition/absorption did not follow similar patterns at the different carcass locations evaluated.

In the second experiment (Table 5), we determined the influence of a dietary fat supplement high in palm oil (40 percent) and sex type on lipid composition (Lough *et al.*, 1990). Lambs averaged 36 kg body weight at the onset of the experiment and were fed ad libitum for 60 days. The longissimus muscle and corresponding subcutaneous adipose tissue were used for chemical analysis. Lambs fed palm oil were (38 percent) fatter than controls and contained more SFA as a result of an increase in palmitic acid. Diet had no

Table 2. Lipid composition of lamb[a] longissimus lean and subcutaneous fat fed alternate dietary protein sources.

Component	Lean tissue longissimus (LM)[b]				Subcutaneous fat over LM[b]			
	Rapeseed Meal	Whole Rapeseed	Soybean Meal	SEM	Rapeseed Meal	Whole Rapeseed	Soybean Meal	SEM
Total lipid, g/100g	4.97	5.15	4.85	.57	74.20^{cd}	77.04^{c}	69.93^{d}	3.78
Total SFA, %	41.12	40.96	41.09	.68	38.77^{cd}	41.18^{c}	37.90^{d}	2.18
Total MUFA, %	43.49	43.76	43.48	.73	39.99^{d}	42.01^{c}	40.54^{cd}	3.31
Total PUFA, %	6.05	5.39	5.88	.53	3.57	3.74	3.79	.38
Cholesterol, mg/100g	70.4^{d}	74.7^{cd}	77.2^{c}	4.0	72.6^{e}	83.4^{d}	89.7^{c}	5.4
Palmitic acid, %	24.14^{c}	22.05^{d}	22.71^{d}	.41	21.48	20.65	20.29	1.11
Stearic acid, %	15.08^{d}	17.17^{c}	16.58^{cd}	.69	9.17^{d}	13.78^{d}	9.75^{d}	1.55

[a]Ram lambs.
[b]LM= loin.
[c,d,e]Means with different superscripts within a row by group are different (P< .01).

Table 3. Lipid composition of lamb[a] semimembranosus lean and subcutaneous fat fed alternate dietary protein sources.

Component	Lean tissue semimembranosus (SM)[b]				Subcutaneous fat over SM[b]			
	Rapeseed Mean	Whole Rapeseed	Soybean Mean	SEM	Rapeseed Meal	Whole Rapeseed	Soybean Meal	SEM
Total lipid, g/100g	4.02	4.06	4.41	.32	79.80^{c}	76.78^{d}	78.23^{cd}	2.59
Total SFA, %	38.80	39.32	38.78	.92	37.21^{d}	43.44^{c}	39.15^{d}	3.03
Total MUFA, %	43.54	43.79	43.15	.97	40.41	42.93	42.64	3.52
Total PUFA, %	7.20	6.46	6.96	.58	3.72	3.40	4.41	.35
Cholesterol, mg/100g	71.7^{d}	74.2^{cd}	78.1^{c}	3.3	72.2^{d}	83.0^{c}	82.2^{c}	4.7
Palmitic acid, %	22.75	21.30	21.38	.32	21.69	20.28	20.83	1.41
Stearic acid, %	14.27^{d}	16.30^{c}	15.74^{cd}	.74	7.39^{d}	16.69^{c}	10.64^{d}	2.53

[a]Ram lambs. [b]SM = leg. [c,d]Means with different superscripts within a row by group are different (P< .01).

Table 4. Lipid composition of lamb[a] triceps brachii lean and subcutaneous fat fed alternate dietary protein sources.

Component	Lean tissue triceps brachii (TB)[b]				Subcutaneous fat over TB[b]			
	Rapeseed Meal	Whole Rapeseed	Soybean Meal	SEM	Rapeseed Meal	Whole Rapeseed	Soybean Meal	SEM
Total lipid, g/100g	4.66	5.36	5.13	.43	57.88[d]	61.68[cd]	63.85[c]	5.29
Total SFA, %	37.96	39.16	38.09	.79	38.70[d]	45.73[c]	44.22[c]	2.59
Total MUFA, %	43.82	44.78	45.21	.91	42.88	41.72	42.99	3.44
Total PUFA, %	6.79	6.08	6.62	.51	3.38	3.36	4.30	.38
Cholesterol, mg/100g	71.0[d]	71.9[d]	76.3[c]	4.4	85.3[d]	95.3[c]	86.4[d]	5.8
Palmitic acid, %	22.32[c]	21.30[d]	21.27[d]	.23	23.06	21.81	23.31	.97
Stearic acid, %	13.75[d]	15.92[c]	15.00[cd]	.75	8.19[e]	17.61[c]	13.86[d]	2.28

[a]Ram lambs.
[b]TB = shoulder.
[c,d,e]Means with different superscripts within a row by group are different ($P < .01$).

effect on cholesterol content of lean tissue. However, the subcutaneous fat from lambs fed palm oil contained less cholesterol (64.8 vs 89.7 mg/100g) compared to lambs without the palm oil supplementation. Ewes were fatter (63 percent) than rams and contained more MUFA in their lean tissue. Rams had a greater cholesterol content in the subcutaneous adipose tissue than ewes. Sex type had no significant effect on cholesterol content of lean tissue. These results indicate that high amounts of dietary palm oil fed to growing lambs caused changes in fat deposition and cholesterol metabolism and would be a useful investigative tool to study lipid metabolism in growing ruminants.

Porcine Somatotropin (pST) Administration

With greater emphasis on lean tissue deposition and less lipid, the optimal genetic potential for protein deposition of an animal is a very important concept in that this potential, or ceiling, defines the protein requirement of the animal. Underlying efforts at the Beltsville Agricultural Research Center was the thought of defining the optimal/genetic potential for protein deposition. Somatotropin (growth hormone) was used as a tool, to maximize genetic potential for protein accretion. Table 6 summarizes the lipid composition results of pST administration (100 μg/kg body weight) to pigs (barrows) for 33 days (Solomon *et al.*, 1990b). The lipid content from pST treated pigs was 27 percent less in the lean tissue and 23 percent less in the subcutaneous fat compared to controls. The administration of pST resulted in lean tissue containing 40 percent less SFA, 37 percent less MUFA and no difference in PUFA compared to controls. The subcutaneous fat from pST treated pigs contained 33 percent less SFA, 24 percent less MUFA and 9 percent less PUFA than controls. There was no difference in cholesterol content in lean tissue from pST treated pigs compared to controls. However, cholesterol content in the subcutaneous fat from pST pigs was higher (88.8 vs 79.4 mg/100g) than from controls. These results indicate that significant reductions in total lipid and all three classes of fatty acids can be achieved using pST. This represents a favorable change in regard to human dietary guidelines.

Transgenic Pigs Expressing a Bovine Growth Hormone Gene

The technology for introducing recombinant genes into laboratory animals has been available only since 1980 (reviewed by Brinster and Palmiter, 1986). Gene transfer into farm animals is

Table 5. Lipid composition of lean and subcutaneous fat from ewe and ram lambs fed a diet high in palm oil.

Component	Treatments[a]				SEM	Significance		
	Ewe-NPO	EWE-PO	Ram-NPO	Ram-PO		Sex (S)	Diet (D)	S*D
				Lean tissue				
Total lipid, g/100g	4.14	3.98	2.95	3.82	.32	.05	.27	.12
Total SFA, %	34.07	36.76	33.15	37.21	.86	.79	.01	.43
Total MUFA, %	35.82	36.48	30.99	33.22	1.14	.01	.22	.50
Total PUFA, %	7.29	6.40	7.17	6.91	.41	.64	.17	.44
Cholesterol, mg/100g	68.2	67.2	64.7	66.6	1.14	.09	.69	.22
Palmitic acid, %	19.05	21.68	17.69	21.86	.51	.25	.01	.14
				Subcutaneous adipose tissue				
Total lipid, g/100g	85.87	88.54	76.66	81.32	2.03	.01	.08	.63
Total SFA, %	44.19	49.82	41.72	49.35	1.36	.29	.01	.47
Total MUFA, %	40.67	39.66	39.99	40.15	.77	.91	.59	.46
Total PUFA, %	4.95	3.30	3.96	3.49	.25	.12	.01	.03
Cholesterol, mg/100g	80.6	56.9	98.8	72.7	4.05	.01	.01	.78
Palmitic acid, %	22.33	27.65	22.31	29.40	.68	.21	.01	.20

[a]NPO = no palm oil; PO = palm oil. Diet consisted of 77% alfalfa, 23% concentrate, 13.5% CP and 2.15 Mcal ME/kg.

Table 6. Lipid composition of porcine longissimus lean and subcutaneous fat by pST treatment.

Component	Lean tissue (longissimus)		Subcutaneous fat over longissimus	
	0	100[c]	0	100
Total lipid, g/100g	4.05[a]	2.95[b]	78.40[a]	60.58[b]
Total SFA, g/100g	1.35[a]	.81[b]	28.33	18.88
Total MUFA, g/100g	1.45[a]	.91[b]	27.79	21.17
Total PUFA, g/100g	.44	.45	13.87	12.68
PUFA/SFA	.33[b]	.56[a]	.49[b]	.67[a]
Cholesterol, mg/100g	57.8	57.2	79.4[b]	88.8[a]

[a,b]Means with different superscripts within a row and within tissue type are different ($P < .05$).
[c]pST administration : 0 = Control; 100 υg/kg/body weight = Treated.

Table 7. Lipid composition of carcass ground tissue from transgenic pigs.

Component	Weight group (W), kg								Significance[b]	
	Transgenic[a] (T)				Control					
	14	26	48	88	14	26	48	88	T	W
Total lipid, g/100g	6.19	7.62	7.54	3.27	10.03	12.32	15.50	19.55	*	*
Total SFA, g/100g	1.90	2.19	2.26	.81	3.17	3.64	4.92	6.59	*	*
Total MUFA, g/100g	1.99	2.47	2.35	.81	3.74	4.69	6.48	7.73	*	*
Total PUFA, g/100g	1.08	1.81	1.75	.93	1.50	2.38	2.54	2.51	*	*
PUFA/SFA	.57	.83	.77	1.23	.47	.65	.52	.38	*	*
Cholesterol, mg/100g	106.5	100.0	85.6	75.5	100.9	95.0	85.6	75.1	NS	*

[a]Transgenic pigs expressing a bovine growth hormone gene.
[b]* = $P < .05$; NS = not significant ($P > .05$).

required to evaluate its potential for improving production efficiency and disease resistance of livestock. Little is known regarding body/carcass composition and lipid composition of transgenic animals. This study was designed to determine what effect the introduction of recombinant bovine growth hormone gene into pigs had on lipid and carcass composition. Lipid composition of carcass tissue from 16 transgenic (T) pigs expressing a bovine growth hormone gene (Pursel *et al.*, 1989; Solomon and Pursel, 1990) and 16 control (C) pigs were compared. Pigs were slaughtered at four different weight groups: 14, 26, 48 and 88 kg.

At 14 kg slaughter weight, carcasses from T-pigs contained 38 percent less fat (T = 6.19 and C = 10.03 percent), 44 percent less SFA, 48 percent less MUFA and 38 percent less PUFA compared to C-pigs. At 26 kg the T-pigs had 38 percent less total carcass fat (T = 7.62 and C = 12.32 percent), 42 percent less SFA, 46 percent less MUFA and 24 percent less PUFA compared to C-pigs. At 48 kg slaughter weight, T-pigs contained 51 percent less carcass fat (T = 7.54 and C = 15.50 percent), 55 percent less SFA, 63 percent less MUFA and 33 percent less PUFA compared to C-pigs. At 88 kg carcasses from T-pigs contained 83 percent less carcass fat (T = 3.27 and C = 19.55 percent), 89 percent less SFA, 88 percent less MUFA and 55 percent less PUFA compared to C-pigs. Cholesterol content was not different between T- and C-pigs at any of the slaughter weights. However, the trend was for cholesterol content to decrease from 14 to the 88 kg weight group. These results suggested a dilution effect of fatty acids in carcass tissue from T-pigs with increasing slaughter weight. Furthermore, carcass fat follows a different pattern of deposition in T-pigs compared to C-pigs. As body weight increased in T-pigs, carcass fat decreased, whereas carcass fat increased in C-pigs as body weight increased.

Conclusions

Potential for manipulation of growth and composition of farm animals has never been greater than at present with the wide array of strategies for altering the balance between lean and adipose tissue. The need to improve the composition of the foods in our diets is evident and the industries are responding to the consumers' needs. Although progress has been made, much more needs to be accomplished. Eating quality and safety of meats must not be sacrificed as leaner animals are developed.

References

Brinster, R.L. and Palmiter, R.D. 1986. Introduction of genes into germ line of animals. Harvey Lect. 80: 1.

Keys, A., Anderson, J. and Grande, F. 1965. Serum cholesterol response to changes in the diet. IV. Particular saturated fatty acids in the diet. Metabolism 14: 776.

Lough, D.S., Solomon, M.B., Lynch, G.P., Rumsey, T.S. and Slyter, L.L. 1990. Influence of dietary palm oil supplementation on carcass characteristics and cholesterol content of tissues from growing ram and ewe lambs. J. Anim. Sci. (Suppl. 1) 68: 336.

Lynch, G.P. and Solomon, M.B. 1989. Growth and carcass characteristics of young ram lambs fed to attain market weight in 120 days from birth. A Review. SID Res. J. 5: 18.

National Research Council. 1988. Designing Foods. Animal product options in the market place. National Academy Press. Washington, D.C.

Pursel, V.G., Rexroad, C.E., Jr. and Bolt, D.J. 1989. Gene transfer for enhanced growth of livestock. In "Animal Growth Regulation." Campion, D.R., Hausman, G.J. and Martin, R.J. (Eds.). Plenum Press, New York, New York.

Solomon, M.B., Lynch, G.P. and Berry, B.W. 1986. Influence of animal diet and carcass electrical stimulation on the quality of meat from youthful ram lambs. J. Anim. Sci. 62: 139.

Solomon, M.B. and Lynch, G.P. 1988. Biochemical, histochemical and palatability characteristics of young ram lambs as affected by diet and electrical stimulation. J. Anim. Sci. 66: 1955.

Solomon, M.B., Lynch, G.P., Paroczay, E. and Norton, S. 1989. Influence of rapeseed meal, whole rapeseed and soybean meal on fatty acid profiles and cholesterol of tissues from ram lambs. J. Anim. Sci. (Suppl. 2) 67: 222.

Solomon, M.B., Lynch, G.P., Ono, K. and Paroczay, E. 1990a. Lipid composition of muscle and adipose tissue from crossbred ram, wether and cryptorchid lambs. J. Anim. Sci. 68: 137.

Solomon, M.B., Caperna, T.J. and Steele, N.C. 1990b. Lipid composition of muscle and adipose tissue from pigs treated with exogenous porcine somatotropin (pST). J. Anim. Sci. (Suppl. 1) 68: 217.

Solomon, M.B. and Pursel, V.G. 1990. Lipid composition of carcass tissue from transgenic pigs expressing a bovine growth hormone gene. J. Anim. Sci. (Suppl. 1) 68: 339.

USDHHS. 1988. USDHHS: The Surgeon General's Report on Nutrition and Health. Public Health Service, USDHHS, PHS, NIH, Publ. No. 88-50210. Washington, D.C.

Biological Conversion of Inedible Biomass to Food

K.J. Senecal, M. Mandels, and D.L. Kaplan
U.S. Army Natick RD & E Center
Natick, MA 01760

The issue of conversion of non-nutritional lignocellulose substrates to produce edible materials is under study. The research emphasis is on substrates (e.g. cornstover, bran, corn cobs and wood chips), enzyme systems, and organisms needed for the conversions in small scale systems with minimal processing requirements. *Trichoderma reesei, Chrysosporium pruinosum* and *Lentinus edodes* are the organisms being studied for their potential to break down lignocellulose to basic food components. End products of the conversions such as protein, glucose and/or microbial biomass are being evaluated. Solid substrate and shake flask fermentation methods are being used to obtain the edible products. A corollary study on algal photosynthesis for food production using *Spirulina maxima* and *Spirulina platensis* is also underway in the laboratory. A review of the recent literature as well as preliminary results found in our investigations will be presented.

Introduction

Large scale biomass conversion has been considered for energy production (i.e., methanol) and production of feedstocks (i.e., single cell protein (SCP)). The production of low cost protein and energy from renewable resources has sparked a great deal of research and development although much of the potential of these technologies has yet to be realized. Many reasons can be identified for this failure to meet expectations, including the transient low cost of petroleum, the high cost of enzyme production, end-product inhibition, and the need for improved pretreatments to disrupt lignocellulose.

Problems With Lignocellulose Conversions

One of the principle sources of biomass is plant lignocellulose, a product of photosynthesis and often available as a result of forestry or agricultural processes (i.e., straw, bark, leaves, wood, roots). Lignocellulose is the most abundant renewable carbon source. Plant biomass is composed of lignin-hemicellulose-cellulose complexes that cannot be readily hydrolyzed by enzymes or acids to liberate fermentable sugars. Lignin is a phenolic heteropolymer that protects cellulose and hemicellulose from rapid chemical or enzymatic hydrolysis (Litchfield, 1983). It is the strong association between these three major components that results in both the unusual and useful properties of wood materials and the difficulty in their conversion into useful products through biomass conversion. Unless the components of the biomass are separated, converison efficiences are low and few organisms are capable of their complete utilization. Once separated, the lignin provides no nutritional benefit for animals and is only slowly digested by some fungi (basidomycetes). The cellulose and hemicellulose are only digested by ruminant animals, fungi and bacteria. Once the hemicellulose is digested to monomer sugars (i.e., pentoses such as xylose and arabinose) these are not metabolized by most animals but can be utilized by some microorganisms. The microbial enzyme glucose isomerase (=xylose isomerase) can be used to convert xylose to xylulose which is metabolized by animals (Pigman, 1957). Unlike the hemicellulose, glucose derived from cellulose depolymerization can provide nutritional sustenance for humans and microorganisms.

A variety of conversion processes to break down lignocellulose can be conceived including: direct growth of mushrooms, fermentation to edible biomass (i.e., SCP) and hydrolysis (acids or enzymatic) for conversion of carbohydrate polymers to simple sugars. The conversion of plant materials to edible products usually requires a pretreatment step to improve surface area and increase accessibility of the usable portions of the plant to further attack by enzymes or acids. Weakening or disintegration of the cell walls are necessary in processing SCP; and physical/chemical methods such as grinding, steam explosion, ball milling, alkaline swelling and acid treatments are used. Vallander and Eriksson (1985) achieved up to 650 g of sugar per kg steam-exploded wheat straw.

Chemical and Enzymatic Hydrolysis Studies

A recent study conducted by Koba *et al.* (1990), on the chemical hydrolysis (acid and alkali) of lignocellulose, reported over 90 percent digestion of the total carbohydrate of palm fiber. The fiber was delignified using 3 percent NaOH, and glucose and xylose were detected as the two main sugars. Xylose, glucose, and arabinose were detected as the main sugars after acid hydrolysis (10 percent sulfuric acid, 1 hr, and 100C). Xylose represented about 90 percent of the total sugar in both hydrolysates. Two of the problems with acid hydrolysis are low yield and the formation of impurities. The acid can degrade sugar monomers such as glucose and xylose to other products, such as furfural, that are toxic to yeasts.

The enzymatic conversion of cellulose materials to sugars has been reviewed (Marsden and Gray, 1985; Mandels, 1981; Mandels and Kaplan, 1986). These enzymes need to be extracellular or at least located on the outside of the producing microorganism in order to gain acess to the substrate. One of the first steps in an enzymatic reaction involves the adsorption of enzyme to the substrate. Reese (1978) has stated that adsorption of the enzyme is a function of: (1) the amount of enzyme present; (2) availability of surface area; (3) physical properties of the enzyme (charge, size and solubility); and (4) the environment (pH, salt concentration, temperature).

Bacteria or fungi harboring the appropriate enzymes could be grown on pretreated, or even untreated plant materials for conversion to edible bacterial or fungal biomass. Branches, bark, wood, leaves, grasses and other materials, agricultural waste, and solid waste such as paper could be used in this process. Many species of fungi and a few bacteria are capable of producing cellulases that degrade insoluble cellulose *in vitro.* Examples of fungal species include *Trichoderma reesei, T. viride, T. konigii, T. lignorum, Penicillium funiculosum, Aspergillus wentii, Sporotrichum pulverlentum, Fusarium solani,* and *Sclerotium rolfsii.* Bacterial species include *Cellulomonas thermocellum* (Bisaria and Ghose, 1981), *Clostridium thermocellum* (Robson and Chambliss, 1989) and *Thermomonospora fusca* (Wilson, 1988). These latter two bacteria are thermophilic, which offers potential benefits in terms of commercialization.

The enzymatic conversion of starch to glucose has been commercialized using the enzyme alpha-amylase from *Bacillus subtilis* and glucoamylase from *Aspergillus niger.* The success of this enzymatic process has displaced the acid hydrolysis process. Similiar success has not been realized for cellulose conversion processes, primarily due to the high cost of enzyme production due to the low specific activity of cellulase and the resulting high enzyme protein requirement. The fungus *T. reesei* is one of the best sources of cellulases. This fungus produces a complete cellulase (endo-and exo-beta-glucanases and beta-glucosidase (cellobiase)) as well as hemicellulases to convert the pentosans. For the production of edible protein from cellulose, a remarkable yield of extracellular protein was found during the growth of *T. reesei* on cellulose (Mandels, 1981). Yields of better than 20 grams per liter representing a 33 percent yield of extracellular protein from the cellulose substrate were attained and productivities of 150 mg of protein per liter hour were achieved in pilot scale runs. This protein is colorless, odorless, freely soluble in water, and has a good amino acid composition (Table 1). A great deal of research has focused on developing hyperproducers or constituitive mutants of cellulase (Reese and Mandels, 1984). Recently the focus has shifted towards cloning cellulase genes to modify expression and regulation and enhance activity (Tanaka *et al.*,1990).

Solid Substrate Fermentation Studies

Fermentation of plant biomass by filamentous fungi can be carried out in the solid state, in submerged liquid culture, or in a composting-like intermediary environment. Extracellular enzymes depolymerize the polysaccharides and the fungal biomass; secreted fungal enzymes (protein) and residual plant biomass can serve as a satisfactory animal food (Peitersen and Andersen, 1978). SCP production using thermophilic actinomycetes (*Thermomonospora* sp.) was actively studied in the 1970's (Bellamy, 1974; Humphrey et al., 1977). In general, cellulolytic bacteria are slow growing and protein yields are low (i.e., *Cellulomonas* sp. and *Alcaligenes* sp.). Although yeast are the most common sources of SCP they do not grow on cellulose or hemicellulose; they only grown on simple sugars. *Fusarium graminiearum* has been used in Britain to convert glucose syrups derived from starch into fungal biomass (myco-protein) (Litchfield, 1983). This fungus was chosen rather than a yeast or bacterium because its fibrous structure matches the texture of meat. Laboratory

Table 1. Composition of soluble proteins from *Trichoderma reesei* grown on cellulose.

Amino Acid	Soluble Protein mg AA/g Protein	Mycelial Protein mg AA/g Protein
Trytophan	21	ND
Lysine	23	33
Histidine	9	8
Ammonia	16	10
Arginine	22	20
Aspartic Acid	106	54
Threonine	76	18
Serine	78	24
Glutamic Acid	74	51
Proline	41	18
Glycine	61	27
Alanine	47	29
Cystine	33	5
Valine	25	9
Methionine	7	3
Isoleucine	14	6
Leucine	40	19
Tyrosine	56	73
Phenyl Alanine	25	12

Soluble Protein = Cellulase, extracellular protein.
Mycelial Protein = Solids from a 1 percent cellulose culture.
ND = not determined

Adapted from Andreotti *et al.* (1978).

animal feeding studies and short-term human feeding studies with this fungal product have shown no ill effects (Tuse, 1983).

A number of different processes can be used for the production of SCP from plant biomass. These include: 1) solid substrate conversion to SCP with axenic or a group of symbiotic microorganisms, and 2) conversion of soluble breakdown products after the substrate has been chemically or enzymatically hydrolyzed in a pretreatment step. According to Hesseltine (1972), solid-state fermentation (SSF) relates to the growth of microorganisms on solid substrates in the absence of any free water. The term SSF is not applied to fermentation on solid materials suspended in a liquid phase. Solid substrate fermentation is used to produce enzymes, fermented foods and mycotoxins. The advantages of SSF as described by Hesseltine (1972) are: 1) operation under non-aseptic conditions,

2) high productivity per unit volume of reactor, 3) use of raw materials as substrates since pretreatments are not required, 4) reduced energy requirements, 5) low waste water output, and 6) low capital investment. Iadoo *et al.* (1982) described drawbacks with SSF including: 1) difficulties in fermentation control, 2) lack of knowledge relating to growth and physiology of converting microorganism(s), 3) the types of microorganism that can be used are limited to those which can grow at reduced moisture levels, and 4) problems in scaleup. A disadvantage to fermentation of biomass in submerged liquid culture is the dilution effect of desirable products which must then be recovered through a concentration step.

Trichoderma Sp.

Trichoderma viride and *Trichoderma reesei* have been the focus of much research because of their cellulase enzyme complexes (Mandels, 1982). The enzyme complexes of these fungal organisms are extracellular, capable of hydrolyzing crystalline cellulose and obtained in relatively high yields. The cellulase enzyme system includes: endoglucanases, cellobiohydrolases, and B-exoglucosidases. The relative activities of these components depend on the source of the enzyme (Coughlan and Ljungdahl, 1988; Mandels and Reese, 1965).

The parent strain of *T. reesei* QM6a was isolated from a deteriorated shelter from Bougaineville Island at the end of World War II. In the early 1950's, *T. reesei* was found to be an excellent source of the enzymes required to hydrolyze cellulose (Mandels, 1982). This strain was originally identified as *T. viride*; but in 1977 it was recognized as a new species. Many enhanced cellulase mutants have been developed from this strain which differ in levels of cellulase prodution. The composition and properties of the enzyme complex are similar for all strains, and the relative proportions of endo- and exo-B-glucanases are similiar. *T. reesei* grows on a wide variety of carbon sources as well as inorganic salts, including ammonium as a nitrogen source (Ryu *et al.*, 1980). End product inhibition is a recurring problem with these reactions. From the original isolate, *T. reesei* QM6a, several mutations programs were initiated to increase cellulase production, at the U.S. Army Natick Laboratories, Rutgers State University and elsewhere (Figure 1, Table 2). The differences between the mutants are only quantitative; the mutants produce higher

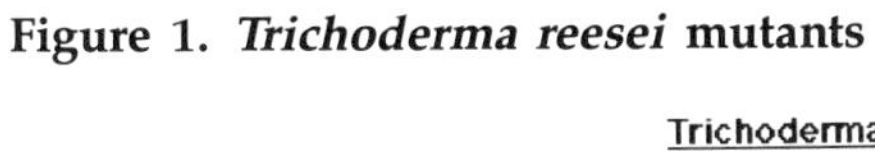

Figure 1. ***Trichoderma reesei*** **mutants**

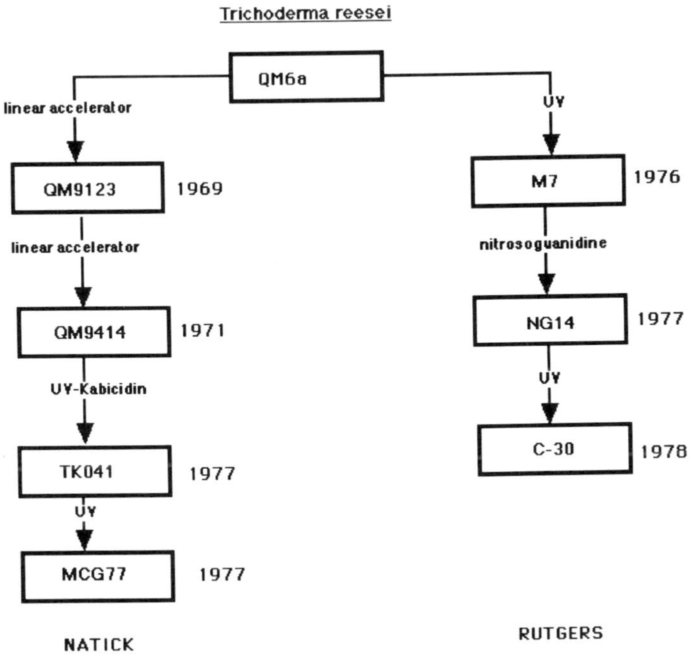

Table 2. Effect of strain and substrate on cellulase production by *T. reesei*.

Strain	Substrate 6 %	Soluble Protein mg/ml
Qm6a	Lactose	0.9
	Ball Milled Pulp	0.8
	FB Cotton	7.4
QM9414	Lactose	0.9
	Ball Milled Pulp	8.7
	FB Cotton	13.6
MCG77	Lactose	3.7
	Ball Milled Pulp	9.6
	FB Cotton	16.2
Rutgers NG14	Ball Milled Pulp	14.4
	FB Cotton	21.2
Rutgers C30	Lactose	3.5
	Ball Milled Pulp	17.8
	FB Cotton	20.6

Adapted from Andreotti *et al.* (1981)

levels of cellulase than the parent strain. The protein was produced in high yield from cellulose, plant biomass and municipal trash such as paper, with an inorganic ammonium nitrogen source.

In a recent study by Liaw and Penner (1990), the influence of substrate and enzyme concentrations on the rate of saccharification of two cellulose substrates, Avicel (FMC Corp. Philadelphia, Pa) and Solka-Floc (James River Co., Berlin, NH), by the cellulase enzyme system of *T. viride* was examined. The authors reported comparable maximum rates of saccharification and similar Km values for the two substrates, although at high substrate concentrations different inhibition properties were found for the substrates. Deschamps *et al.* (1985) studied cellulase production by growing *T. harizianum* on lignocellulosic materials in solid-state cultivation by static and mixed techniques. They concluded that although it is possible to obtain higher enzyme activity under submerged fermentation conditions, the requirement for strict aseptic conditions works against you by increasing energy costs and the cost of production of the enzymes. Cellulase production by solid substrate fermentation offers advantages because it avoids the need for aseptic conditions and the expensive steps of extraction and purification of the enzymes.

Toyama and Ogawa (1974) concluded that hemicellulytic, celluloytic, and lignolytic enzymes should be produced by wood degrading fungi during their growth. Of these components, hemicellulose such as xylan, was the most readily degraded substrate. The authors used cellulase activity as a measure of comparison between wood-degrading activity of different fungal species. Both edible and inedible fungi were included in the study. They found that *T. viride* cellulase was unable to degrade sawdust, rice straw, and bagasse without chemical pretreatment to effect delignification. Since oxidizing enzymes were not detected with the exception of catalase, the authors concluded that oxidizing enzymes play an important role in the degradation of lignin. Acebal *et al.*, (1986), grew *T. reesei* QM9114 on wheat straw as the sole carbon source. The straw was pretreated by physical and chemical methods. The authors found maximal growth on alkali-pretreated straw, whereas cellulase production was optimal when physically pretreated straw was used as the substrate. This implies that delignification does not appear necessary for the production of cellulases when optimal contact between *T. reesei* and cellulosic substrate occurs. Toyama and Ogawa, in 1975, examined enzymatic saccharification of rice straw, bagasse, and sawdust with the intention of producing sugar at low cost. The cellulase was derived from *T. viride,* and saccharification of the delignified substrates at various substrate concentrations resulted in the formation of 5 percent to 10 percent sugar solutions after incubation at pH 5.0, 45 C for 48 hr (Table 3).

T. reesei has already been approved by the U.S. Food and Drug Administration for inclusion on the GRAS list based on an application from Novo industries. This company plans to market cellulase for food applications. The amino acid composition is complete and without deficiencies in any essential amino acids.

Table 3. Thick sugar solution attainable with cellulase preparations using cellulosic wastes.

Substrate	Enzyme conc. percent	Incubation 96 Hr.	
		Sugar %	Decom. %
Rice Straw	3	21.89	65.7
	1	15.92	47.8
Bagasse	3	21.06	63.2
	1	14.90	44.7

Powdered substrates were delignified by boiling with 1 percent NaOh for 3 hr. Substrate conc. 25 percent, pH 5.0, 45C.
Enzyme: Meicelase CEP-233

Adapted from Toyama *et al.* (1979)

Chrysosporium pruinosum

Phanerochaete chrysosporium, a white-rot wood-decaying basidiomycete, produces a potent lignin-degrading system that oxidizes lignin completely to CO_2. Considerable research has been done with *P. chrysosporium* (the perfect form of *Chrysosporium pruinosum)* in Sweden and the United States. Physiological, biochemical and genetic studies on lignin breakdown in *P. chrysosporium* have indicated that: 1) lignolytic activity appears in cultures with or without the presence of lignin; 2) lignolytic activity results due to nutrient starvation; 3) activity is not induced by lignin and; 4) lignin breakdown is regulated as a secondary metabolic event

(Ander, Hatakka and Eriksson, 1980). At least some white rot fungi do not use lignin as a carbon source, using cellulose as a co-metabolite for energy for lignin degradation (Ander, Hatakka and Eriksson, 1980).

Previous studies on *P. chrysosporium* have shown that the ligninolytic system is synthesized in the absence of lignin in response to nitrogen starvation, and that a supplemental growth substrate is required for lignin degradation (Keyser, Kirk, and Zeikus, 1978). Restricting the carbohydrate source led to the appearance of ligninolytic activity earlier than in nitrogen starved cultures while phosphorus limitation did not trigger the early onset of ligninolytic activity. However, in cultures containing a low nitrogen level, excess celliobiose and limiting phosphorus, delayed the appearance of ligninolytic activity by one day. Carbohydrate or phosphorus limitation may be more important than nitrogen starvation in stimulating ligninolytic activity (Jeffries, Choi and Kirk, 1981). For total degradation of ligninocellulosic materials, the cellulose-hemicellulose structure must also be degraded; however, excess glucose represses cellulose-degrading enzymes. This interaction resulted in 50–80 percent total lignin degradation in the absence of excess glucose compared with 25–55 percent degradation when glucose was present (Leisola, Ulmer and Feichter, 1984). *P. chrysosporium*, secretes a haem-protein, which in the presence of hydrogen peroxide and oxygen, degrades lignin model compounds (Tein and Kirck, 1983). Kirk *et al.* (1986) recently purified and characterized this lignase, showing that it catalyses several different oxidative reactions in lignin model compounds. In a chemically defined medium using growth conditions conducive to high ligninolytic activity, the enzyme is produced only at the specific activity level of approximately 20 units per liter. Prior attempts to characterize the enzyme, as measured by the oxidation of veratryl alchohol have been difficult due to the low levels of the enzyme produced by the fungus.

Increased ligninase activity was reported in cultures of *P. chrysosporium* ME446 when veratryl alcohol, a secondary metabolite was added (Faison and Kirk, 1985). Kirk *et al.* (1986) also reported an increase in total ligninase activity by the addition of certain trace metals. The results of this study showed that either Cu^{2+} or Mn^{2+} causes an increase in activity equal to the complete trace element solution. The increase in enzyme activity is not due to just one protein, since high performance chromatography profiles showed increases in several peaks exhibiting activity. Leatham (1986) reported that ligninolytic activity in *P. chrysosporium* develops after vegetative growth ceases, and requires both oxygen and an exogenous carbon source. Certain inorganic elements including Fe^{2+}, Ca^{2+}, and Mo^{6+} were found to stimulate lignolytic activity, with calcium addition at 40 mg/liter giving the highest activity.

Eriksson and Larsson (1975) studies submerged cultures of *Sporotrichum pulverulentum* using lignin-containing waste fibers from newsprint as the only carbon source. The influence of different nitrogen sources on growth was also investigated. The study showed that the protein content decreased in the residual substance when the crytstallinity of the cellulose carbon source increased. Results indicated that the more complex the carbon source, the more difficult it is to digest and the more enzyme has to be produced for its degradation. Mudgett and Paradis (1985) characterized a strain of white rot fungus, *P. chrysosporium Burds.* ME446, that was used to convert birch lignin by solid-substrate fermentation. They found that moisture content, inoculum density, nitrogen supplementation and autoclaving of birch solids significantly affected lignin conversion rates and yields in 20 day fermentations. They also reported that partial pressure of oxygen at 1.0 atmosphere favored lignin conversion while oxygen pressures of 2.0 atmosphere severely inhibited lignin conversion rates.

Lentinus edodes

The shiitake mushroom *Lentinus edodes* fruits on the wood of dead deciduous trees and is widely cultivated in the Orient (Declaire, 1978). Suitable species for growth of *L. edodes* are oaks *(Quercus serata; Q. acustissima; Q. monogolica* var *grossoserrata; Cyclobalanopsis acuta; C. glauca;);* Shii tree *(Castanopsis cuspidata; C. cuspidata* var *Shieboldii);* chestnut *(Castanea crenata);* and hornbeans *(Carpinus tschonoskii; C. laxiflora; C. Japonica)* (Ito, 1978).

Leatham (1985) addressed the problem of growth and development on oakwood-oatmeal medium for the commercially important mushroom *L. edodes*. The study determined the components of the lignocellulose degraded during a 150 day incubation, vegetative growth patterns of the fungus, the likely growth-limiting nutrient, and extracellular enzyme activity. All major components of the oakwood-oatmeal medium were degraded by *Lentinus*, including the lignin component. The vegetative growth rate of *Lentinus* was highest during the first 90 days and nitrogen was found to be the limiting growth factor. Certain enzyme activities were identified including cellulase, hemicellulases, a ligninolytic system, amylase, pectinase, acid protease, cell wall lytic enzymes, acid phosphatase, and laccase. Ligninolytic activity was initiated and maintained from day 15 onward and postively correlated with glucan degradation. The author suggested that *L. edodes,* like *P. chrysoporium* must utilize an alternate substrate as a cometabolite to support lignin degradation.

In another study by Leatham (1986), the ability of *L. edodes* to degrade lignin and the factors controlling the ligninolytic activity are described. Ligninolytic activity was found to require oxygen and a carbon source other than lignin. *L. edodes* showed maximal degradation activity during the vegetative growth period. In contrast to *P. chyrsosporium,* ligninolytic activity was not stimulated by Ca^{2+} and Mn^{2+} was required, with 10 mg/liter Mn^{2+} giving optimal activity. When oxygen concentration was increased from 20.9 percent to 100 percent in a standard liquid medium the rate of lignin degradation increased by at least two-fold.

Algae

Algae have been grown and harvested for centuries as food in different regions of the world. Microalgae such as *Spirulina* have been extensively used as a food source due to its high protein content (Table 4). *Nostoc* is also consumed extensively. Algae, as with all green plants, have the capability to use the energy of sunlight to convert water and carbon dioxide into carbohydrates. *Spirulina* contain all the essential amino acids, fatty acids, vitamins, and minerals of any plant; however, they grow faster owing to the fact that energy is not spent synthesizing support tissues. *Spirulina* grown in alkaline lakes have been consumed in regions of Africa and Mexico for years as a major source of protein (Litchfield, 1979).

Spirulina, a blue green algae, is easy to digest and contains up to 70 percent protein. Additional benefits to consumption of microalgae are: they serve as a source of beta-carotene (food coloring and potential application in cancer treatment therapy) and eicosapentaenoic acid (a polyunsaturated fatty acid with potential to reduce serum cholesterol). *Nostoc, Chlorella, Dunaliella* and *Scenedesmus* are other algae that have been extensively studied (Brown and Frape, 1988).

Table 4. Gross chemical composition of *Spirulina platensis* and *Spirulina maxima* percent dry matter.

Alga	Protein	Lipids	Carbohydrates	Nucleic Acids
Spirulina platensis	46-50	4-9	8-14	2-5
Spirulina maxima	60-70	6-7	13-16	2.9-4.5

Adapted from Becker (1988).

Environmental conditions, growth and yields for many algae have been studied (Litchfield, 1979). Most research on algal food sources with regard to nutritional composition and toxic component removal has been conducted with the green algal organisms. Blue green algae have been studied for nitrogen fixation and agricultural applications.

Advantages to the use of algae as a food source are: 1) high photosynthetic activity and 2) minimal need for carbon and nitrogen in the growth medium. Disadvantages to consuming untreated and unpurified algal biomass include nondigestible cell wall components, excess nucleic acids, toxins, poor flavor and acceptability. Methods such as supercritical carbon dioxide extraction have been used to improve removal of algal lipids and pigments (Nakhost and Krukonis, 1987). Heat treatments have been developed to reduce nucleic acid levels. *Spirulina* studies have shown that this blue-green algae is advantaeous due to its size (500 μm), a digestible cell, and resistance to contamination because of the alkaline medium in which it is grown. The protein in *Spirulina* is of high quality and no toxicity problems have been reported. Blue-green algae have also been shown to produce extracellular amino acids and peptides, serving as possible food supplements (Craig and Reichelt, 1987).

Current Research

The focus of our current work on biomass conversion has been on small scale processes for the conversion of inedible environmental substrates to edible foods. The processes we are studying are limited in scope since they are geared towards producting foods *in situ* in survival scenarios. Therefore we are focusing on simple processes and characterizing the yields, nutritional contents, and overall feasibility of these processes. Our initial efforts involved consideration of all potential options for generating edible biomass in survival scenarios, including conversion of structural materials. No attempt will be made here to summarize all of the options explored and the conclusions reached since this was detailed in a recent (Kaplan *et al.*, 1988). However, the outcome of the study was to focus on two processes: (1) biomass conversion of lignocellulose using enzymatic conversion and (2) algal photosynthesis for the production of edible biomass. We will briefly describe our approaches and some of the findings from our ongoing work.

Materials and Methods

Shake flask fermentations. Substrates: The substrates used in the shake flask fermentation studies were: BW200 (pure cellulose pulp); NEP40 (uninked newsprint that has been hammermilled to 40 mesh) containing 30–40 percent lignin + ash, and 60–70 percent carbohydrate; shredded corn stover (leaves and stems after harvesting) containing approximately 35 percent lignin + ash, 35 percent glucose, and 30 percent pentosan.

Organisms: The organisms used in all experiments were *T. reesei* QM6a wild strain, *T. reesei* QM 9414 enhanced cellulase mutant, and *C. pruinosum* QM 826. Both QM6a and QM 9414 are cellulolytic; QM 826 is an organism which is both cellulolytic and ligninolytic.

Growth: The organisms were grown in 50 ml *T. reesei* salts medium (Mandels and Weber, 1969), with 1 percent substrate as the carbon source, and shaking at 29 C. In experiment 1 the flasks contained NEP 40, experiments 2 and 3 contained BW 200 as the substrate. After the initial growth phase, sodium azide was added to the culture to stop growth (but not inhibit enzyme activity) and citrate buffer was added to adjust the pH to 4.8 for optimal enzyme activity. More substrate was then added either in the form of NEP 40 or shredded corn stover. Total substrate concentration was brought up to 5 percent. The flasks were then allowed to incubate 2 to 3 days at 40 C or 50 C unshaken.

Analysis: Dry weight was analyzed by filtering the culture on tared glass filter paper and saving both filtrate and solid cake for sugar and protein analyses. The cake was washed and dried at 80 C. Sugar analysis was performed by the DNS procedure using glucose as a standard (Miller, 1959). The protein was measured in the filtrate by the Lowry method (Lowry and Rosebrough, 1951) after precipitation with 10 percent trichloroacetic acid. The protein in the filtrate is a measure of the soluble protein (secreted protein). The collective protein (mycelial protein, adsorbed protein, and substrate protein) remaining in the solid product was extracted after saccharification with 0.1N NaOH for 2 hours (room temperature), then filtered and precipitated with 10 percent trichloroacetic acid.

Solid Substrate Fermentations.

Substrates: The substrates used in this experiment were powdered cellulose (Schleicher and Schuell Inc. Keene, NH) and shredded cornstover.

Organisms: *T. reesei* QM 6a.

Growth: The fungus in the initial growth phase was grown *T. reesei* salt medium (50 ml) with 0.2 percent lactose, for 7 days at 29 C on a reciprocal shaker. After the growth phase, a 1 ml aliquot of culture with 1 ml of sterile MQ water was inoculated on 1 gram substrate. The solid substrate fermentation was incubated for 7 days at 29 C.

Analysis: Mycelial and plant proteins were extracted with 1 N NaOH for 16 hours at room temperature. The supernatant was collected to be read for protein content by the BioRad Assay (BioRad, CA).

Algal Fermentations.

Organisms: *Spirulina maxima* and *Spirulina platensis.*

Growth: The algae were grown on 50 ml of Allens medium or Spirulina medium (Starr and Zeikus, 1988) for 14 days at 600 foot candles of light.

Analysis: Dry weight was obtained by centrifuging cultures and repeated rinsing with MQ water to remove salts. The biomass obtained was then dried in a 80 C oven.

Results

Shake Flask Fermentation (Table 5). In experiment 1, both cultures QM 826 and QM 9414 grew fairly well on the NEP 40 in the initial growth phase. Any soluble sugars produced during this phase were consumed by the organisms. Both organisms saccharified NEP 40 substrate; yields ranged from 7 to 16 mg/ml of reducing sugar. Cornstover was more resistant, yielding only 3 to 4 mg/ml of reducing sugar. Protein yields from both organisms on NEP 40 substrate were

Table 5. Shake flask fermentation results.

	Organism	Sustrate 5 percent	Protein(mg/ml)	RS(mg/ml)
Exp. 1	QM 826	NEP 40	0.83	7.44
		Cornstover	1.00	3.88
	QM 9414	NEP 40	1.47	16.20
		Cornstover	1.66	3.58
Exp. 2	QM 826	NEP 40	0.42	1.35
		Cornstover	1.22	0.82
	QM 6a	NEP 40	0.79	6.96
		Cornstover	1.38	3.04
Exp. 3	QM 826	NEP 40	0.51	3.90
		Cornstover	3.41	2.58
	QM 6a	NEP 40	0.54	2.48
		Cornstover	3.29	2.78

RS = Reducing sugars

0.8 to 1.4 mg/ml. Cornstover protein yields ranged from 1.0 to 1.6 mg/ml. Although differences in protein yield between organisms were not detectable, reducing sugar yields on the NEP 40 substrate differed significantly. QM 9414 produced about twice the amount of sugar.

In experiment 2, QM 826 and QM 6a also grew well on the BW 200 substrate in the initial growth phase. Reducing sugar yields on NEP 40 ranged from 1.35 mg/ml for QM 826 to 6.96 mg/ml for QM6a. Cornstover yields 0.8 mg/ml for QM 826 to 3.04 mg/ml for QM 6a. Protein yields were 0.42 mg/ml for QM 826 and 0.79 mg/ml for QM6a.

Experiment 3 results showed increased protein yields on the cornstover substrate over that of experiments 1 and 2. Protein yields on the NEP 40 substrate were down slightly from experiment 2, but differed significantly from experiment 1. Reducing sugar yields were similiar in experiments 2 and 3.

Solid Substrate Fermentations. Solid substrate fermentation protein yields are low (Table 6). The powdered cellulose substrate yielded approximately 4.6 mg protein/gram of substrate, whereas the cornstover fermentation yielded approximately 30 mg protein/gram. Cornstover yielded more protein/gram indicating that protein was also extracted. Low proteins yields overall correlate with poor growth

Table 6. Solid substrate fermentation results.

Organism	Substrate	Protein yield mg/gram substrate
Trichoderma reesei	milled cellulose	4.6
	cornstover	30.0

of the fungus. Observations of fermentations have shown that spore formation is prevalent, and moisture availability may be the problem with SSF. Additional observations have shown that by increasing the ratio of water to substrate, spore formation was minimized and improved yields of protein should be realized.

Algal Fermentations.

Algal results obtained were 2 mg to 3 mg per culture in a 50 ml flask. The two species did not differ in yields.

Discussion

With the goal of developing small scale simplified systems for biomass conversion in survival scenarios, we report relatively low yields of edible products from lignocellulose substrates during our preliminary studies. The edible nutritional products from the conversion process include protein, sugar and fungal biomass. Yields of proteins were around 83 mg total protein in the liquid cultures and 30 mg in the SSF where cornstover was used as the carbon source. The most optimal yield of sugar was 972 mg total in liquid culture (7 days) where NEP40 was used as the carbon source. Based on caloric needs for an individual (2900 calories/day), we can extrapolate the yield combining protein and sugar data to 0.57 calories/day, 0.02 percent of the daily caloric requirement. Algal biomass production under the conditions of this study were also very low. To increase the yields of protein, the use of lignocellulosic substrates, instead of cellulosic substrates as the initial carbon source, might increase enzyme production and secretion in the growth phase, especially pertaining to the SSF studies where protein yields were very low. Another thought might be using fungal organisms together as a synergistic approach to increase enzymatic hydrolysis, although problems might arise with inhibitors. Moisture availability is a problem with solid substrate fermentation. *T. reesei* used in this study sporulated correlating with the lack of moisture. Perhaps incremental additions of liquid over the course of the fermentation would increase growth and improve enzymatic activity. Some of the above options are currently being evaluated.

References

Acebal, C., Castillon, M.P., Estrada, P., Mata, I., Costa, E., Aguado, J., Romero, D. and Jimenez., F. 1986. Enhanced cellulase production from *Trichoderma reesei* QM 9414 on physically treated wheat straw. Appl. Microbiol. Biotechnol. 24: 218-223.

Aidoo, K.E., Hendry, R. and Wood, J.B. 1982. Solid substrate fermentations. Adv Appl. Microbiol. 28: 201-235.

Ander, P. Hatakka, A. and Eriksson, K.E. 1980. Degradation of lignin and lignin-related substances by *Sporotrichum pulverulentum (Phanaerochaete chrysoporium)*, In "Lignin Biodegradation: Microbiology, Chemistry, and Potential Applications," Vol. 2. CRC Press, Boca Raton Florida.

Andreotti, R.E., Medieros, J.E., Roche, C.R. and Mandels, M. 1981. Effects of strain and substrate on production of cellulase by *Trichoderma reesei*. Proceedings 2nd International Symposium on Bioconversion. Vol. 1 p. 353–372.

Becker, E.W. 1988. Micro-algae for human and animal consumption. In "Micro-algal Biotechnology," p. 222-256. Cambridge Univ. Press, London.

Bisaria, V.S. and Ghose, J.K. 1981. Biodegradation of cellulosic materials; substrates, enzymes, microorganisms and products. Enzyme Microb. Technol. 3: 90-104.

Brown, D.E. and Frape, D.L. 1988. Algal Culture Technology and the Agricultural Economy of Jordon-UK Research Groups - Collaborative Study Jan./Feb. Intern. Industrial Biotechnol. 8: 1.

Coughlan, M.P., Ljungdahl, L.G. 1988. Comparative biochemistry of fungal and bacterial celluloytic enzyme system. In "Biochemistry and Genetics of Cellulose Degradation." p. 11-30. Academic Press, Inc. (London), Ltd., London, England.

Craig, R. and Reichelt, B.Y. 1987. Genetic engineering in algal biotechnology. Trends in Biotech. 4(11): 280-285.

Delcaire, J. 1978. Economics of cultivated mushrooms. In "The Biology and Cultivation of Edible Mushrooms." pp. 727-793. Academic Press, New York.

Deschamps, F., Guiliano, G., Asther, M., Huet, M.C. and Rousoos, S. 1985. Cellulase production by *Trichoderma harzianum* in static and mixed solid-state fermentation reactors under nonaseptic conditions. Biotechnology and Bioengineering. Vol. 27: 1385-1388.

Eriksson, K.E. and Larsson, K. 1975. Fermentation of waste mechanical fibers from a newsprint mill by the rot fungus *Sportrichum pulverulentum.* Biotechnol. Bioeng. 17: 327-348.

Faison, B.D. and Kirk, T.K. 1985. Factors involved in the regulation of a ligninase activity in *Phanerochaete chrysosporium.* Appl. Environ. Microbiol. 49: 299-304.

Forage, A.J. 1985. Utilization of agricultural and food processing wastes containing carbohydrates. Res. Conserv. Novel Biol. Processes. Part III. p. 309-314. Genetic Technology News. Feb. 2–3.

Hedenskog, G. 1978. Properties and composition of single-cell protein, influence of processing. Proceedings of the FEBS. 44: 78–88.

Ito, T., Cultivation of *Lentinus.* 1978. In "The Biology and Cultivation of Edible Mushrooms." Acad. Press. New York, San Francisco, London.

Jefferies, T.E. Choi, S. and Kirk, T.K. 1981. Nutritional Regulation of Lignin Degradation by *Phanerochaete chrysosporium.* Appl. and Env. Microbiol. p. 290-296.

Keyser, P., Kirk, T.K. and Zeikus, J.G. 1978. Ligninolytic enzyme system of *Phanerochaete chrysosporium:* Synthesized in the absence of lignin in response to nitrogen starvation. J. Bacteriol. 135: 790-797.

Kim, J.-M., Kong, I.-S., Yu, J.-H. 1987. Molecular cloning of an endoglucanase gene from an alkalophilic *Bacillus* sp. and its expression in *Escherichia coli.* Appl. Env. Microbiol. 53(11): 2656-2659.

Kirk, T.K., Croan, S. and Tien, M. 1986. Production of multiple ligninases by *Phanerochaete chrysosporium:* Effect of selected growth conditions and use of a mutant strain. Enz. Microb. Tech. 8: 27-32.

Koba, Y., and Ishizaki, A. 1990. Chemical composition of palm fiber and its feasibility as cellulosic raw material for sugar production. Agric. Biol. Chem. 54(5): 1183-1187.

Leathum, G.F. 1986. The ligninolytic activities of *Lentinus edodes* and *Phanerochaete chrysosporium.* Appl. Microbiol. Biotechnol. 24: 51-58.

Leathum, G.F. and Kirk, T.K. 1983. Regulation of ligninolytic activity by nutrient nitrogen in white-rot basidiomycetes. FEMS Microbiol Lett. 16(5): 65-67.

Leatham, G.F. and Stahmann, J. 1983. Stimulatory effect of nickel or tin or fruiting of *Lentinus edodes*. Trans. Br. Mycol. Soc. 83(3): 513-516.

Leisola, M., Ulmer, D.C. and Fiechter. A. 1984. Factors affecting lignin degradation in lignocellulose by *Phanerochaete chrysosporium*, Arch Microbiol. 137: 171-175.

Liaw, ET., Penner, M.H. 1990. Substrate - Velocity Relationships for the *Trichoderma viride* Cellulase-Catalyzed Hydrolysis of Cellulose, Appl. Environ. Micro. 8: 2311-2318.

Litchfield, J.H. 1979. Production of Single-Cell Protein for Use in Food or Feed. Microbiol. Tech 2nd Ed., Vol. 1. Academic Press 93-155.

Litchfield, J.H. 1983. Single cell proteins. Science. 219: 740-746.

Lobarzewski, J. and Paszczynski, A. 1985. Lignocellulose biotransformation with immobilized cellulase, D-glucose oxidase and fungal peroxidases. Enz. Microb. Technol. 7: 564-566.

Lowry, O.H. and Rosebrough, N.J. 1951. Protein measurement with the folin phenol reagent. J. Biol. Chem. 193: 265-275.

Mandels, M. 1981. Enzymatic hydrolysis of cellulose to glucose. A Report on the Natick Program. Final Report.

Mandels, M. 1982. Cellulases. Ann. Rep. Ferment. Process. 5: 35-78.

Mandels, M. and Kaplan, D.L. 1987. Conversion of inedible wheat biomass to edible foods for space missions. Final Report to NASA.

Mandels, M. and Reese E.T. 1965. Inhibition of cellulase. Annu. Rev. Phytopathol. 3: 85-102.

Mandels, M. and Weber, J. 1969. The production of cellulases. Adv. Chem. Series. 95: 391-414.

Marsden, W.L., and Gray, P.P. 1985. Enzymatic hydrolysis of cellulose in lignocellulosic materials. CRC Critical Reviews in Biotechnology, Vol. 3(3): 235-276.

Miller, G.L. 1959. Use of dinitrosalicylic acid reagent for determination of reducing sugar. Analy. Chem. 31: 426-428.

Mudgett, R.E. and Paradis, A.J. 1985. Solid-state fermentation of natural birch lignin by *phanerochaete chrysosporium*. Enzyme Microb. Techol. 7: 150-154.

Nakhost, Z. K., and Krukonis, V.J. 1987. Non-conventional approaches to food processing in CELSS. I. algal proteins; characterization and process optimizing. In "NASA Conference Publication 2480. Controlled Ecological Life Support System in Space." p. 27-36.

Pigman, W. 1957. "The Carbohydrates, Chemistry, Biochemistry and Physiology." p. 795-796. Academic Press, New York.

Reese, E.T. and Mandels, M. 1984. Rolling with the times: production and application of *T. reesei cellulase.* Ann Reports on Ferment. Proc. 7: 1-20.

Robson, L.M., and Chambliss. 1989. Cellulase of bacterial origin. Enzyme Microbial Technol. 11(10): 626-644.

Ryu, D.D.Y., Andreotti, R., Medeiros, J. and Mandels, M. 1980. Comparative quantitative physiology of high cellulase producing strains of *Trichoderma reesei.* Enzyme Engineering 5: 33-40.

Tanaka, M. M., R. and Converse, A.O. 1990. n-Butlyamine and aid-steam explosion pretreatments of rice straw and hardwood: effects on substrate structure and enzymatic hydrolysis. Enzyme and Microbial. Tech. 12(3): 190-195.

Tien, M. and Kirk, T.K. 1983. Lignin-degrading enzyme from the hymenomycete *Phanerochaete chyrsosporium Burds.* Science 221: 661-663.

Toyama, N., and Ogawa, K. 1974. Comparative studies on celluloytic and enzyme activities of edible and inedible wood rotters. Mushroom Science IX (Part I) Proceedings of the Ninth International Scientific Congress on the Cultivation of Edible Fungi, Tokyo, Japan.

Toyama, N. and Ogawa, K. 1975. Sugar production from agricultural woody wastes by saccharification with *Trichoderma viride.* Biotechnol. and Bioeng. Symp. 5: 225-244.

Toyama, T. Ogawa, K., and Toyama, H. 1979. Production of sugar from cellulose with *Trichoderma* cellulase. In "Biosynthesis and Biodegradation of Cell Wall Components." ACS/CSJ Chemical Congress, Honolulu, HA.

Tuse, D. 1983. Single cell protein current status and future prospects. CRC Critical Reviews in Food Science 19(4): 273-325.

Vallander, L. and Eriksson, K.E. 1985. Enzymatic saccharification of pretreated wheat straw. Biotech. Bioeng. 27: 650-659.

Wilson, D.B. 1988. Cellulases of *Thermomonospora fusca.* In "Methods in Enzymology." p. 314-323. Academic Press, Inc. Harcourt, Brace, Janovich Publishers. San Diego, CA.

CARBOHYDRATE NUTRITION—NEEDS AND MOLECULAR APPROACHES

Industry Trends and Nutritional Issues for Food Uses of Starch: The Impact of Biotechnology on Future Opportunities

Robert B. Friedman
Research Department
American Maize-Products Company
1100 Indianapolis Blvd.
Hammond, Indiana 46320-1094

Starch biopolymers constitute the major energy storage polymer in plants. Over the years they have been used by human society as a direct source of nutrition; as societies have become more developed, starches have been employed as a source of animal nutrition as well. Even primitive people have recognized, however, that starch polymers possess functionalities as food ingredients in addition to their nutritive properties. Only recently, through developments in polymer and analytical chemistry, has it become possible to correlate function with structure in food systems. Biotechnological advancements in biochemical genetics have made feasible the control of structural characteristics of biopolymers through genetic manipulation. This permits the preparation of biopolymers which require no further chemical modification, or minimal chemical treatment. This genetic control has the potential to be achieved either through classical breeding methods or through genetic engineering. Ramifications of these routes will be discussed.

Introduction

Starch is the major energy storage polymer in the vegetable kingdom. Its presence and biological availability in seed kernels permits the successful plant to grow and compete in its environment (Rowe and Garwood, 1978). Members of the animal kingdom have long ago recognized the nutritional importance of starch-containing plants; carbohydrates derived from these plants constitute the major source of calories in the human diet.

Primitive human society used these vegetable nutrients as direct nutritional sources. As human civilization has "progressed," more and more of the dietary nutritional contribution of starchy plants have appeared as a secondary nutrient (CIMMYT, 1984). That is to say, more starch is being used as animal feed than as human food.

While the relative proportion of starch consumption as a nutritive component in the human diet has decreased, its importance as an ingredient to enhance food quality has steadily increased. From the first primitive baker who discovered that chewing corn meal before mixing a dough increased its sweetness to the contemporary food scientist who adds maltodextrins to prepare a "genuine imitation fat-free mayonnaise," starches and starch derivatives have been playing an increasingly important role in food quality enhancement.

In modern times the improvement to the quality of a food system by an "all natural" food starch has sometimes proved to be inadequate. The most simple means available to control the behavior of starch biopolymers has been to modify the starches chemically to provide the desired type and level of functionality. A less frequently employed route has been to utilize different starches from alternate plant sources whose functionalities were closer to those desired. It is well known to practitioners of the art, for example, that tapioca, potato, wheat, and maize starches all have different viscosity behavior patterns. Decisions on the most appropriate route to control starch behavior in a commercial product have usually been based on cost and availability of the desired starch.

Recent advances in the analysis of starch oligosaccharides have made possible a clearer appreciation of their structures. Indeed, an improved concept of starch structure has permitted the use of an approach similar to that employed in the field of synthetic polymer chemistry (Slade and Levin, 1987). This has empowered establishing a working approach towards understanding and applying starch structure-function relationships.

Other areas of science showing great advances are biochemistry and more particularly, biochemical genetics. Significant strides have been made towards identifying different biosynthetic pathways in starch formation (Preis *et al.*, 1985). Different biosynthetic routes operating in concert confer diversity to structures of starch biopolymers. These biosynthetic mechanisms are controlled genetically (Nelson, 1985). As the balance of contributions from different biosynthetic pathways becomes better understood and controlled, it will become increasingly practical to direct the biosynthetic course of production of a specific starch biopolymer structure. The value of this process will increase immeasurably as the structure-function relationships of starch biopolymers become better understood.

Benefits from understanding starch structure, however, will not be limited to enhancements in their structural functionality in food systems. In the same way that different ratios of biosynthetic enzyme systems will yield biopolymers with different structures, so, too, will different biopolymers be depolymerized through diverse catabolic pathways. The significance of starch biopolymer structure for human nutrition is now only beginning to be unraveled. Many major issues remain as questions. Is there a meaningful difference in metabolism between so called "complex" carbohydrates and "simple" carbohydrates (Crapo, 1983)? Does resistant starch (Jane and Robyt, 1984) have any nutritional significance or is it an artifact of analysis? Does the amylose content of a food product have any importance for diabetics (Behall *et al.*, 1988)? These and other questions will have to be resolved by thorough and effective scientific research.

Our discussion of starches in foods has returned full circle to nutrition. Actually, it is more a helix than a circle, for each turn brings us further along towards an enhanced understanding of the roles played by starch biopolymers in food systems. As we advance along this path we will, hopefully, find a greater relationship between the function of starch biopolymers in food systems and their nutritional qualities.

The purpose of today's discussion is to review strategies using biotechnology for controlling functionality as well as nutritional aspects of starch biopolymers in food systems. Included in this review will be perspectives on starch functionalities themselves, in addition to regulatory and public concerns. In this presentation the term "biotechnology" will be used in a broad sense. It will be employed to encompass the wide range of biological processes which can be applied in the food and agriculture industries.

Starch in Nutrition

Even in today's modern society starch is an important dietary element. According to the current recommended dietary allowance tables (NRC, 1989) approximately 55-60 percent of the caloric content of our diet should be derived from carbohydrates. The difference is that, as the understanding of catabolic processes has become sophisticated, it is now possible to select appropriate starch biopolymer structures to meet specific needs. A few examples of this are in order.

Diabetes is a disease of great concern (Kolata, 1983) in our aging society. As a result of its widespread nature, nutritional means to help manage the disease are quite desirable. Early literature (Crapo, 1983) has suggested that "complex" carbohydrates, that is to say larger polysaccharide structures, would be nutritionally preferable as energy sources since they might be metabolized more slowly. More sophisticated analysis has indicated, however, that since digestive enzymes are sufficiently efficient, there really is little difference in availability between the glucose of a dextrose solution and the glucose found in starch (Jenkins *et al.*, 1982).

This general observation on the metabolic utility of starch seems to hold true for "common" starch. Other structural configurations of starch, though, might not yield the same results. It has been known for some time that starch structures may be divided into two general categories. One group of starch biopolymers, called amylose, has mostly linear chains with minimal amounts of short branching. The molecular weights of this group tend to be smaller. The other group of biopolymers is called amylopectin. Here the degree of branching and the size of the branches are quite large; the molecular weights of amylopectins tend to be much larger than those of amyloses. Common, or normal, starches have roughly 25 percent of their composition as amylose while the remainder is amylopectin.

Other commercial varieties of starch with varying ratios of amylose/amylopectin are available; it has been recognized that metabolic differences, as a reflection of starch structural variations, might exist. Indeed, it was noted that high amylose-containing starch was somewhat resistant to the action of amylase (Sandstedt *et al.*, 1962); distinctions were reported between high amylose-containing starches and normal starches when used as animal feed (Ackerson, 1961; Borchers, 1962, Preston *et al.*, 1964) shortly after these types of maize starch were reported in the literature. It has even been recommended that high amylose-containing barley be avoided as feedstock for brewing (Ellis, 1976). It did not take long before these biopolymers were evaluated in terms of human nutrition as well (Wolf *et al.*, 1977). The effects of high amylose starches from rice (Goddard, 1984) and maize (Behall *et al.*, 1988; 1989) on insulin response were soon reported. The nature of the effect is still being studied to document the actual dimensions of the response. It is intriguing to speculate, however, if the metabolic differences are a reflection of structural differences or whether they result from the inability of the catabolic enzymes to depolymerize the linear amylose rapidly.

If, in fact, the latter situation is valid, one can then raise the issue of whether a biopolymer which is resistant to certain enzymes *in vitro* has significance as a distinct class of materials *in vivo*. In recent years a number of different laboratories around the world have recognized the existence of starches which are resistant to the action of amylolytic enzymes (Asp *et al.*, 1986; Jane and Robyt, 1984; Ring *et al.*, 1988; Sievert and Pomeranz, 1989) *in vitro*. The bulk of this literature is concerned with the nature of the resistant starch, analytical methodology, and whether it is appropriate to classify this material as a fiber. It remains to be seen whether there is a relationship between the reported value of high amylose starches for diabetic diets and the analysis of resistant starches.

Starch Functionality

The functionality of starches in food systems is diverse. It is impossible to deal with the subject in adequate depth within the framework of a short chapter such as this. The reader is, thus, referred to current reviews on the subject (Moore *et al.*, 1984). For purposes of this discussion, however, it is important to touch on several of the key interactions of starch biopolymers in food systems.

Starch interactions may be divided arbitrarily between those with low molecular weight compounds and those with large molecular weight substances. Low molecular weight interactions might include associations with water and lower molecular weight lipids while larger molecular weight interactions might include associations with proteins and other polysaccharide molecules. This latter situation includes polysaccharides such as vegetable gums or even other starch molecules. Such intramolecular relationships between starch molecules are especially important in starch gel formation, in retrogradation and, possibly, in staling.

Among the low molecular weight interactions, water interactions are especially important in food systems. Their significance becomes readily apparent in low water systems when viscosity and mouthfeel are important. Associations of water with polymers are a direct consequence of their polymeric structures; highly crystalline polymers, for example, have little room for associating with water molecules. Crystallinity is usually associated with amylopectin structure (Pfannemuller, 1987), however, and is attributed to formation of double helix structures on amylopectin branches (Gidley and Bulpin, 1987; Ring, 1987). Starch interactions with volatile fragrance and flavor compounds are well known (Reineccius, 1989). Structural differences have been shown to be important for controlled release of low molecular weight fragrance compounds (Wienen *et al.*, 1989).

Starch interactions with higher molecular weight materials are important in food functionalities as well. For example, lipids with hydrocarbon-like chains readily form stable complexes with high amylose-containing starches (Holm *et al.*, 1983). High amylose starches are known to form very tough intermolecular aggregates on retrogradation. These aggregates might be a useful bulking agent if they are not metabolized effectively (Jane and Robyt, 1984). What is being emphasized here, however, is that interactions with both high and low molecular weight materials are significantly affected by the structure of the starch biopolymers involved.

Controlling Starch Function

The main point of all the preceding discussion is that both intermolecular and intramolecular associations are especially sensitive to starch biopolymer structure. Modification of the biopolymer structure, even very subtle changes, can have very substantial effects on starch behavior. Contemporary starch biopolymer function control is carried out, on the whole, through chemical modification. Chemically modified starches have proven themselves to be safe, suitable, and economical food ingredients. Their uses have enhanced the quality of life in contemporary society. For example, intermolecular associations can be controlled effectively through chemical crosslinking while retrogradation can be restrained by etherification. The reader is referred to a recent book on chemical modification of starches (Wurzburg, 1986) for a more detailed review of the subject.

A specific and effective means for changing starch functionality, especially when new forms of functionality are realized, is the employment of specific enzyme activities (Saha and Zeikus, 1987). The use of a special glucosyl transferase enzyme activity results in the formation of the family of cyclodextrin compounds (Friedman and Hedges, 1989). Specific amylolytic enzymes have been used in the production of oligosaccharides of distinct properties such as chain length (Inglett, 1988). Enzymatic depolymerization of a starch so that its amylopectin portion has been reduced in size is reported to yield a thermally reversible gel which may have utility when used in a reduced fat system (Bulpin *et al.*, 1984; Schierbaum *et al.*, 1986). Biotechnology has a great deal of potential in this regard since, among other capabilities, it has the capacity for incorporating specific desirable enzyme traits in a target organism (Kennedy *et al.*, 1988; Sharma, 1989).

An alternative method for changing functionality is to seek out other polymers which possess the desired functionality. These might be selected from a list of polymers containing, for example, starches from different plant sources, vegetable gums, microbial polysaccharides, hard-shelled animal derivatives, or synthetic polymers. The difference in behavior between all these materials is grounded in their unique structures; distinctive function is a reflection of unique structural features. Recognizing and characterizing those structural features can confer a level of predictive skill on the user. This is an area where biotechnology will prove most useful.

Role of Biotechnology in Starch Polymer Function Control

Biotechnology today is a buzzword which covers a multitude of sins. Encompassed within that term are genetic engineering, aspects of biochemistry and enzymology, and a more enlightened approach to conventional breeding (Whitehouse, 1977). The latter term is used here to describe selective breeding for a specific biochemical trait. In the past, most mutants were identified by a distinctive phenotypic characteristic. Most biochemical differences were not recognized until after the mutations were expressed in a tangible form. Much progress has been made, especially in maize research, in identifying those biosynthetic components which contribute to the biosynthesis of starch polymer and granule structures (Shannon and Garwood, 1984). The effects of different mutations have been recognized in different starch structures. For example, maize kernels with a waxy phenotype (waxy, *wx*, mutation) were observed to produce a starch which structurally was a form of amylopectin (Powell, 1973). Maize kernels with a dull phenotype (*du* mutation), on the other hand, produce starches with a higher amylose content (Kramer *et al.*, 1958). It must be recognized that these genetic effects are often controlled by multiple factors which might be located on several different genes. In other words, the genetic impact on starch structure is most likely not carried out by a single mutation, but rather by the balance of biosynthetic enzymes which are controlled by different genetic loci. Starch structure, then, might be called a "biochemical phenotype."

The implication of this latter concept is that, in the case of starch polymers, functional differences do not arise only between amylopectin and amylose. In amylopectin structures there are differences of functionality which result from frequency of branches, size of branches, and differences in molecular weights which are just as broad as the differences between an all-amylopectin starch and amylose-containing biopolymers (Wienen *et al.*, 1990; Friedman *et al.*, 1988). As these structure-function relationships continue to be elucidated, the selection of specific biochemical factors becomes feasible. Because of the delicate nature of this biosynthetic balance, the use of an enlightened conventional breeding program becomes a more feasible strategy for producing starches with specific structural characteristics. It would be difficult to produce a complete starch biopolymer using genetic engineering alone, at least with the state of today's art. A similar strategy has been employed for the improvement of cotton fibers (Triplett, 1990).

Regulatory and Public Concerns

The state of technology, however, is not the sole factor in the development of such strategies. Since the quality of the food supply is such a central concern of both the general population and its elected representatives, it becomes important that the population's concerns must be taken into account. A recent review (Salvage, 1990) reveals that the public is concerned about its perception of the quality, healthfulness, and convenience of the food supply. Minimal processing certainly relates to the first two characteristics. The public's perception of what is desirable is usually reflected subsequently in the government's regulatory stance. New products are generally developed with those guideposts in mind.

Genetic engineering is a powerful and useful tool for developing new food products. Its ability to incorporate specific qualities within a normal biological structure is beyond question. As a result of the activities of some activists in addition to the public's general suspicion of science, there appears to be some public uneasiness about the rapid utilization of food products which have been developed through genetic engineering. There is no doubt that products will be forthcoming; their introduction rate will be slow and deliberate, however, and probably will come first through products which will be tangential to the food chain. Animal growth hormones and food treatment enzymes come to mind as possible first entry products. Genetic engineering has permitted the development of tomatoes with reduced polygalacturonase (Sheehy *et al.*, 1988; Smith *et al.*, 1988) which is thought to permit vine ripening of tomatoes without softening. Safety will have to be documented for each developmental step. For example, questions of the safety of some disease resistant plants have been raised recently (Fenwick *et al.*, 1990). Regulatory agencies within our government are monitoring developments within the food sector of biotechnology (FDA, 1989) and guidelines will be established (McNamara, 1986).

Meanwhile, other approaches within the framework of biotechnology will be able to proceed unhampered by regulatory constraints and negative public concern as long as they do not step outside established and accepted procedures. These biotechnology methods will undoubtedly provide the public with those desired food products which show minimal processing, enhanced convenience, and extended stability without sacrificing healthfulness and quality. As the trend towards reduced reliance on chemical modification continues, both as a result of public demand and regulatory restrictions, the need for biopolymers from natural sources will continue to escalate.

Summary

1. The uses of starch biopolymers in food systems were reviewed.
2. These uses include nutritional and functional qualities.
3. Methods of control of starch properties in food systems were discussed.
4. These methods include conventional chemical modification as well as structural developments through biotechnology.
5. Finally, future roles for different aspects of biotechnology in food systems were discussed.

References

Ackerson, C.W. 1961. High amylose corn for day-old chicks. Feed Age: 25.

Asp, N.G., Bjorck, I., Holm, J., Nyman, M. and Siljestrom, M. 1986. Enzyme resistant fractions and dietary fibre. Scand. J. Gastroenterol. Suppl. 129: 29.

Behall, K.M., Scholfield, D.J., Yuhaniak, I. and Canary, J. 1989. Diets containing high amylose vs amylopectin starch: effects on metabolic variables in human subjects. Am. J. Clin. Nutr. 49: 337-44.

Behall, K. M., Scholfield, D. J. and Canary, J. 1988. Effect of starch structure on glucose and insulin responses in adults. Am. J. Clin. Nutr. 47: 428–32.

Borchers, R. 1962. Note on the digestibility of the starch of high-amylose corn by rats. Cereal Chem.: 39.

Bulpin, P.V., Cutler, A.N. and Dea, I.C.M. 1984. Thermally-reversible gels from low DE maltodextrins. In "Gums and Stabilisers For The Food Industry 2." Glyn O. Phillips, David J. Wedlock and Peter A. Williams (Ed.). p. 475. Pergamon Press, Oxford, England.

CIMMYT Maize Facts and Trends 1984 Report Two: An analysis of changes in Third World food and feed uses of maize. p. 2.

Crapo, P. 1983. Food fallacies and blood sugar. New Eng. J. Med. 44.

Ellis, R.D. 1976. The use of high amylose barley for the production of whiskey malt. J. Inst. Brew. 82: 280-281.

F.D.A. 1989. "Biotechnology and the U.S. Food Industry." Technomic Publishing Co., Lancaster, PA.

Fenwick, G.R., Johnson, I.T. and Hedley, C.L. 1990. Toxicity of disease-resistant plant strains. Trends in Food Sci. Technol. 1990: 23.

Friedman, R.B. and Hedges, A.R. 1989. Recent advances in the use of cyclodextrins in food systems. In "Frontiers in Carbohydrate Research - 1: Food Applications." R.P. Millane, J.N. BeMiller and R. Chandrasekaran (Eds.) p. 74. Elsevier Applied Science, New York.

Friedman, R.B., Gottneid, D.J., Faron, E.J., Pustek, F.J. and Katz, F.R. 1988. Foodstuffs containing starch of a dull waxy genotype. U.S. Patent 4,789,557, December 6.

Gidley, M.J. and Bulpin, P.V. 1987. Crystallization of malto-oligosaccharides as models of the crystalline forms of starch: minimum chain-length requirements for the formation of double helices. Carbohydr. Res. 161: 291-300.

Goddard, M.S., Young, G. and Marcus, R. 1984. The effect of amylose content on insulin and glucose responses to ingested rice. Am. J. Clin. Nutr. 39: 388-392.

Holm, J., Bjorck, I., Ostrowska, S., Eliasson, A.C., Asp, N.G., Larsson, K. and Lundquist, I. 1983. Digestibility of amylose-lipid complexes *in-vitro* and *in-vivo*. Starch/Starke 35: 294-297.

Inglett, G.E. 1988. Novel malto-oligosaccharides and their manufacture from starch with alpha-amylase. U.S. Patent Application 189,093. October 15. (C.A. 111:55855y).

Jane, J.L. and Robyt, J.F. 1984. Structure studies of amylose-V, amylose-V complexes, and retrograded amylose by action of alpha amylases, and a new method for preparing amylodextrins. Carbohydr. Res. 132: 105-118.

Jenkins, D.J.A., Taylor, R.H. and Wolever, T.M.S. 1982. The diabetic diet, dietary carbohydrate and differences in digestibility. Diabetologia. 23: 477-484.

Kennedy, J.F., Cabalda, V.M. and White, C.A. 1988. Enzymic starch utilization and genetic engineering. Trends in Biotechnol. 8: 184-189.

Kolata, G. 1983. Dietary dogma disapproved. Science 220: 487-488.

Kramer, H.H., Pfahler, P.L. and Whistler, R.L. 1958. Gene interactions in maize affecting endosperm properties. Agron. J. 50: 207.

McNamara, S. 1986. Regulatory issues in the food biotechnology arena. In "Biotechnology In Food Processing." S.K. Harlander and T. P. Labuza (Ed.). p. 15. Noyes Publications, Park Ridge, NJ.

Moore, C.O., Tuschhoff, J.V., Hastings, C.W. and Schanefelt, R.V. 1984. Applications of starches in foods. In "Starch: Chemistry and Technology." R.L. Whistler, J.N. BeMiller and E.F. Paschall (Eds.). p. 575. Academic Press, Inc.

Nelson, O.E. 1985. Genetic control of starch synthesis in maize endosperms — a review. In "New Approaches to Research on Cereal Carbohydrates." R.D. Hill and L. Munck (Eds.). p. 19. Elsevier, New York.

N.R.C. 1989. "Recommended Dietary Allowances" Food and Nutrition Board. National Academy Press, Washington, D.C.

Pfannemuller, B. 1987. Influence of chain length of short monodisperse amyloses on the formation of A- and B-type X-ray diffraction patterns. Int. J. Biol. Macromol. 9: 106.

Powell, E.L. 1973. Starch amylopectin (waxy corn and waxy sorghum). In "Industrial Gums Polysaccharides and Their Derivatives." R.L. Whistler and J.N. BeMiller (Eds.). p. 567. Academic Press, New York.

Preis, J., MacDonald, F.D., Singh, B.K., Robinson, N. and McNamara, K. 1985. Various aspects in the regulation of starch biosynthesis. In "New Approaches to Research on Cereal Carbohydrates." R. D. Hill and L. Munck (Eds.), p. 1. Elsevier, New York.

Preston, R.L., Zuber, M.S. and Pfander, W.H. 1964. High-amylose corn for lambs. J. Animal Sci. 23: 1182.

Radosta, S. and Schierbaum, F. 1990. Polymer-water interaction of maltodextrins Part III: non-freezable water in malto-dextrin solutions and gels. Starch/Starke. 42: 142-147.

Reineccius, G.A. 1989. Flavor encapsulation. Food Revs. Intl. 1989: 147-176.

Ring, S.G., Gee, J.M., Whittam, M., Orford, P. and Johnson, I. T. 1988. Resistant starch: its chemical form in foodstuffs and effect on digestibility *in vitro*. Food Chem. 28: 97-109.

Ring, S.G. 1987. Molecular interactions in aqueous solutions of the starch polysaccharides: a review. Food Hydrocolloids. 1 (5/6): 449-454.

Rowe, D.E. and Garwood, D.L. 1978. Effects of four maize endosperm mutants on kernel vigor. Crop Science 18: 709.

Saha, B.C. and Zeikus, J.G. 1987. Biotechnology of maltose syrup production. Proc. Biochem. 22(3): 78-82.

Salvage, B. 1990. Charting the course of R & D top industry researchers put their heads together for a new decade. Food Business. 32.

Sandstedt, R.M., Strahan, D., Ueda, S. and Abbot, R.C. 1962. The digestibility of high-amylose corn starches compared to that of other starches. The apparent effect of the ae gene on susceptibility to amylase action. Cereal Chem. 39: 123.

Schierbaum, F., Vorwerg, B., Kettlutz, B. and Reuther, F. 1986. Interaction of linear and branched polysaccharides in starch gelling. Die Nahrung 30: 1047-1049.

Shannon, J.C. and Garwood, D.L. Genetics and physiology of starch development. In "Starch: Chemistry and Technology." R.L. Whistler, J.N. BeMiller and E.F. Paschall (Eds.). p. 26. Academic Press, Inc., New York.

Sharma, B.P. 1989. Genetic modification of enzymes used in food processing. In "Biotechnology and Food Quality." S.D. Kung, D.D. Bills and R. Quatrano (Eds.). pp. 287-305. Butterworth-Heinemann, Boston, Massachusetts.

Sheehy, R.E., Kramer, M., and Hiatt, W.R. 1988. Reduction of polygalacturonase activity in tomato fruit by antisense RNA. Proc. Natl. Acad, Sci. 85: 8805-8809.

Sievert, D. and Pomeranz, Y. 1989. Enzyme-resistant starch. I. Characterization and evaluation by enzymatic, thermoanalytical and microscopic methods. Cereal Chem. 66(4): 342-347.

Slade, L. and Levine, H. 1987. Recent advances in starch retrogradation. In "Industrial Polysaccharides—The Impact of Biotechnology and Advanced Methodologies," S.S. Stivala, V. Crescenzi and I.C.M. Dea (Eds.). p. 387. Gordon and Breach Science Publishers, New York, New York.

Smith, C.J.S., Watson, C.F., Ray, J., Bird, C.R., Morris, P.C., Schuch, W. and Grierson, D. 1988. Antisense RNA inhibition of polyfalacturonase gene expression in transgenic tomatoes. Nature 334: 724-726.

Triplett, B.A. 1990. Evaluation of fiber and yarn from three cotton fiber mutant lines. Text. Res. J. 60: 143-148.

Whitehouse, R.N.H. 1977. Cereal breeding and its future trends. Proc. Nutr. Soc. 36: 127.

Wienen, W.J., Delgado, G.A. and Friedman, R.B. 1990. Studies of branching in major starches containing the waxy gene. Intl. Symp. Cereal and Other Plant Carb. 1990. p. 12

Wienen, W.J., Tenbarge, F.L. and Friedman, R.B. 1989. Encapsulation and controlled release of flavors by corn starches. Proc. Intern. Symp. Control Rel. Bioact. Mater. 16: 283.

Wolf, M.J., Khoo, U. and Inglett, G.E. 1977. Partial digestibility of cooked amylomaize starch in humans and mice. Die Starke 12: 401-405.

Wurzberg, O.B. 1986. "Modified Starches - Properties and Uses." CRC Press, Boca Raton, Florida.

Human Physiological Responses to Dietary Fiber

Barbara O. Schneeman
Department of Nutrition
University of California
Davis, CA 95616

Dietary fiber refers to the components of plant foods that are not digestible by the endogenous enzymes in the mammalian digestive system. The major components include nonstarch polysaccharides and lignin. This definition includes a diverse array of polysaccharides, including: cellulose, hemicelluloses, β-glucans, pectins, and gums. Three physiological responses associated with fiber consumption in human studies are providing fecal bulk, lowering plasma cholesterol, and blunting the plasma glycemic response. The properties that appear most important to understand the physiological responses to dietary fiber are waterholding capacity (WHC), viscosity, bile acid binding capacity (BABC), and microbial degradation. An increase in fecal bulk can be due to increasing the WHC of the fecal residue or supporting increased microbial growth. A high viscosity and BABC are both associated with the ability of a fiber source to lower plasma cholesterol, and high viscosity and WHC contribute to blunting the glycemic response. Improvements in plants that are sources of dietary fiber should consider the functional properties of the fibers as well as the chemical composition.

Introduction

Dietary fiber refers to the components of plant foods that are not digestible by the endogenous enzymes in the mammalian digestive system. The major components are the non-starch polysaccharides, such as cellulose, hemicelluloses, β-glucans, pectins, and gums; in addition, lignin is a nonpolysaccharide included in most definitions of dietary fiber. Plants are a unique source of dietary fiber because of the polysaccharides associated with the plant cell wall and the secretion and storage of polysaccharides in plant cells and seeds. The current interest in dietary fiber is largely associated with the efforts of Burkitt, Painter, Walker and Trowell (Trowell and Burkitt, 1975), who published their observations that certain chronic disorders such as obesity, large bowel cancers, heart disease, diverticulosis, and type 2 diabetes that are common in countries like England or the United States are rarely observed in rural African populations. Their observations led to the hypothesis that fiber consumption is a protective factor against these gastrointestinal and metabolic disorders. An unequivocal role for fiber in preventing these disorders is difficult to establish primarily because of the multifactorial nature of their incidence. For example, genetics can be a contributing factor to the incidence of chronic diseases as well as other dietary factors such as fat intake. However, as a result of the potential association of fiber intake and a lower incidence of certain chronic diseases, numerous clinical and experimental studies have been conducted with dietary fiber to determine its nutritional importance. In this paper the chemical and physical properties of dietary fiber that are related to gastrointestinal function and other physiological responses associated with fiber consumption in human studies will be reviewed.

Properties of Dietary Fiber

The clinical and experimental research conducted with various sources of dietary fiber have demonstrated its potential importance for normal gastrointestinal function, lowering plasma cholesterol, and blunting glycemic response and insulin release. The properties of fiber that are most important in understanding these responses are viscosity, water-holding capacity (WHC), binding of organic molecules, and microbial degradation. Table 1 summarizes these properties that are discussed below.

Viscosity refers to the ability of a fiber source or polysaccharide to thicken a solution so that mobility in the solution is reduced. The ability to increase viscosity is related to the chemical nature of the polysaccharide. For example in pectins the molecular weight and methyl ester content are determinants of viscosity. In oat products the hydrolysis of the beta 1->3 linkages in beta-glucans will decrease viscosity (Tietyen *et al.*, 1990). The consumption of viscous polysaccharides has been associated with the ability of certain fiber sources to decrease plasma cholesterol and blunt the increase in plasma glucose and insulin following a carbohydrate load. Preliminary data from our laboratory have indicated that hydrolysis of the beta-glucan fraction in oats, reducing the viscosity of the oat products, diminishes the ability of oat bran to lower plasma and liver lipid levels in rats.

Property	Physiological Consequences	Types of Fiber Involved	Nutritional Implication
Microbial degradation	• Breakdown of polysaccharide structure in the large intestine • Production of short chain fatty acids (SCFA) and other microbial metabolites • Growth of microflora	• Polysaccharides • Extent of microbial action is dependent on solubility	• Fecal bulking of fiber is dependent on extent of microbial degradation • Increase in stool weight due to residual polysaccharides and/or an increase in microbial cells • SCFAs provide energy to cells and may have metabolic effects • Reduction in the pH of colon contents
Water-holding capacity (WHC)	• Swelling with water in the gut contents • Increased viscosity of gastrointestinal contents • Affects the microbial breakdown of fiber	• Pectins • Gums • β-glucans • Some hemicelluloses	• Increased viscosity slows gastric emptying and the digestion and absorption of nutrients • Increased viscosity interferes with mixing in the intestinal contents • High solubility allows greater microbial degradation
Adsorption/binding of organic molecules	• Interaction with bile acids and digestive enzymes in the intestine	• Pectins • Gums • Lignin • Non-purified fiber sources (e.g. cereal brans, legumes)	• Increased fecal excretion of bile acids • Slows the rate of digestion in the small intestine
Particle size	• Determines surface area exposed • Degree of cell wall disruption due to grinding	• Primarily important for non-purified fiber sources	• Increasing surface area and disrupting cell walls will enhance exposure to microbial action and digestive enzymes

Viscosity has also been associated with slowing gastric emptying, disrupting mixing in the small intestine contents, and thickening the unstirred layer in the small intestine (Schwartz *et al.*, 1982; Edwards, 1990). The results of all 3 of these actions will slow digestion and absorption of nutrients from foods, if viscous polysaccharides are present. Sources of viscous polysaccharides have been shown to delay the disappearance of lipids and carbohydrates from the small intestine, indicating that the ability of certain polysaccharides to increase viscosity *in vitro* undoubtedly has physiological significance (Schneeman, 1990). Tinker and Scheneeman (1989) demonstrated that in rats fed a viscous polysaccharid (guar gum) the disappearance of starch from the small intestine was slower than from animals fed a fiber free diet. Adiotomre *et al.* (1990) demonstrated the importance of viscosity *in vitro* by showing that highly viscous polysaccharides such as guar gum retard the flow of glucose from a dialysis bag whereas a nonvisous material such as wheat bran has a minimal effect on the dialysis of glucose. We and other investigators have proposed that the effect of viscosity on slowing the rate of nutrient absorption as well as shifting the site of absorption to more distal regions of the small intestine is important in explaining the hypocholesterolemic and hypoglycemic responses to some sources of dietary fiber (Schneeman, 1990; Jenkins *et al.* 1978).

Water-holding Capacity (WHO) is a measure of the ability of a fiber source to hold water and is related to the solubility and viscosity of the polysaccharides. For example, certain pectins and many gums are soluble in water and have a relatively high WHC. The ability of these polysaccharides to become viscous in the gut is associated with a relatively high solubility and WHC. In contrast cellulose is an example of a polysaccharide that is relatively insoluble in water and has a low WHC and viscosity.

WHC is important in understanding fecal bulking due to consumption of fiber. However, the relationship between WHC and fecal bulking ability is complex. A high WHC allows greater microbial penetration and subsequent breakdown of the polysaccharide structure. On the other hand, WHC of the fecal residue, which contains undegraded polysaccharides, contributes to stool bulk. Consequently measures of WHC made *in vitro* cannot be used to predict the fecal bulking ability of a fiber source. Adiotomre *et al.* (1990) reported that to predict the ability of certain fiber sources to increase fecal bulk, the WHC of the fiber residue, the dry weight of fiber residue following incubation with fecal microflora, and the ability to produce SCFAs must be considered.

Microbial Degradation is based on the ability of the microflora normally present in the large intestine to degrade the polysaccharides associated with dietary fiber. In healthy human subjects consuming a mixed diet, 70-80 percent of the fiber disappears due to transit through the gut (Cummings, 1986). Microbial action is important for determining the stool bulking ability of sources of dietary fiber, altering the colonic environment, and potentially inducing other metabolic responses. As a result of microbial action carbon dioxide, hydrogen, methane, and short chain fatty acids, of which acetate, butyrate, and propionate are the major anions, are produced in the large intestine.

In the human large intestine, the primary carbohydrate source for the microflora are the nonstarch polysaccharides (NSP). Cummings (1986) estimates that about 20-60 g of carbohydrate enters the human colon daily, which includes some nondigested starch (referred to as resistant starch). Plant foods are the primary source of these carbohydrates. The extent to which these polysaccharides are degraded by the microflora is dependent upon their solubility, water-holding capacity, and chemical composition. Sources of insoluble polysaccharides such as wheat bran and cellulose are not degraded extensively by the microflora, whereas soluble polysaccharides such as pectins or gums tend to be more completely degraded. The interaction of a fiber source with the microflora is a critically important factor in determining the mechanisms by which it increases fecal weight. Nyman *et al.* (1986) published data demonstrating the relationship between microbial breakdown and fecal bulking ability (Table 2).

Table 2. Correlation between recovery of fecal neutral sugars and increment in fecal dry weight.

	% Recovery of fecal sugars	Increase fecal weight (g) per g fiber
Wheat	66 ± 3	1.18 ± 0.09
Apple	19 ± 11	0.35 ± 0.08
Cabbage	10 ± 3	0.46 ± 0.04
Carrot	25 ± 8	0.50 ± 0.06
Guar gum	13 ± 1	0.29 ± 0.14

Nyman *et al.*, 1986

When wheat bran was fed, the recovery of neutral sugars in the fecal collections was relatively high (66 percent) and was associated with the largest increment in fecal weight. The high recovery of neutral sugars indicates a relatively low microbial degradation of wheat bran. On the other hand, a low recovery of neutral sugars (e.g. guar gum), indicating a more complete degradation of the polysaccharide, resulted in a smaller increment in fecal weight. The greater recovery of neutral sugars in the fecal residue in subjects fed wheat bran suggests two mechanisms whereby wheat bran increases fecal bulk: directly by adding bulk to the stool and indirectly by increasing the water-holding capacity of the fecal residue. The fiber source degraded to a greater extent primarily increased fecal bulk by increasing the microbial mass in the stool contents.

In addition to the influence of microbial action on fecal bulking ability, this property is also associated with the effects of fiber on the overall large intestinal environment and production of metabolites during the transit through the large bowel. Polysaccharides that are fermented extensively are more likely to produce SCFAs and lower the pH of the large intestinal contents. The cells in the large intestine can use SCFAs as an energy source, and butyrate may be a primary energy source for cells in the more distal colon (Cummings, 1986). Investigators are now interested in the potential role that butyrate has in the metabolism of colon cells. Experimental evidence obtained with cultured cells suggests that butyrate may be protective against colon cancer. However, animal studies have indicated that highly fermentable polysaccharides increase tumor yield in animals given a chemical carcinogen (Jacobs, 1990). This effect in rodent models is undoubtedly due to the ability of metabolites produced during the fermentation to stimulate mucosa cell growth.

The other two major anions produced during microbial degradation are propionate and acetate and metabolic roles have been proposed for these compounds as well. Propionate appears in the portal circulation and is cleared by the liver. Nishina and Freedland (1991) reported that incubation of isolated hepatocytes from rats with propionate inhibits the incorporation of tritiated water into fatty acids but not cholesterol. It is not known whether this response occurs in intact animals, but may be involved in the effects of certain fiber sources on lipid metabolism. Acetate reaches the peripheral tissues where it can be metabolized to CO_2.

Our current state of knowledge about the microbial actions on dietary fiber in the large intestine are inadequate for us to fully understand the physiological importance of this property. However, when one considers the size and diversity of the colonic microflora, the need for these microflora to adapt readily to different substrates, and the fact that the polysaccharides in dietary fiber are their primary carbohydrate source, it is evident that this is an important area for future research activity—especially if we are to understand the overall importance of dietary fiber for the health of the large intestine.

The *Bile Acid Binding Capacity* of dietary fibers has been associated with their ability to lower plasma cholesterol. The original hypothesis states that by increasing bile acid excretion from the small intestine, fibers that interact with bile acids increase the conversion of cholesterol to bile acids, thus enhancing cholesterol removal from the body. In testing this hypothesis it is evident that bile acid excretion alone is not sufficient or consistent enough to explain completely the hypocholesterolemic effects of certain fiber sources (Pilch, 1987). However, within the small intestine the interaction of fiber with bile acids can also contribute to slowing the rate of lipid absorption since bile acids are important for micelle formation in the gut. Table 3 contains the ratio of bile acids and phospholipids in the aqueous phase of the small intestine contents to the total intestine contents. This ratio is a measure of bile acid or phospholipid binding in the small intestine (Gallaher and Schneeman, 1986).

Table 3. Ratio of bile acids and phospholipids in the aqueous phase of the small intestine contents.

Diet Treatment	Bile acid Binding Ratio	Phospholipid Binding Ratio
Fiber free	0.326±0.038	0.170±0.035
Cellulose	0.336±0.018	0.167±0.017
Cholestyramine	0.134±0.021[ab]	0.071±0.010[ab]
Lignin	0.166±0.015[ab]	0.132±0.030
Chitosan	0.288±0.006[b]	0.085±0.006[ab]
Wheat bran	0.244±0.026[b]	0.120±0.020
Oat bran	0.394±0.035	0.146±0.045
Guar gum	0.636±0.049[ab]	0.145±0.016
Konjac mannan	0.695±0.016[ab]	0.196±0.007

[a]Indicates a significant difference ($p<0.05$) from the fiber free group.
[b]Indicates a significant difference ($p<0.05$) from the cellulose group.
Gallaher and Schneeman, 1986; Ebihara and Schneeman, 1989.

In these experiments the control treatments are the fiber free and cellulose groups, since cellulose does not differ from the fiber free treatment. A ratio less than either of the control groups indicates that the fiber or drug treatment is insoluble in the intestine contents and binds bile acids or phospholipids, removing them from the aqueous phase; whereas, a ratio higher than either control groups indicates that the fiber is soluble in the intestinal contents and interacts with the bile acids or phospholipids in the aqueous phase. The bile acid binding resin, cholestyramine, was included as a positive control to demonstrate the reduction in the ratio that can be achieved with a potent bile acid sequestering agent. The results indicate that lignin, chitosan, and wheat bran significantly lower the ratio, thus binding bile acids in the undigested or insoluble phase of the intestinal contents. Additionally, guar gum and konjac mannan are soluble polysaccharides and interact with bile acids in the aqueous phase. Only cholestyramine and chitosan affected the phospholipid ratio, indicating that the interactions are specific between the bile acids and the other fibers, not a general entrapment of micelles. Among the fiber treatments only guar gum and konjac mannan have been shown to slow the disappearance of triglyceride and cholesterol from the small intestine. Thus the interaction in the aqueous phase is undoubtedly important during lipid digestion and absorption. Cholestyramine, chitosan, oat bran, guar gum, and knojac mannam have been shown to lower plasma cholesterol whereas wheat bran and cellulose have not. These results indicate that interaction with bile acids is not the only factor that predicts the effect of a fiber source on fecal bile acid excretion or plasma cholesterol.

In addition to the potential importance for lipid absorption, the interaction of fibers with bile acids illustrates that the presence of fiber in the lumen of the small intestine can have significant effects on the utilization of nutrients from foods that contain fibers since they will alter the absorptive process.

Nutritional Responses

In healthy individuals, fiber is needed in the diet for normal gastrointestinal function. Several investigators have proposed that the adequacy of fiber intake can be assessed by the amount of fiber needed to maintain wet stool weight at 160-200 g per day and transit time at less than two days (Spiller, 1986; Pilch, 1987). Various estimates indicate that no less than 20 g and up to 45 g per day of fiber are needed to maintain the stool output. Because of the variation in the properties of fibers, as discussed above, the daily fiber intake should come from a mixture of sources. Table 4 shows the increase in stool weight per gram of fiber fed using data that Cummings has compiled from various sources.

Table 4. Increase in stool weight per gram of fiber fed.

Fiber Source	Grams
Wheat bran	5.7±0.5
Fruits & Vegetables	4.9±0.9
Oat products	3.9±1.5

Cummings, 1986 (Reproduced with permission).

These data demonstrate that coarse wheat bran is one of the most effective fiber sources to increase stool bulk. This effect is probably due to the various properties of wheat bran that allow it to be partially degraded by colonic microflora but not completely; thus it increases stool weight directly by increasing the water-holding capacity of the fecal residue. Because fruits, vegetables and oat products contain a higher proportion of soluble polysaccharides, they are more completely degraded by the microflora. Hence a greater percentage of the increase in fecal mass may be due to an increase in microbial cells. For example when cabbage fiber was fed to human subjects, fecal weight increased about 70 percent but the fiber itself was completely degraded and did not appear in the stool. Isolated polysaccharides also increase stool weight; however, the increase is generally less than that calculated for the nonpurified food sources of fiber.

Several food manufacturers are interested in producing foods that contain fibers that have been shown to lower plasma cholesterol levels. The effect of fiber on blood lipids involves a complex set of dietary interactions. In part, foods that provide dietary fiber are foods that are lower in total fat and saturated fatty acid content (e.g. cereals, fruits, vegetables, and legumes). However, the experimental evidence indicates that certain fibers have a hypocholesterolemic effect that is independent of lowering fat intake. It is important that we consider methods to increase the availability and likelihood of consuming these sources of fiber, but without adding additional fat to the diet. A review of the clinical and experimental studies that have been conducted indicates that wheat bran and cellulose, sources of non-viscous, insoluble polysaccharides, do not lower plasma cholesterol; whereas pectin, guar gum, oat bran, psyllium husk, beans (legumes) and fruits and vegetables lower plasma cholesterol and specifically LDL-cholesterol levels (Pilch, 1987). These fiber sources can be characterized as containing viscous, soluble polysaccharides. The exact reasons why these properties are important are, as yet, not well understood. Viscous polysacchrides can slow cholesterol and triglyceride absorption from the small intestine, increase fecal bile acid excretion, and increase the production of SCFA in the large intestine. Each of these factors probably contributes to the hypocholesterolemic response but the relative importance is unclear.

Conclusions

The carbohydrates included in the definition of dietary fiber are a nutritionally important part of plant foods because of their importance for maintaining bulk in the large intestine and a normal transit time, providing substrates for the microflora, and regulating the rate and site of nutrient absorption from the small intestine. These attributes of fiber are important in normal, healthy individuals. The properties of dietary fiber that determine its contribution to these essential functions are microbial degradation, viscosity, water holding capacity, and the interaction with bile acids. These properties are related to the presence of cell wall polysaccharides in foods. Modifications of plant foods to improve crop value need to maintain these attributes.

In addition to the importance of fiber in healthy individuals, clinical evidence indicates that polysaccharides that are soluble and viscous may have therapeutic use in the treatment of hypercholesterolemia and hyperglycemia.

Acknowledgements

Research reported in this paper has been supported in part by NIH grant DK20446. The author is grateful for the research contributions from Janet Tietyen, Dan Gallaher, Ph.D., and Dr. Kiyoshi Ebihara.

References

Adiotomre, J., Eastwood, M.A., Edwards, C.A. and Brydon, W.G. 1990. Dietary fiber: *in vitro* methods that anticipate nutrition and metabolic activity. Am. J. Clin. Nutr. 52: 128.

Cummings, J.H. 1985. Cancer of the large bowel. In "Dietary Fibre, Fibre-Depleted Foods and Disease." p. 161. Academic Press, London.

Cummings, J.H. 1986. The effect of dietary fiber on fecal weight and composition. In "CRC Handbook of Dietary Fiber in Human Nutrition." p. 211. CRC Press, Boca Raton, FL.

Ebihara, K. & Schneeman, B.O. 1989. Interaction of bile acids, phopholipids, cholesterol, and triglyceride with dietary fibers in the small intestine of rats. J. Nutr. 119: 1100.

Edwards, C.A. 1990. Physiological effects of fiber. In "Dietary Fiber: Chemistry, Physiology, and Health Effects." p. 167. Plenum Press, New York.

Gallaher, D. and Schneeman, B.O. 1986. Intestinal interaction of bile acids, phospholipids, dietary fibers, and cholestyramine. Am. J. Physiol. 250: G420.

Jacobs, L.R. 1990. Influences of soluble fibers on experimental colon carcinogenesis. In "Dietary Fiber. Chemistry, Physiology, and Health Effects." p. 389. Plenum Press, New York.

Jenkins, D.J.A., Wolever, T.M.S., Leeds, A.R., Gassull, M.A., Haisman, P., Dilawari, J., Goff, D.V., Metz, G.L. and Albertti, K.G.M.M. 1978. Dietary fibres, fibre analogues, and glucose tolerance: importance of viscosity. Br. Med. J. 1: 1392.

Nishina, P.M. and Freedland, R.A. 1991. Effects of propionate on lipid biosynthesis in isolated rat hepatocytes. J. Nutr. (In press).

Nyman, M., Asp, N-G., Cummings, J.H. and Wiggins, H. 1986. Fermentation of dietary fibre in the intestinal tract: comparison between man and rat. Br. J. Nutr. 55: 487.

Pilch, S.M. (editor) 1987. "Physiological effects and health consequences of dietary fiber" (FDA 223-84-2059). Federation of American Societies for Experimental Biology, Bethesda, MD.

Schneeman, B.O. 1990. Macronutrient absorption. In "Dietary Fiber. Chemistry, Physiology, and Health Effects." p. 157. Plenum Press, New York.

Schwartz, S.E., Levine, R.A., Singh, A., Scheidecker, J.R. and Track, N.S. 1982. Sustained pectin ingestion delays gastric emptying. Gastroenterology 83: 812.

Spiller, G.A. 1986. Suggestions for a basis on which to determine a desirable intake of dietary fiber. In "CRC Handbood of Dietary Fiber in Human Nutrition." p. 281. CRC Press, Boca Raton, FL.

Tietyen, J.L., Nevins, D.J. and Schneeman, B.O. 1990. Characterization of the hypocholesterolemic potential of oat bran. FASEB J 4: A527.

Tinker, L.F. and Schneeman, B.O. 1989. The effect of guar gum and wheat bran on the disappearance of ^{14}C-labled starch from the rat gastrointestinal tract. J. Nutr. 119: 403.

Trowel, H.C. and Burkitt, D.P. 1975. "Refined Carbohydrate Foods and Disease." Academic Press, London.

Molecular Strategies to Optimize Forage and Cereal Digestion by Ruminants

Cecil Forsberg
University of Guelph
Guelph, Ontario, Canada
K.-J. Cheng
Agriculture Canada Research Station
Lethbridge, Alberta, Canada

Introduction

The beef and dairy industries are economically important and indeed, essential sectors of national economies (Table 1). Sheep production is a minor component, but may increase as farming strategies change. The present trend in agricultural practice is to reduce the dependence on chemical inputs which lead to serious pollution problems and which, in some instances, are not realizing the productivity they once did (MacRae *et al.*, 1990; National Research Council, 1989). This new ecological approach to agricultural practice is frequently called sustainable, or alternative agriculture, and it relies upon crop rotations, crop residues and animal manure (Liebhardt *et al.*, 1990; National Research Council, 1989), and puts into practice techniques employed in developing countries (Bradford, 1989). This strategy depends upon an increase in legumes and grasses to maintain and improve soil fertility, and increases their availability for cattle feed. The associated change in feeding practices probably will result in more decentralized animal production, and because

Table 1. Value of the American and Canadian beef, dairy and sheep industries

Product	Cash Receipts ($Billions)	
	United States[a]-1988	Canada[b]-1989
Cattle and calves	36.3	3.9
Dairy products	17.7	3.1
Sheep	0.5	.003
TOTAL	54.5	7.0

[a] Economic Indicators of the Farm Sector, National Financial Summary, 1988. USDA.ECIFS 8–1

[b] Statistics Canada.

cereal grain production will be reduced, a greater dependence on forage for high productivity in ruminant animals can be expected. Pretreatment of feed and the feeding of growth promotants will be restricted due to cost and pollution problems, thus there will be a greater dependence on the genetic potential of forage crops, and the digestive capacity of rumen fermentation. Since cereal grains are widely fed to ruminants to maintain high productivity, it is also necessary to focus on efficient digestion of cereals by ruminants. Because of these requirements, further research on strategies to optimize microbial digestion of forages in the rumen has been recommended (National Research Council, 1989) and, since cereal grains are widely fed to ruminants to maintain high productivity, similar research is necessary for the cereals.

The major factors regulating ruminant feed digestion include: (i) plant structure which regulates bacterial access to nutrients; (ii) animal factors which increase the availability of nutrients through mastication and salivation, and provide a stable environment; (iii) microbial factors which control adhesion and hydrolysis by complexes of oriented hydrolytic enzymes of the adherent microorganisms (Cheng *et al.*, 1990; Forsberg and Cheng, 1990).

The objective is to optimize digestive processes in order to accelerate digestion of forages and other fibrous materials such as straw, while slowing the utilization of readily digestible feeds such as cereals and alfalfa which frequently cause acidosis and bloat, respectively. Strategies used to maximize the efficiency of rumen fermentation include: (i) biological and/or chemical pretreatment of feed, (ii) the inclusion of feed additives, (iii) plant breeding, and

(iv) genetic manipulation of rumen microorganisms. The combination of methods employed depends upon the circumstances that prevail, although feed pretreatment and additives are only short term solutions to the problem. Before examining the various strategies for optimizing digestion, we will review the structure of the plant cell wall of forages and cereal grains, and the state of research on the biodegradation of feed by rumen microorganisms.

Structure of Plant Cell Walls and Cereal Starch

Forages and cereals differ qualitatively and quantitatively in composition (Table 2). Forages contain a large proportion of cell wall material which is high in cellulose and hemicellulose. In contrast, cereals are low in these polymers but contain high concentrations of starch.

The plant cell walls of grasses share common structural features. Cellulose microfibrils in a typical primary cell wall are elliptical in cross section, with axes 0.005 to 0.300 nm, and occupy some 15 percent or more of the wall. The microfibrils result from the association of individual molecules of cellulose, a hydrogen-bonded β-1,4-linked D-glucan, in crystalline or near crystalline arrays (Figure 1). In some primary cell walls the layers of microfibrils are at right angles to each other (Chafe, 1970).

The hemicellulose fraction is a major component of the cell walls of both stems and leaves of legumes and grasses (monocotyledonous plants) with the ratio of cellulose to hemicellulose ranging from

Table 2. Chemical composition of grasses and cereals

Plant Material	Percent of dry weight					
	Cell Walls	Cellulose	Hemi-cellulose	Lignin	Protein	Starch
Grasses and legumes[a]	40–77	24–32	8–38	3–8	7–17	–[c]
Cereals[b]	9–19	2–5	6–12	1–2	11–14	60–80

[a]Van Soest, 1982.

[b]United States - Canadian tables of feed composition.

[c]Not available.

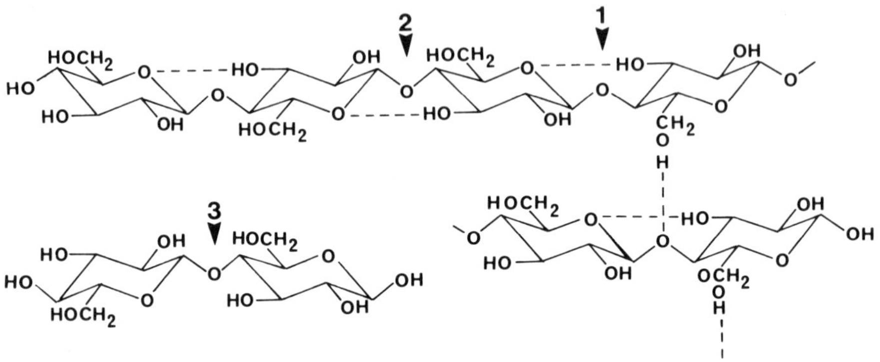

Figure 1. Major linkages in the cellulose fibrils of the plant cell wall. Dashed lines illustrate intra-strand and inter-strand hydrogen bonding between glucose residues. Sites of cleavage by enzymes are illustrated by numbers: 1, endoglucanase cleaves randomly within a glucan strand; 2, cellobiohydrolase (or cellobiosidase) cleaves disaccharides from the nonreducing end of the glucan strand; 3, cellobiase (β-glucosidase) cleaves the dissacharide.

0.8–1.6:1 (Wilkie, 1979). The hemicellulose fraction is composed mainly of xylans with a backbone structure of β-1,4-linked xylose residues (Figure 2). The strands of xylan are hydrogen bonded to cellulose fibrils. In grasses, 25 to 53 percent of the xylose residues are acetylated at the O-2 and/or O-3 positions (Chesson *et al.*, 1983). Single L-arabinofuranosyl groups are attached to O-3 atoms of the main chain D-xylose residues, and D-glucuronopyranosyl uronic acid, or its 4-methyl ether, or both, are attached to O-2 atoms of other main-chain D-xylosyl residues. In grass and cereal plant non-endosperm cell wall tissue, arabinose and glucuronic acid are present in approximately equimolar concentrations of 1.6–3.6 percent each (Waite and Gorrod, 1959) and occupy 12 percent of xylose residues (Wilkie, 1979). In grasses, 40 to 50 percent of the arabinose residues linked to xylan are themselves substituted in the O-5 position (Chesson *et al.*, 1983). In barley straw, 1 in 15 arabinose residues is esterified at the O-5 position with a ferulic acid residue while 1 in 31 is esterified with coumaric acid (Mueller-Harvey *et al.*, 1986). Some of the ferulic acid and coumaric acid residues linked to xylan form cyclodimeric (Figure 2) and dihydroxytruxillic acid interstrand cross-links (Hartley and Ford, 1989; Hartley and Jones, 1976). Data

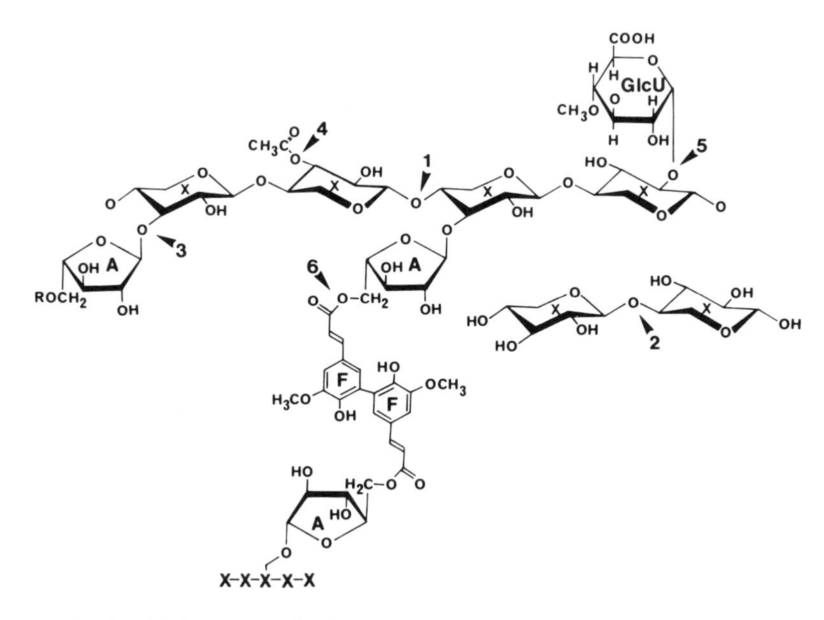

Figure 2. **Major linkages in the hemicellulose polymers of the plant cell wall. The proportions of sugars are not representative of cell walls. Abbreviations: X, xylose; A, arabinose; GlcU, glucuronic acid or 4-O-methylglucuronic acid; F, ferulic acid. The ferulic acid may be replaced by coumaric acid. The phenolic acids may form dimers crosslinking xylan strands; a dehydrodiferulic acid crosslink is illustrated. Sites of cleavage by enzymes are illustrated by numbers: 1, xylanase - cleaves the xylan backbone in a random fashion; 2, xylobiase - cleaves xylobiose; 3, arabinofuranosidase; 4, acetylxylan esterase; 5, α-glucuronidase; 6, ferulic acid esterase.**

from other studies demonstrate that ferulic and coumaric residues serve as linkages between arabinoglucuronoxylan and lignin (Jung, 1989; Scalbert *et al.*, 1985). It has been proposed that if all ether linked ferulic acid residues in *Zea mays* (maize) were involved in bridging between lignin and polysaccharide, these bridges would be 5–10 times more abundant than the diferulic acid bridges between polysaccharides (Iiyama *et al.*, 1990).

Structural proteins called extensins are commonly found in dicotyledonous cell walls (Cassab and Varner, 1988). They are basic flexuous rod-shaped proteins rich in hydroxyproline residues glycosylated with arabinose. They frequently form intermolecular isodityrosine cross-links (Fry, 1986) and, as a consequence, entrap other polymers within the wall. Similar extensin-like proteins have recently been isolated from *Zea mays* (Kieliszewski *et al.*, 1990) and from *Beta vulgaris* (sugar beet) (Li *et al.*, 1990). The presence and role of these types of proteins in the grasses remain to be determined. It is possible that in some grasses their presence may be an important factor in determining the rate and extent of microbial forage digestion.

The surface cuticle of leaves, stems, and grain is a potent barrier to microbial penetration (Forsberg and Cheng, 1990; Nordin and Campling, 1976), but is given comparatively little consideration because of extensive mastication by ruminants (Bailey, 1962; DeBoever *et al.*, 1990).

A somewhat different worty layer on the surface of barley straw also restricts access by rumen microorganisms (Engels and Brice, 1985). The presence of these layers, which prevent entry of microorganisms except at damaged sites, results in an "inside-out" digestive process (Cheng *et al.*, 1990) where microbial degradation proceeds on the inside leaving the resistant shell largely intact (Akin, 1989).

The composition of cereal grains differs dramatically from the stems and leaves of forages and cereals (Table 2). Cereal grains are complex in structure. The interior contains endosperm cells filled with starch granules which are embedded in a protein matrix (Rooney and Pflugfelder, 1986). Endosperm cell walls represent only a minor component of the starch granule-laden cells and are composed of hemicellulosic polymers. In wheat endosperm cell walls, arabinoxylan predominates, while in barley a β-D-glucan predominates (Selvendran, 1983).

The highly organized starch granules range in size from submicron to 200 μm in diameter. They are held together by hydrogen bonding, contain highly organized crystalline as well as relatively unorganized amorphous regions, and are surrounded by a protein matrix which in wheat consists of a 30,000 dalton highly basic protein (Gallard, 1983). Starch is a glucan composed of amylose and amylopectin. Amylose is a linear polymer of α-1,4-linked D-glucose units and usually accounts for 20 to 30 percent of normal cereal starch, while little or none is present in waxy starches. Amylopectin accounts for the balance of the starch. It is composed of linear chains of α-1,4-linked D-glucose with α-1,6-branch points every 20 to 25 glucose residues. Amylopectin is the only starch in waxy genotypes of corn, sorghum, barley, rice and millet.

Major Fibrolytic Organisms, Their Interactions, and Adhesion to Fiber

In the rumen, fibrous plant materials are degraded by a limited number of bacteria, fungi and protozoa. The bacteria always predominate numerically and appear to be responsible for the bulk of forage digestion (Figure 3; Akin and Benner, 1988; Cheng *et al.*, 1990). The major fibrolytic bacteria include *Fibrobacter succinogenes*, *Ruminococcus flavefaciens*, and *R. albus* (Cheng *et al.*, 1990). *Butyrivibro fibrisolvens* appears to have a role only under restricted conditions such as in the high arctic Svalbard reindeer (Orpin *et al.*, 1985). Generally speaking, the *Ruminococcus* species tend to predominate during the degradation of more readily available forms of cellulose, while *F. succinogenes* increases in number when mature forages and other recalcitrant forms of cellulose, such as barley and wheat straw, are consumed by the ruminant (Cheng *et al.*, 1990). Fibrolytic bacteria tend to degrade the more readily digestible structures such as the mesophyll cells, although *F. succinogenes* digests parenchyma bundle sheaths, epidermal cell walls and leaf sclerenchyma (Akin, 1989). *F. succinogenes* interacts synergistically with non-cellulolytic bacteria during forage digestion. Examples of these synergistic actions include a doubling in the utilization of orchard grass hemicellulose and pectin by co-culture with the hemicellulolytic bacterium, *Bacteroides ruminicola*, over that utilized by either organism alone (Osborne and Dehority, 1989) and a 15 to 20 percent increase in barley straw digestion by co-culture with the non-cellulolytic bacterium, *Treponema bryantii* (Kudo *et al.*, 1987).

The fungi have received less attention than the bacteria, primarily because they were comparatively recently identified (Figure 4; Orpin and Joblin, 1988). The fungi are universally present in the rumen and have been reported to account for as much as 8 percent of the microbial biomass (Orpin and Joblin, 1988). Their removal results in an increase in the propionic acid concentration in rumen fluid and a reduction in barley straw degradation during incubation of the material *in vivo* in nylon bags, thereby suggesting that they have a significant role in the rumen (Elliot *et al.*, 1987). The rumen fungi have two unique attributes. First, they are able to penetrate both the cuticle and the cell wall of lignified tissues (Akin *et al.*, 1986), which may suggest that they possess cutinase activity (Kolattukudy, 1984). Second, they have the ability to degrade more recalcitrant cell wall materials, including the sclerenchyma and vascular tissue (Akin,

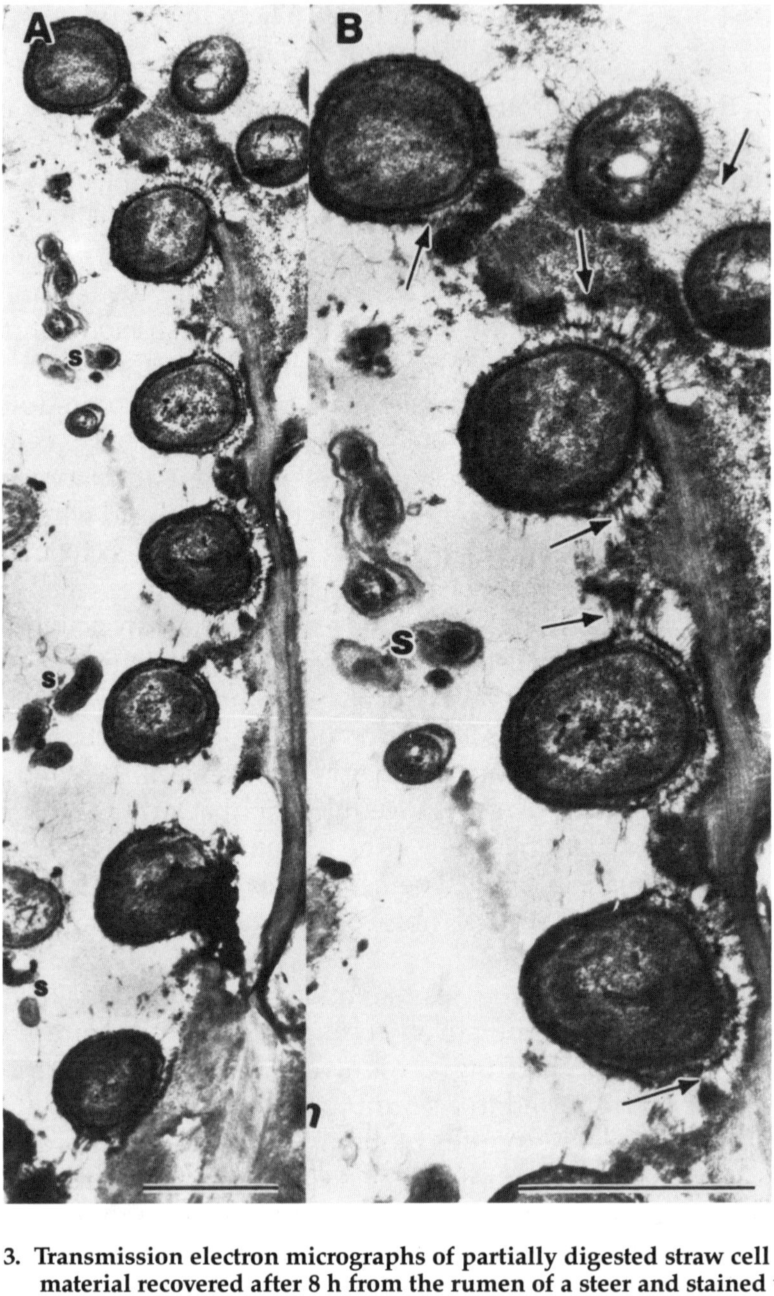

Figure 3. Transmission electron micrographs of partially digested straw cell wall material recovered after 8 h from the rumen of a steer and stained with ruthenium red during fixation. Arrows indicate glycocalyx fibers (collapsed during dehydration) by which the Gram-positive cocci (*Ruminococcus* spp.) are specifically adhered to the straw. 'S' indicates spiral cells of a *Treponema* species often seen to be associated with the monolayer of cellulolytic bacteria which are the primary colonizers of the cellulosic substratum. Bars indicate 1.0 μm. (Reproduced with permission from Cheng *et al.*, 1990; Copyright 1990 Academic Press, Inc.)

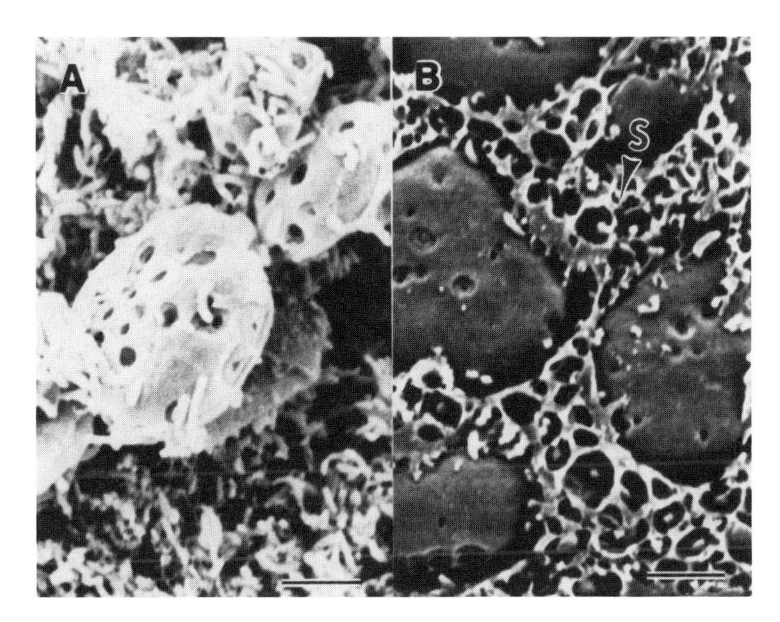

Figure 4. Scanning electron micrographs of hammer-milled barley after incubation with rumen microorganisms for 24 h. A. Untreated barley - the protein matrix in which the starch granules are embedded has been digested away, exposing the granules to microbial attack and eventual digestion. B. Formaldehyde-treated barley - the protein matrix (see arrow) is largely undigested, which restricts access of digestive bacteria to the starch granules (S) thereby slowing the overall rate of barley starch digestion. Bars indicate 5.0 μm. (Reproduced with permission from McAllister *et al.*, 1990a; Copyright 1990 Canadian Journal of Animal Science.)

1989). For example, they degrade 37 to 50 percent of barley straw while the rumen bacteria digest only 14 to 25 percent (Joblin *et al.*, 1989). The fibrolytic activity of the fungi which includes both cellulase and hemicellulase activities is enhanced by growth with hydrogen utilizing methanogens (Joblin *et al.*, 1989; 1990) which decrease the repressive effect of hydrogen (Orpin and Joblin, 1988).

The fact that fungi do not tend to predominate may be due to several factors. First, they have slower growth rates than the bacteria, with doubling times of 6 to 9 hours with simple sugars as the sources of carbohydrate (Lowe *et al.*, 1987), as compared to 0.5 to 3.5 hours for the bacteria (Huang and Forsberg, 1990; Varel and Jung, 1986). Second, their growth is repressed by culture with some bacteria, for example, *Ruminococcus* sp. (Bernalier *et al.*, 1988; Joblin *et al.*, 1990). Research to elucidate the interactions of the bacteria and the fungi may result in the development of methods to circumvent this growth repression, thereby enabling enhanced fungal fibrolytic activity in the rumen.

In vitro studies have suggested that 19-28 percent of total cellulase activity is attributed to the protozoa (Gijzen *et al.*, 1988), although digestion appears to be limited to very susceptible tissue such as mesophyll cells (Akin, 1989). Defaunation to remove the protozoa inhibits fiber digestion. However, in the absence of protozoa there is an increased requirement for non-protein nitrogen because of an increase in the bacterial population; therefore a shortage of nitrogen may account, at least in part, for the reduction in fiber digestion (Ushida and Jouany, 1990). The protozoa also reduce the tendency for acidosis in animals receiving a high cereal ration because they engulf starch granules and ferment the starch (Williams and Coleman, 1988) more slowly than do the bacteria.

Adhesion of the fibrolytic bacteria and fungi appears to be essential for digestion (Kudo *et al.*, 1987; Cheng *et al.*, 1990). Of the rumen fibrolytic bacteria, *F. succinogenes* binds the most tenaciously (Miron *et al.*, 1989). Adhesion is reduced by treatment of cells with trypsin, thereby suggesting that binding involves a surface protein(s) (Gong and Forsberg, 1989). Mutants of *F. succinogenes* unable to bind to cellulose have been isolated and were separated into different classes because of different growth patterns on cellobiose, amorphous cellulose and crystalline cellulose (Gong and Forsberg, 1989). We do not know how many proteins are involved in adhesion, or whether any of them are cellulase enzymes. A role for cellulase enzymes in adhesion would seem reasonable because at least one possesses a cellulose-binding domain (McGavin and Forsberg, 1989).

Hydrolytic Enzymes of Rumen Microorganisms

The rumen bacteria possess a vast array of hydrolytic enzymes including cellulases and hemicellulases, but they lack ligninases (Cheng *et al.*, 1990; Forsberg and Cheng, 1990). Most research has been conducted on *F. succinogenes* and this genus has been shown to possess all of the enzymes illustrated in Figures 1 and 2. Enzymes purified to homogeneity and characterized include three endoglucanases (McGavin and Forsberg, 1988; McGavin *et al.*, 1989), two cellobiosidases (Huang *et al.*, 1988; Huang and Forsberg, 1988), two xylanases (Matte and Forsberg, unpublished data) and an acetylxylan esterase (McDermid *et al.*, 1990a). Other enzymes known to be present include a β-glucosidase (Groleau and Forsberg, 1981), an arabinofuranosidase, a ferulic acid esterase (McDermid *et al.*,

1990b) and an α-glucuronidase (Smith and Forsberg, unpublished data). Erfle *et al.* (1988) have also characterized a cloned mixed linkage glucanase (β-1,3;1–4–β–D-glucanase) from this bacterium. In addition, gene cloning of the endoglucanases has revealed the presence of at least six endoglucanases (Crosby *et al.*, 1984); therefore we will probably be able to purify more of these enzymes from *Fibrobacter.* Similarly we know that there is at least one more xylanase in *Fibrobacter* based on enzymological data and gene cloning. A detailed discussion of the cellulase enzymes has been presented elsewhere (Forsberg and Cheng, 1990); however, it is worth noting that the studies have progressed to structural details, for example, we have found that the endoglucanase 2 has a high affinity for crystalline cellulose and that the enzyme is composed of two domains, a catalytic domain and a cellulose binding domain (McGavin and Forsberg, 1989). Endoglucanase 2, like many other endoglucanases, exhibits catalytic activity on amorphous cellulose, short chain cellooligosaccharides and, in addition, both lichenin and barley β-glucan (McGavin and Forsberg, 1988). The xylanases that were purified have rather unique properties as well. One xylanase hydrolyzes arabinoxylan, commonly found in rye and wheat, by first cleaving off a large proportion of the arabinose residues from the xylan backbone, then hydrolyzing the β-1,4-linkages of the main chain. The second xylanase exhibits endoglucanase activity as well as xylanase activity (Matte and Forsberg, unpublished data).

Ruminococcus albus and *R. flavefaciens* also possess a variety of cellulases and hemicellulases. Enzymes purified from *R. albus* include an endoglucanase (Ohmiya *et al.*, 1987), a cellobiosidase (Ohmiya *et al.*, 1982), a β-glucosidase (Ohmiya *et al.*, 1985), a xylanase (Greve *et al.*, 1984a), and an arabinofuranosidase (Greve *et al.*, 1984b). Only one enzyme, a cellobiohydrolase, has been purified from *R. flavefaciens* (Gardner *et al.*, 1987), although numerous enzymes including endoglucanases and xylanases are produced by this bacterium (Doerner and White, 1990).

The presence of a broad array of fibrolytic enzymes helps explain why these bacteria are able to compete successfully in the rumen environment. The synergistic interaction of an arabinofuranosidase and a xylanase from *R. albus* (Greve *et al.*, 1984a) documents cooperativity required for efficient cleavage of the complex array of wall polymers. Additional studies using this approach are essential; however, for most of the bacteria, difficulty with obtaining large quantities of purified enzymes has precluded these more detailed studies. Amplification of cloned genes will eventually solve the problem.

The anaerobic rumen fungi, as well, possess a broad range of fibrolytic enzymes including cellulases and xylanases (Borneman *et al.*, 1989; Cheng *et al.*, 1990; Williams and Orpin, 1987a and 1987b). One fungus, *Neocallimastix frontalis*, has been documented as having the highest cellulolytic activity of any organism ever reported in the literature (Wood *et al.*, 1986), and a *Neocallimastix* sp. reportedly degrades straw more extensively than the aerobic fungus *Trichoderma reesei* (Lowe *et al.*, 1987) which has been acclaimed as the prototype for industrial fermentation (Wood *et al.*, 1986). Recently, Borneman *et al.* (1990) demonstrated the presence of both ferulic and coumaric acid esterase activities in two monocentric and three polycentric fungi. Whether the presence of both of these enzymes is a factor in their high fibrolytic activity remains to be determined. To date only three fungal glycosidases have been purified and characterized (Hebraud and Fevre, 1988).

The protozoa possess cellulases, xylanases and a broad range of glycosidases although none have been purified and characterized in detail (Cheng *et al.*, 1990).

Because the cell wall is composed of a matrix of intertwined polymers, it is difficult to pick out one specific linkage as being primarily responsible for the slow rate of wall degradation (Ford and Elliot, 1987). However, present evidence points to cross-links formed between xylan chains by esterified ferulic and p-coumaric acids, and to links between lignin and hemicellulose by the same phenolic acids ether linked to lignin, as the components limiting the rate and extent of cell wall digestion (Bohn and Fales, 1989; Hartley and Ford, 1989; Iiyama *et al.*, 1990). The location of the phenolic compounds is also important. For example, in bermuda grass stems the epidermis, sclerenchyma ring and xylem contain high concentrations of phenolic acids as demonstrated by use of specific histochemical stains and microspectrophotometry. These tissues are highly resistant to rumen microbial degradation (Hartley *et al.*, 1990).

Starch Digestion

The role of rumen microorganisms in the digestion of raw starch has received surprisingly little attention. Rumen bacteria with high amylolytic activity include *Streptococcus bovis* JB1, *Ruminobacter* amylophilus H18, *Butyrivibrio fibrisolvens* A38, and *Bacteroides ruminicola* 23 and B_14 (Cotta, 1988). When grown on starch, these bacteria produced α-amylase which is largely cell-associated. Other organisms lacking α-amylase, including *B. fibrisolvens* D1, *Selenomonas ruminantium* D, and *Succinovibrio dextrinosolvens* 22-B, also grew on starch, and it was suggested that they possess β-amylase which releases maltose as the product of hydrolysis that could be readily used as a carbon source. When amylolytic bacteria were grown on milled barley, wheat, and maize, different digestion patterns were observed (McAllister *et al.*, 1990b). *S. bovis* digested starch in wheat to a greater extent than that in maize or barley, and colonized endosperm randomly. *R. amylophilus* was more active on barley than maize or wheat, and colonized starch granules preferentially over protein. In contrast, *B. fibrisolvens* was shown to preferentially colonize endosperm cell wall material, and utilized protein when growing on wheat. The overriding feature influencing starch digestion was the association between starch and protein within the endosperm. Therefore it is not surprising that the major amylolytic bacteria also exhibit high proteolytic activity (Wallace and Cotta, 1988). It can be concluded from this data that the contribution of any one species to digestion of starch and protein in cereal grains is dependent upon the grain fed.

Selected species of fungi (Williams and Orpin, 1987a) and protozoa (Prins and Van Hoven, 1977) also utilize starch, but their respective hydrolytic enzymes have not been studied in detail. The fungi are proteolytic (Wallace and Joblin, 1985) as well as being amylolytic, therefore they have suitable attributes for cereal grain colonization.

Amylose digestion begins in amorphous regions and proceeds to crystalline regions, where digestion is much slower (Rooney and Pflugfelder, 1986). Digestibility is generally inversely proportional to amylose content. As indicated earlier, cereal starch granules are completely embedded in a protein matrix which protects the starch and decreases the rate of starch hydrolysis. The greatest inhibition occurs in sorghum and the least in barley and wheat. To reduce the rate of starch digestion in barley by the rumen microflora, treatment with formaldehyde to cross-link proteins has proven successful (McAllister *et al.*, 1990a). This appears to inhibit access of bacteria to underlying starch granules presumably because it blocks access to the granule. The cross-linking through free amino groups on the protein presumably interferes with hydrolysis of the surface protein by proteases, and the protein net has too small a mesh for bacterial entry.

Genes Coding for Fibrolytic Enzymes

The opportunity to genetically manipulate rumen microorganisms has served as a catalyst for the cloning of genes coding for fibrolytic enzymes. At this time at least 34 genes coding for fibrolytic enzymes of rumen bacteria have been cloned, and seven have been sequenced (Table 3). This has contributed to a valuable gene pool for biotechnological studies, but more importantly at the present time helps develop a greater understanding of the range of different enzymes possessed by the rumen bacteria. It is immediately apparent from Table 3 that several of the bacteria possess a multiplicity of genes coding for different endoglucanases and xylanases. For example, as mentioned earlier, *F. succinogenes* reportedly has six different genes which code for endoglucanase activity (Crosby *et al.*, 1984). We have identified a similar number of genes from a collection consisting of the Crosby clones and other clones from plasmid and λ-Dash gene libraries (Malburg and Forsberg, unpublished data). Another cellobiosidase has been cloned (Schellhorn, Malburg and Forsberg, unpublished data) in addition to that previously reported by Gong *et al.* (1989). We have also cloned another xylanase gene, and have discovered restriction fragment polymorphism within each of the xylanase genes (Wang, Smith, Lee and Forsberg, unpublished data) which in reality means four separate genes. As indicated previously we have purified two xylanases, at least one of which is not coded for by these genes, therefore, there probably are at least three gene families which code for distinctly different xylanases. It is notable that a large number of cellulase clones have been isolated from *R. albus* and *R. flavefaciens* (Table 3). The multiplicity of cellulases and xylanases is also common among bacteria in other environments (Beguin, 1990).

Many of the enzymes although specific for β-1,4-linkages do not differentiate between related polymers. Good examples include the *end1* gene from *Butyrivibrio fibrisolvens* (Berger *et al.*, 1989) and the *cel-3* gene from *F. succinogenes* (McGavin *et al.*, 1989). Both genes code for enzymes active on CMC, lichenin and p-nitrophenyl–β–D-cellobioside. It is fascinating to see that clones from *R. flavefaciens,* lambda CM404, and lambda CM407 degrade crystalline cellulose (Huang *et al.*, 1989). From studies with non-rumen bacteria, it has been found that a minimum of two proteins is necessary for cellulose hydrolysis (Wu *et al.*, 1988; Shoseyov and Doi, 1990).

A similar feature common to practically all the genes sequenced to date was that they exhibited promoter consensus patterns for regulation of transcription, translation, and processing characteristic of those in *Escherichia coli*. This is not necessarily unexpected since they are expressed in *E. coli*. An outstanding question is whether a lack of this consensus may be the reason for not being able to clone genes coding for other enzymes known to be present in the rumen bacteria of interest.

The large number of cellulases and xylanases possessed by each bacterium, some of which exhibit a broad specificity for β-1,4-linkages, provide each of the major fibrolytic bacteria with a potent battery of hydrolytic enzymes for cleavage of the diversity of β-1,4-linkages constituting the complex matrix of the plant cell wall. However, we do not know whether they are all necessary for the degradation of plant cell walls.

Genetic Diversity

The genetic diversity of the rumen microflora is much greater than formerly expected. Heterogeneity of the bacterial population has been demonstrated by use of a variety of molecular techniques including 16S r-RNA analysis, G+C content, DNA-DNA hybridization and restriction enzyme analysis of DNA, and the combination of polyacrylamide gel electrophoresis and zymogram analysis. A comparison of 16S r-RNA sequences showed that strains classified as *Bacteroides succinogenes* were not closely related to other species of *Bacteroides* and as a consequence were renamed *Fibrobacter* (Montgomery *et al.*, 1988). *Fibrobacter* was divided into two species, *succinogenes* and *intestinalis*. Greater complexity has been reported among rumen strains of *Butyrivibrio fibrisolvens*. Hybridization and G+C content has been used to separate them into five separate species encompassing 20 different strains. Nineteen other strains of *Butyrivibrio* were not closely related to any other strains (Mannarelli, 1988). Similar research on a smaller group of *Butyrivibrio fibrisolvens* isolates also resulted in a conclusion of limited relationship between strains (Hudman and Gregg, 1989). Analysis of ten recently isolated strains of *B. ruminicola* subsp. *brevis* revealed characteristics suggesting three to four distinct subgroups (Hudman and Gregg, 1989). Similar diversity has been noted in strains of *R. albus* (Ware *et al.*, 1989) and *R. flavefaciens* (Hudman and Gregg, 1989). Using polyacrylamide gel electrophoresis and zymogram analysis

Table 3. Genes from rumen bacteria coding for cellulase and hemicellulase enzymes

Bacterium	Clone/Gene Designation	Enzyme	Substrates[a] Cleaved	Protein (Da)[b]	Ref.[c]
B. ruminicola subsp. *brevis* B_14	-[d]	Endoglucanase	CMC	40,481/88,000	9
B. ruminicola subsp. *brevis* AR20	pJW4	Endoglucanase	CMC, ASC, xylan	–	19
B. ruminicola 23	pRX26	Xylanase	Xylan	–	18
B. fibrisolvens GS113	*xy1B*	Xylosidase	X_2-X_5 pNPX	–	13
B. fibrisolvens 49	*xyn A*	Xylanase	Xylan	/46,664	8
B. fibrisolvens H17c	*end1*	Endoglucanase	CMC, lichenin, pNPC	/61,000	1
B. fibrisolvens	–	Amylase	–	–	2
F. succinogenes S85	*cel-3*	Endoglucanase	CMC, lichenin barley β-glucan pNPC	73,432/118,000	10
F. succinogenes S85	pCB5	Cellodextrinase	pNPC	-/50,000	4
F. succinogenes S85	pBX6	Xylanase	Xylan	–	14
F. succinogenes S85	–	Lichenase	Lichenin, oat β–glucan	35,168/32,200 37,200	15
F. succinogenes 135	pFSX02	Xylanase	Xylan	–	7
R. albus F-40	*Eg1*	Endoglucanase	CMC	40,848/50,000	11
R. albus AR67	pTC1	Endoglucanase	CMC, ASC, xylan	–	16
AR68	pMH1	Endoglucanase	CMC, ASC, xylan	–	
R.albus SY3	pRLA1	Endoglucanase	CMC, MUC, xylan lichenin	–	12
	pRLA2	Endoglucanase	CMC		
R. albus F-40	pMU1	β-Glucosidase	pNPG, pNPC	–	5

Bacterium	Clone/Gene Designation	Enzyme	Substrates[a] Cleaved	Protein (Da)[b]	Ref.[c]
R. flavefaciens FD-1	*cel A*	Cellodextrinase	pNPC, G_4-G_6 CMC	39,400/	17
R. flavefaciens 186	CM201	Endoglucanase	CMC	–	6
	λCM301	Endoglucanase	CMC	–	
	λCM401	Endoglucanase	CMC	–	
	λCM404	Endoglucanase	CMC,pNPC, Avicel xylan	–	
	λCM407	Endoglucanase	CMC,pNPC, Avicel	–	
	λCM901	Endoglucanase	CMC, pNPC	–	
	λCM902	Endoglucanase	CMC, pNPG, cellobiose	–	
	λCM903	Endoglucanase	CMC, pNPC, pNPG, Avicel	–	
R. flavefaciens 17	L9	Xylanase	Xylan, lichenin, MUX	–	3
	X10	Xylanase	Xylan	–	
	X2	Xylanase	Xylan	–	
	X4	Xylanase	Xylan	–	
	L1	Endoglucanase	CMC, lichenin	–	

[a] Abbreviations: CMC, carboxymethylcellulose; ASC, acid swollen cellulose; pNPC, p-nitrophenyl-β–D-cellobioside; MUC, 4-methylumbelliferyl-β–D-cellobiose; MUX, 4-methylumbelliferyl-β–D-xyloside.

[b] The molecular weight of a protein derived from the DNA sequence is presented before the slash, and that of enzyme purified from either the *E. coli* host or the original rumen bacterium after the slash.

[c] References: 1, Berger *et al.*, 1989; 2, Clark *et al.*, 1990; 3, Flint *et al.*, 1989; 4, Gong *et al.*, 1989; 5, Honda *et al.*, 1988; 6, Huang *et al.*, 1989; 7, Hu *et al.*, 1990; 8, Mannarelli *et al.*, 1990; 9, Matsushita *et al.*, 1990; 10, McGavin *et al.*, 1989; 11, Ohmiya *et al.*, 1989b; 12, Romaniec *et al.*, 1989; 13, Sewell *et al.*, 1989; 14, Sipat *et al.*, 1987; 15, Teather and Erfle, 1990; 16, Wang and Thompson, 1990; 17, Ware *et al.*, 1989; 18, Whitehead and Hespell, 1989; 19, Woods *et al.*, 1989.

[d] Not available.

to detect differences, considerable phenotypic diversity has also been demonstrated in *Selenomonas ruminantium* (Flint and Bisset, 1990). This information is very important because it helps to determine groups within which gene transfer is possible.

Diversity also exists at the level of the gene. Flint *et al.* (1990) reported that a cloned DNA fragment specifying an endoglucanase from *Bacteroides succinogenes* strain BL2 hybridized with different *BamH1* fragments of chromosomal DNA from each of five rumen strains of *B. succinogenes,* and the degree of binding ranged from 16 to 42 percent of homologous binding. Gong *et al.* (1989) isolated two cellobiosidase clones which coded for antigenically identical enzymes, but the DNA inserts differed in one restriction site. As indicated earlier, we have also detected restriction fragment polymorphism in the two xylanase genes cloned from *F. succinogenes* S85 (Wang *et al.*, unpublished data). Taken together, this data shows that intra-strain divergence occurs, and this eventually exhibits itself as divergence between strains and eventually shows up as interspecies differences.

Plasmids from Rumen Bacteria

Plasmids have been isolated from some of the major species of rumen bacteria (Table 4); however, only two of these reportedly have antibiotic resistance markers: pRR14, a 19.5 kb plasmid from *B. ruminicola* 223/M2/7 (Flint *et al.*, 1988) and pRAC, a 15.3 kb plasmid from *R. albus* F-40 (Ohmiya *et al.*, 1989a). The *B. ruminicola* plasmid codes for tetracycline resistance while that from *R. albus* codes for *β*-lactamase which gives rise to penicillin resistance. A second 7.4 kb plasmid from *R. albus* strain F-40 was unique in that it was detected only in cellulose-grown cells. Champion *et al.* (1988) were unable to isolate plasmids from 17 strains of *R. albus* and *R. flavefaciens* although they did detect resistance to tetracycline and to erythromycin, which may be plasmid mediated. In the case of *Selenomonas ruminantium*, one plasmid pSR1 was reported by Martin and Dean (1989). They observed that after the plasmid was cloned into pBR322, it was cleaved by a number of hexanucleotide-cleaving restriction enzymes which were inactive on the original plasmid, thereby indicating that the DNA was modified in *E. coli*. Asmundson and Kelly (1987) reported the detection of two plasmids in *R. flavefaciens*, one of which (pRf186) they characterized in more detail. Thus, with the exception of *F. succinogenes*, for which there are no reports

Table 4. Plasmids from rumen bacteria

Source	Designation Clone/Gene	Size (kb)	Selection Factor	Host Range	Ref.[a]
B. ruminicola 223/M2/7	pRRI4	19.5	Tetracycline	*B. ruminicola* subsp. *brevis*	2
	–[b]	3.0–3.5	–	–	
	–	7.0	–	–	
B. ruminicola 23	pRRI7	9.0	–	–	3
B. ruminicola 46/5(2)	p*RRI1*	*12*	–	–	3
B. fibrisolvens Bu49	pOM1	2.8	–	–	4
B. fibrisolvens ATCC 19171	–	–	–	–	7
R. albus F-40	pRAB	7.4	Cellulose	–	6
	pRAC	15.3	Ampicillin	–	
R. albus RA8	–	7.4	–	–	1
R. flavefaciens 168	pRf 186	5.2	–	–	
S. ruminantium HD4	pSR1	4.8	–	–	5

[a] References: 1, Asmundson and Kelly, 1987; 2, Flint *et al.*, 1988; 3, Flint and Stewart, 1988; 4, Mann *et al.*, 1986; 5, Martin and Dean, 1989; 6, Ohmiya *et al.*, 1989a; 7, Teather, 1982.

[b] Not available.

of plasmids having been found, these major rumen bacterial species are known to possess a range of plasmids with potential for the construction of vectors for gene transfer into other members of the rumen population.

Gene Transfer

Gene transfer in bacteria can occur via transformation, conjugation, transfection, protoplast fusion, and electroporation.

Natural transformation has been tested in several rumen bacteria. Hazlewood *et al.* (1987) reported the unsuccessful transformation of a fatty acid auxotroph of *Butyrivibrio* S29 to prototrophy with chromosomal DNA extracted from six wild type strains of *B. fibrisolvens*. In a similar vein Lee, Forsberg and Gibbins (unpublished data) attempted the transformation of *F. succinogenes* S85 to lactose prototrophy using DNA from a lactose-positive mutant of the same organism (Javorsky *et al.*, 1990). In this case the recipient was treated with $CaCl_2$ to induce competence, but no transformants were detected. Hazlewood *et al.* (1987) also reported the construction of a series of *E. coli/B. fibrisolvens* hybrid plasmids which comprised the B. fibrisolvens plasmid pOM1 inserted into each of pBR325, pA153 and pHV33. These plasmids replicated in *E. coli*, but again there was no evidence of transfer into either L-phase cells or protoplasts of *Butyrivibrio* strains.

Transfer of plasmids via conjugation has been successful in several instances, including the transfer of pRP4 from *E. coli* into *B. fibrisolvens* (Teather, 1985) and of the colonic *Bacteroides* shuttle vector pE5-2 from *B. fragilis* into *B. ruminicola* B_14 (Russell and Wilson, 1988), however, in the latter instance the additional protein synthesis as a consequence of plasmid function significantly reduced the growth rate of the bacterium. The colonic bacterium shuttle vector pVAL-1 developed for conjugative transfer between *E. coli* and *Bacteroides* was used successfully for the transfer of a *B. ruminicola* 23 xylanase gene from *E. coli* to *B. fragilis* and *B. uniformis* (Whitehead and Hespell, 1990). Surprisingly, the enzyme activity of the xylanase was increased 1400-fold and 1600-fold, respectively, in the colonic bacteria. As indicated by Whitehead and Hespell (1990) attempts to introduce *B. fragilis* vectors into *B. ruminicola* by either conjugation or electroporation were unsuccessful.

The plasmid pRRI4 from *B. ruminicola* 223/M2/7 has been successfully transferred into *B. ruminicola* strain F101 by conjugation (Flint *et al.*, 1988). *B. ruminicola* F101 was also successfully transformed with pRRI4 by electroporation with an efficiency of 10^5 transformants per μg DNA (Thomson and Flint, 1989). Attempts to introduce it into three other strains of *B. ruminicola* (118B, M384, GA33) by electroporation were unsuccessful, which may reflect stringent restriction barriers.

The use of a temperate bacteriophage for the development of a transfer system was reported for *S. ruminantium* (Lockington *et al.*, 1988). In this approach, DNA isolated from the temperate bacteriophage was used to transfect the host bacterium by both electroporation and polyethylene glycol-mediated DNA uptake. As indicated by the authors, bacteriophage DNA is an ideal vector for testing transformability in bacteria lacking endogenous plasmids carrying antibiotic resistance markers, and provides the basic components for development of a transfer system.

Intergeneric protoplast fusion is an alternate approach that has been successfully used for the transfer of the *R. albus Eg1* gene on pBR322 in *E. coli* HB101 to a *Fusobacterium* recombinant able to degrade lignin-related compounds (Chen *et al.*, 1988). Southern hybridization experiments suggested that pBR322 with the *Eg1* gene was transferred and existed autonomously in the recombinant.

This information clearly shows that an aggressive research program will be essential to develop transfer systems for rumen bacteria before expression of heterologous genes can be tested.

Manipulation of Fiber Digestion

Pretreatment

Although it is desirable to devise alternative technologies for enhancing fiber digestion, chemical pretreatment has potential under unique circumstances and provides a temporary solution for improving fiber digestion. Variations of the alkaline hydrogen peroxide method remain a favorite for enhancing digestion of recalcitrant plant fiber such as wheat straw (Cecava *et al.*, 1990). Ammonia treatment has also shown potential and has been developed for enhancing the nutritive value of mature graminaceous straws (Mason *et al.*, 1990).

Fungal pretreatment of straw to remove the lignin is another approach to enhancing fiber utilization. Reid (1989) has exhaustively reviewed fungal treatment processes, and has described a process using the white rot fungus *Phlebia tremellosa* which reportedly has excellent potential, although it necessitates sterility which makes it expensive.

In the United States and Canada these processes have not been implemented to an appreciable extent.

Feed Additives

A variety of antibiotics, especially ionophores, chemicals (Van Nevel and Demeyer, 1988), and probiotics (Parker, 1990) are either used or have been tested as potential feed additives.

The ionophores have consistently brought about improved efficiency and/or rate of gain of animals receiving either high concentrate or high fiber diets. Monensin, the most commonly used ionophore, increases propionate production, reduces methane synthesis, reduces ruminal degradation of preformed protein and leads to improved mineral absorption. Monensin does not interfere with cellulolysis *in vivo* which stems, in part, from the fact that the highly fibrolytic bacterium *F. succinogenes* is not inhibited (Russell and Strobel, 1989).

The fungi tolerate monensin although their numbers may decrease (Grenet *et al.*, 1989). Monensin appears to have minimal effect on protozoa *in vivo* (Dennis and Nagaraja, 1986; Grenet *et al.*, 1989). The ionophores are usually included in the feed, and this has been a shortcoming for animals on pasture. The advent of a monensin ruminal delivery device circumvents the problem (Davenport *et al.*, 1989).

Other agents have been tested for their effect on the rumen fermentation, including antifungal materials (Elliot *et al.*, 1987), antiprotozoal agents (Coleman, 1988; Ushida and Jouany, 1990), buffers and isoacids (Kane *et al.*, 1989). Of all microbial additives tested, the only one claimed to enhance fiber digestion directly is *Aspergillus oryzae*, but the benefit is disputed (Frumholtz *et al.*, 1989). Formaldehyde treatment decreases the rate and extent of starch digestion in the rumen (McAllister *et al.*, 1990a).

Plant Breeding

Plant breeding to improve rumen digestibility of feed has received comparatively little attention. However, mutant strains of maize, sudan grass, and sorghum low in lignin have been identified and shown to have enhanced ruminant digestibility (Cherney *et al.*, 1986). Increased *in vitro* dry matter digestibilities as high as 33 percent were reported for brown midrib mutants of sorghum (Porter *et al.*, 1978). Reduced concentrations of phenolic acids (especially p-coumaric acid) and lower levels of etherylated syringyl moieties in conjunction with reduced concentrations of lignin are important factors in the improved quality of brown midrib mutant line 12 of sorghum (Akin *et al.*, 1986; Cherney *et al.*, 1986). The breeding of forages to improve nutritional quality has lagged behind improvements in agronomically desirable characteristics (Wheeler and Corbett, 1989). The testing of isogenic strains of birdsfoot trefoil differing in tannin content is a move in this direction; however, the tannin content of this grass did not influence the apparent digestibility (Chiquette *et al.*, 1989).

Future successes in plant breeding to enhance forage digestion will undoubtedly rely on a more thorough appreciation of cellular organization and chemical linkages responsible for low digestibility such as diferulic cross-linkages in hemicellulose (Hartley and Ford, 1989) and lignified sclerenchyma and vascular tissues (Akin, 1989; Hartley *et al.*, 1990). The focus must be on the major forage grasses since their enhanced digestibility is a cornerstone to successful alternative agriculture.

Selection of alfalfa (Lucerne) to reduce the potential for bloat has been highly productive with the development of a bloat-safe variety (Kudo *et al.*, 1985). This program of breeding now is in the fourth generation. It has been found that the initial rate of digestion (4–6 hour) is reduced by 25 percent, and this significantly reduces the bloat potential of the forage (Cheng *et al.*, unpublished data).

In the case of cereal grains such as barley, where the need is to slow digestion, it would seem appropriate to develop cultivars with greater protection of the starch granules for the passage through the rumen. The selection for plants producing high concentrations of trypsin and α-amylase inhibitors to reduce matrix coat protein and granule digestion within the rumen might be one approach, since the genes coding for these inhibitors have been cloned (Halford *et al.*, 1988). Alternatively, it may be possible to select cultivars with a modified starch granule coat protein, which is less susceptible to rumen microbial proteases.

Whether the objective is to decrease or increase susceptibility of polymers to degradation by microbial enzymes, it will be essential to know the substrate specificities of the predominant microbial enzymes in order to design specific screening procedures.

Genetic Manipulation

Bacteria provide the most useful system for studying the genetic manipulation of rumen microorganisms, since more is known about them and they are the easiest to grow. There have been numerous reviews on the genetic manipulation of the rumen microflora, including those by Flores, 1989; Forsberg *et al.*, 1986; Hazlewood *et al.*, 1987; Hazlewood and Teather, 1988; Russell and Wilson, 1988. Several proposals have been made for the genetic manipulation of the flora to enhance fiber digestion and some of these proposals have been implemented with limited success.

Russell and Wilson (1988) reported on initial studies to manipulate the common non-cellulolytic bacterium *B. ruminicola* to become highly cellulolytic. This strategy was based on the observation that when ruminant rations are high in cereal grain and low in forage, the rumen pH is usually below 6.0. The major fibrolytic bacteria are sensitive to pH values below 6.0. Therefore, to enable cellulose digestion to continue under these stringent conditions of low pH, Russell and Wilson proposed to introduce an acid-tolerant cellulase into the common rumen bacterium, *B. ruminicola,* because it is able to grow at pH values as low as 5.1. *B. ruminicola,* although not cellulolytic, binds to cellulose and produces an endoglucanase able to degrade amorphous cellulose (Russell and Wilson, 1988). The strategy was to clone a cellulase gene from *Thermomonospora fusca* which codes for a cellulase capable of degrading crystalline cellulose, and to transfer it to *B. ruminicola* using the *Bacteroides* shuttle vector pE5-2. Since the original proposal, they have cloned the endoglucanase from *B. ruminicola* $B_1$4 (Matsushita *et al.*, 1990). The native enzyme hydrolyzed amorphous cellulose, but failed to bind to cellulose. They speculated that the failure of *B. ruminicola* to grow on cellulose may be due to the absence of a cullulose-binding domain on the endoglucanase. Since cellulose-binding domains from other cellulases have been characterized (Ong *et al.*, 1989), it was inferred that they would ligate the cellulose-binding domain nucleotide sequence from another bacterium to the *B. ruminicola* gene which could then be reintroduced into the organism. If the modified cellulase binds to amorphous cellulose and hydrolysis occurs, they will have achieved a major part of their goal.

In collaboration with Dr. A.M. Gibbins and S.F. Lee, we proposed to modify *F. succinogenes* to utilize xylose as described previously by Forsberg *et al.* (1986). We anticipated that this would substantially increase the growth rate of this bacterium on fiber, since it has the capacity to extensively degrade xylan, but is unable to use the xylose derived from it. We have found that the bacterium lacks the enzymes necessary for the uptake and metabolism of xylose to xylulose-5-PO_4 (Matte *et al.*, 1988). The major pitfall of this research has been the problem of developing a transfer system for the introduction of genes into *Fibrobacter.* A basic requirement is an indigenous plasmid which can be used for construction of a vector, but no plasmids have been isolated from this genus. Several of the *Bacteroides* vectors including pE5-2 and pDP1 were tested for conjugal transfer into *F. succinogenes* S85, but no conjugants were detected (Forsberg and Teather, unpublished data). We have since constructed a vector which could be introduced by electroporation (Lee *et al.*, 1990). This construct consisted of the *cel-3* promoter (McGavin *et al.*, 1989) upstream from the promoterless *cat* gene from phage P1 carried on pBR322. *E. coli* harboring the recombinant CAT plasmid was chloramphenicol resistant (8 mg/ml), however, when it was introduced into *F. succinogenes* by electroporation no transformants were detected (S.F. Lee, personal communication). Introduction into the CAT plasmid of the insert from the plasmid pCB5 which codes for a *Fibrobacter* cellodextrinase (Gong *et al.*, 1989), to provide the opportunity for homologous recombination, still did not provide the opportunity for chloramphenicol resistance in *Fibrobacter* (Lee, personal communication). *F. succinogenes* exhibits high nuclease activity (Flint and Thomson, 1990), thus, destruction of the vector before reaching its destination could be one of the reasons for the lack of success in obtaining homologous recombination in *F. succinogenes.*

Molecular Considerations

Numerous genes coding for cellulase and hemicellulase enzymes of the major fibrolytic bacteria have been cloned into *E. coli*. From these data there is the distinct impression that the only major problem slowing these developments is the design of adequate screening methods for detection of desired clones. The situation is not as clear with either the fungi or the protozoa, since no genes cloned from these groups of organisms have been reported, consequently, there is no information on which to predict future prospects.

The characterization of fibrolytic enzymes of rumen microorganisms is not keeping pace with the gene cloning. Research on the substrate specificity of the major fibrolytic enzymes and their amplification to provide large quantities of separate enzymes for studies on plant physiology will aid in the development of plant breeding strategies.

Gene transfer presents a different problem. Numerous plasmids have now been cloned from the major rumen bacteria, with the exception of *F. succinogenes*. However, there still is no efficient gene transfer system for any of the rumen bacteria other than selected strains of *B. ruminicola* (Thomson and Flint, 1989). This need was stressed in a recent review by Hazlewood and Teather (1988) and it remains a significant barrier to future progress. Flint and Thomson (1990) have provided at least one explanation for the lack of transfer, with the finding that many rumen bacteria have high nuclease activity that may destroy heterologous DNA before uptake into the cell can occur. Morrison *et al.* (1990) have found that *R. flavefaciens* FD-1 possesses at least two endonuclease activities while *R. albus* 8 possesses a single activity. *R. albus* was sensitive to *dam* methylation which gives a clue to the nature of the DNA necessary to prevent destruction during transformation. As discussed by Forsberg *et al.* (1986), once plasmids are in the cell they may exert a considerable stress on the cell, thereby slowing growth. This problem may be circumvented by integration of the gene of interest into the host chromosome. However, we know nothing about the mechanism of recombination in the rumen bacteria except that a recombination pathway was demonstrated in the colonic *Bacteroides fragilis* (Goodman *et al.*, 1987). Furthermore, as demonstrated in *E. coli*, the DNA inserts expressed in plasmids may increase the auxotrophic requirements of the host cell (Forsberg *et al.* 1986; Wang, Lee, Smith and Forsberg, unpublished data), thus, some genes may be more appropriate than others for expression in heterologous hosts. Aspects which will require consideration for each host organism include the presence of a recombination system, restriction and modification systems, messenger stability, codon usage, post-translational modification, and cellular location (Reznikoff and Gold, 1986).

Conclusions

The enhancement of fiber digestion in the rumen is dependent upon advances in a number of related areas. The genetic manipulation of rumen microorganisms, once touted as the solution to the problem, in essence, provides only one piece of the puzzle. We need to find out which linkages in plant cell walls cause the greatest constraints on degradability. Extensin-like peptides are increasingly being reported in plant cell walls, and this opens the question as to the importance of proteases in microbial forage degradation. It is fascinating that the major fibrolytic bacteria lack proteases, while the fungi which are able to degrade more recalcitrant structures in the plant cell wall are proteolytic (Wallace and Joblin, 1985) and possess both ferulic and coumaric acid esterase. Recent evidence suggests that it is ferulic and coumaric residues ester-linked to hemicellulose, and ether-linked to lignin (Hartley and Ford, 1989; Iiyama *et al.*, 1990) that are cross-link polymers within the plant cell wall, thereby inhibiting enzymic degradation. The separate enzymes of rumen microorganisms involved in fiber digestion need to be characterized in greater detail to allow precise determination of their substrate specificity. The production of large quantities of these enzymes would be valuable for characterization of the important linkages in the plant cell wall. This will also permit the identification of enzymes present in low amounts which may limit digestion. The isolation of mutant bacteria or fungi deleted for each of the fibrolytic enzymes will allow an unambiguous determination of the enzymes of major importance in cell wall digestion. The availability of functional transposons would assist with the identification of genes and the development of transformation systems for the various bacteria would allow the movement of genes. Application of these methodologies would substantially enhance the capacity to develop a thorough knowledge of the roles of each of the fibrolytic enzymes. This information would open the door for the genetic modification of both the microorganisms and the plants in order to maximize ruminant digestion. However, these studies must be approached with an appreciation of the complex microbial interactions that occur within the rumen in order to ensure a beneficial outcome.

Acknowledgements

We express our appreciation to David Smith, Katherine Jakober and Dave Mowat for suggestions to improve the manuscript. Our research is supported by the Natural Sciences and Engineering Research Council of Canada, Agriculture Canada and the Ontario Ministry of Agriculture of Food.

References

Akin, D.E. 1989. Histological and physical factors affecting digesting of forages. Agron. J. 81: 17.

Akin, D.E. and Benner, R. 1988. Degradation of polysaccharides and lignin by ruminal bacteria and fungi. Appl. Environ. Microbiol. 54: 1117.

Akin, D.E., Hanna, W.W., Snook, M.E., Himmelsback, D.S., Barton II, F.E. and Windham, W.R. 1986. Normal-12 and brown midrib-12 sorghum. II. Chemical variations and digestibility. Agron. J. 78: 832.

Asmundson, R.V. and Kelly, W.J. 1987. Isolation and characterization of plasmid DNA from *Ruminococcus*. Curr. Microbiol. 16: 97.

Bailey, C.B. 1962. Rates of digesting of swallowed and unswallowed dried grass in the rumen. Can. J. Anim. Sci. 42: 49.

Beguin, P. 1990. Molecular biology of cellulose degradation. Annu. Rev. Microbiol. In Press.

Berger, E., Jones, W.A., Jones D.T. and Woods D.R. 1989. Cloning and sequencing of an endoglucanase (*end1*) gene from *Butyrivibrio fibrisolvens* H17c. Mol. Gen. Genet. 219: 183.

Bernalier, A., Fonty, G. and Gouet, Ph. 1988. Degradation and fermentation of cellulose by *Neocallimastix* sp. alone or in association with species of rumen bacteria. Repord. Nutr. Develop. 28: 75.

Bohn, P.J. and Fales, S.L. 1989. Cinnamic acid-carbohydrate esters; an evaluation of a model system. J. Sci. Food Agric. 48: 1.

Borneman, W.S., Hartley, R.D., Morrison, W.H., Akin, D.E. and Ljungdahl, L.G. 1990. Feruloyl and para-coumaroyl esterase from anaerobic fungi in relation to plant cell wall degradation. Appl. Microbiol. Biotechnol. 33: 345.

Borneman, W.S., Akin, D.E. and Ljungdahl, L.G. 1989. Fermentation products and plant cell wall-degrading enzymes produced by monocentric and polycentric anaerobic ruminal fungi. Appl. Environ. Microbiol. 55: 1066.

Bradford, E. 1989. Animal agriculture research and development: challenges and opportunities. Can. J. Anim. Sci. 69: 847.

Cassab, G.I. and Varner, J.E. 1988. Cell wall proteins. Annu. Rev. Plant Physiol. Plant Mol. Biol. 39: 321.

Cecava, M.J., Merchen, N.R., Berger, L.L. and Fahey Jr., G.C. 1990. Intestinal supply of amino acids in sheep fed alkaline hydrogen peroxide-treated wheat based diets supplemented with soybean meal or combinations of corn gluten meal and blood meal. J. Anim. Sci. 68: 467.

Chafe, S.C. 1970. The fine structure of the collenchyma cell wall. Planta 90: 12.

Champion, K.M., Helaszek, C.T. and White, B.A. 1988. Analysis of antibiotic susceptibility and extrachromosomal DNA content of *Ruminococcus albus* and *Ruminococcus flavefaciens*. Can. J. Microbiol. 34: 1109.

Chen, W., Ohmiya, K. and Shimizu, S. 1988. *Escherichia coli* spheroplast-mediated transfer of pBR322 carrying the cloned *Ruminococcus albus* cellulase gene into anaerobic mutant strain FEM29 by protoplast fusion. Appl. Environ. Microbiol. 54: 2300.

Cheng, K.-J., Forsberg, C.W., Minato, H. and Costerton, J.W. 1990. Microbial ecology and physiology of feed degradation within the rumen. In "Digestion and Metabolism in Ruminants," Academic Press Inc., Orlando Florida. In press.

Cheng, K.-J., Kudo, H., Jakober, K.D., Duncan, S.H., Mesbah, A., Stewart, C.S., Bernalier, A., Fonty, G. and Gouet, P. 1990. Prevention of fungal colonization and digestion of cellulose by the addition of methylcellulose. Can. J. Microbiol. In Press.

Cherney, J.H., Moore, K.J., Volenec, J.J. and Axtell, J.D. 1986. Rate and extent of digestion of cell wall components of brown-midrib sorghum species. Crop Sci. 26: 1055.

Chesson, A., Gordon, A.H. and Lomax, J.A. 1983. Substituent groups linked by alkali-labile bonds to arabinose and xylose residues of legume, grass and cereal straw cell walls and their fate during digestion by rumen microorganisms. J. Sci. Food Agric. 34: 1330.

Chiquette, J., Cheng, K.J., Rode, L.M. and Milligan, L.P. 1989. Effect of tannin content on two isosynthetic strains of birdsfoot trefoil (*Lotus corni culatus* L.) on feed digestibility and rumen fluid composition in sheep. Can. J. Anim. Sci. 69: 1031.

Clark, R.G., Hu, Y.J., Hynes, M.F. and Cheng, K.J. 1990. Procedure for screening for amylase genes in rumen bacteria. Proc. Can. Soc. Microbiol. GM3.

Coleman, G.S. 1988. The importance of rumen ciliate protozoa in the growth and metabolism of the host ruminant. International J. Anim. Sci. 3: 75.

Cotta, M.A. 1988. Amylolytic activity of selected species of ruminal bacteria. Appl. Environ. Microbiol. 54: 772.

Crosby, B., Collier, B., Thomas, D.Y., Teather, R.M. and Erfle, J.D. 1984. Cloning and expression in *Escherichia coli* of cellulase genes from *Bacteroides succinogenes*. In "Proceedings of the 5th Canadian Bioenergy R&D Seminar," p. 573. Elsevier Applied Science Publishers Ltd., Barking, England.

Davenport, R.W., Galyean, M.L., Branine, M.E. and Hubbert, M.E. 1989. Effects of a monensin ruminal delivery device on daily gain, forage intake and ruminal fermentation of steers grazing irrigated winter wheat pasture. J. Anim. Sci. 67: 2129.

DeBoever, J.L., Andries, J.K., DeBrabander, D.L., Cottyn, B.G. and Buysse, F.X. 1990. Chewing activity of ruminants as a measure of physical structure — A review of factors affecting it. Anim. Feed Sci. Technol. 27: 281.

Dennis, S.M. and Nagaraja, T.G. 1986. Effect of lasalocid, monensin and thiopeptin on rumen protozoa. Res. Vet. Sci. 41: 251.

Doerner, K.C. and White, B.A. 1990. Assessment of the endo-1,4-β-glucanase components of *Ruminococcus flavefaciens* FD-1. Appl. Environ. Microbiol. 56: 1844.

Elliot, R., Ash, A.J., Calderon-Cortes, F. and Norton, B.W. 1987. The influence of anaerobic fungi on rumen volabile fatty acid concentrations *in vivo*. J.Agric. Sci. Camb. 109: 13.

Engels, F.M. and Brice, R.E. 1985. A barrier covering lignified cell walls of barley straw that restricts access by rumen microorganisms. Curr. Microbiol. 12: 217.

Erfle, J.D., Teather, R.M., Wood, P.J. and Irvin, J.E. 1988. Purification and properties of a 1,3-1,4-β-D-glucanase (lichenase, 1,3-1,4-β-D-glucan 4-glucanohydrolase, EC 3.2.1.73) from *Bacteroides succinogenes* cloned in *Escherichia coli*. Biochem. J. 255: 833.

Flint, H.J. and Bisset, J. 1990. Genetic diversity in *Selenomonas ruminantium* isolated from the rumen. FEMS Microbiol. Ecol. 73: 351.

Flint, H.J., Duncan, S.H., Bisset, J. and Stewart, C.S. 1988. The isolation of tetracycline-resistant strains of strictly anaerobic bacteria from the rumen. Lett. Appl. Microbiol. 6: 113.

Flint, H.J., McPherson, C.A., Avgustin, G. and Stewart, C.S. 1990. Use of a cellulase-encoding gene probe to reveal restriction fragment length polymorphisms among ruminal strains of *Bacteroides succinogenes*. Curr. Microbiol. 20: 63.

Flint, H.J., McPherson, C.A. and Bisset, J. 1989. Molecular cloning of genes from *Ruminococcus flavefaciens* encoding xylanase and β(1-3,1-4)-glucanase activities. Appl. Environ. Microbiol. 55: 1230.

Flint, H.J. and Stewart, C.S. 1988. Antibiotic resistance patterns and plasmids of ruminal strains of *Bacteroides ruminicola* and *Bacteroides multiacidus*. Appl. Microbiol. Biotechnol. 26: 450.

Flint, H.J. and Thomson, A.M. 1990. Deoxyribonuclease activity in rumen bacteria. Lett. Appl. Microbiol. 11: 18.
Flores, D.A. 1989. Application of recombinant DNA to rumen microbes for the improvement of low quality feed utilization. J. Biotechnol. 10: 95.
Ford, C.W. and Elliot, R. 1987. Biodegradability of mature grass cell walls in relation to chemical composition and rumen microbial activity. J. Agric. Sci. Camb. 108: 201.
Forsberg, C.W. and Cheng, K.-J. 1990. Integration of rumen microorganisms and their hydrolytic enzymes during the digestion of lignocellulosic materials. In "Microbial and Plant Opportunities to Improve Lignocellulose Utilization by Ruminants," Akin, D.E., Ljungdahl, L.G., Wilson, J.R., and Harris, P.J. (Eds.), Elsevier Science Publishing Co. Inc., New York, pp. 411-423.
Forsberg, C.W., Crosby, B. and Thomas, D.Y. 1986. Potential for manipulation of the rumen fermentation through the use of recombinant DNA techniques. J. Anim. Sci. 63: 310.
Frumholtz, P.P., Newbold, C.J. and Wallace, R.J. 1989. Influence of *Aspergillus oryzae* fermentation extract on the fermentation of a basal ration in the rumen simulation technique (Rusitec). J. Agric. Sci. 113: 169.
Fry, S.C. 1986. Cross-linking of matrix polymers in the growing cell walls of angiosperms. Ann. Rev. Plant Physiol. 37: 165.
Gallard, T. 1983. Starch-lipid complexes and other non-starch components of starch granules in cereal grains. In "Mobilization of Reserves in Germination," Recent Advances in Phytochemistry 17: 111.
Gardner, R.M., Doerner, K.C. and White, B.A. 1987. Purification and characterization of an exo-β-1,4-glucanase from *Ruminococcus flavefaciens* FD-1. J. Bacteriol. 169: 4581.
Gijzen, H.J., Lubberding, H.J., Gerhardus, M.J.T. and Vogels, G.D. 1988. Contribution of rumen protozoa to fiber degradation and cellulase activity *in vitro*. FEMS Microbiol. Lett. 53: 35.
Gong, J. and Forsberg, C.W. 1989. Factors affecting adhesion of *Fibrobacter succinogenes* subsp. *succinogenes* S85 and adherence-defective mutants to cellulose. Appl. Environ. Microbiol. 55: 3039.
Gong, J., Lo, R.Y.C. and Forsberg, C.W. 1989. Molecular cloning and expression in *Escherichia coli* of a cellodextrinase gene from *Bacteroides succinogenes* S85. Appl. Environ. Microbiol. 55: 132.

Goodman, H.J.K., Parker, J.R., Southern, J.A. and Woods, D.R. 1987. Cloning and expression in *Escherichia coli* of a *recA*-like gene from *Bacteroides fragilis.* Gene 58: 265.

Grenet, E., Fonty, G., Jamot, J. and Bonnemoy, F. 1989. Influence of diet and monensin on development of anaerobic fungi in the rumen, duodenum, cecum, and feces of cows. Appl. Environ. Microbiol. 55: 2360.

Greve, L.C., Labavitch, J.M. and Hungate, R.E. 1984a. α–L–arabinofuranosidase from *Ruminococcus albus* 8: Purification and possible role in hydrolysis of alfalfa cell wall. Appl. Environ. Microbiol. 47: 1135.

Greve, L.C., Labavitch, J.M. and Hungate, R.E. 1984b. Xylanase action on alfalfa cell walls. In "Structure, function, and biosynthesis of plant cell walls," p. 150, American Society of Plant Physiologists, Rockville, MD.

Groleau, D. and Forsberg, C.W. 1981. Cellulolytic activity of the rumen bacterium *Bacteroides succinogenes*. Can. J. Microbiol. 27: 517.

Halford, N.G., Morris, N.A., Urwin, P., Williamson, M.S., Kreis, M. and Shewry, P.R. 1988. Molecular cloning of the barley seed protein CMd: a variant member of the α–amylase/trypsin inhibitor family of cereals. Biochim. Biophys. Acta 950: 435.

Hartley, R.D., Akin, D.E. and Himmelsbach, D.S. 1990. Microspectrophotometry of bermudagrass (*Cynodon dactylon*) cell walls in relation to lignification and wall biodegradability. J. Sci. Food Agr. 50: 179.

Hartley, R.D. and Ford, C.W. 1989. Phenolic constituents of plant cell walls and wall degradability. In "Plant Cell Wall Polymers: Biogenesis and Biodegradation," p. 137, American Chemical Society, Washington, D.C.

Hartley, R.D. and Jones, E.C. 1976. Diferulic acid as a component of cell walls of *Lolium multiflorum*. Phytochemistry 15: 1157.

Hazlewood, G.P., Mann, S.P., Orpin, C.G. and Romaniec, M.P.M. 1987. Prospects for the genetic manipulation of rumen microorganisms. In "Recent Advances in Anaerobic Bacteriology," p. 162, Martinus Nijhoff, Boston, Mass.

Hazlewood, G.P. and Teather, R.M. 1988. The genetics of rumen bacteria. In "The Rumen Microbial Ecosystem," p. 323, Elsevier Applied Science, London and New York.

Hebraud, H. and Fevre, M. 1988. Production and purification of glycosidases secreted by the fungus *Neocallimastix frontalis*. Reprod. Nutr. Develop. 28: 71.

Honda, H., Saito, T., Iijima, S. and Kobayashi, T. 1988. Molecular cloning and expression of a β-glucosidase gene from *Ruminococcus albus* in *Escherichia coli*. Enzyme Microb. Technol. 10: 559.

Hu, Y.J., Cheng, K.J. and Forsberg, C.W. 1990. Cloning of *Fibrobacter succinogenes* 135 xylanase gene and its expression in *E. coli*. Proc. Can. Soc. Microbiol. GM1.

Huang, L. and Forsberg, C.W. 1988. Purification and comparison of the periplasmic and extracellular forms of the cellodextrinase from *Bacteroides succinogenes*. Appl. Environ. Microbiol. 54: 1488.

Huang, L. and Forsberg, C.W. 1990. Cellulose digestion and cellulase regulation and distribution in *Fibrobacter succinogenes* subsp. *succinogenes* S85. Appl. Environ. Microbiol. 56: 1221.

Huang, L., Forsberg, C.W. and Thomas, D.Y. 1988. Purification and characterization of a chloride-stimulated cellobiosidase from *Bacteroides succinogenes*. J. Bacteriol. 170: 2923.

Huang, C.M., Kelly, W.J., Asmundson, R.V. and Yu, P.K. 1989. Molecular cloning and expression of multiple cellulase genes of *Ruminococcus flavefaciens* strain 186 in *Escherichia coli*. Appl. Microbiol. Biotechnol. 31: 265.

Hudman, J.F. and Gregg, K. 1989. Genetic diversity among strains of bacteria from the rumen. Curr. Microbiol. 19: 313.

Iiyama, K., Lam, T.B.T. and Stone, B.A. 1990. Phenolic acid bridges between polysaccharides and lignin in wheat internodes. Phytochemistry 29: 733.

Javorsky, P., Lee, S.F., Verrinder-Gibbins, A.M. and Forsberg, C.W. 1990. Extracellular β-galactosidase activity of a *Fibrobacter succinogenes* S85 mutant able to catabolize lactose. Appl. Environ. Microbiol. 56: 3657-3663.

Joblin, K.N., Naylor, G.E. and Williams A.G. 1990. Effect of *Methanobrevibacter smithii* on xylanolytic activity of anaerobic ruminal fungi. Appl. Environ. Microbiol. 56: 2287.

Joblin, K.N., Campbell, G.P., Richardson, A.J. and Stewart, C.S. 1989. Fermentation of barley straw by anaerobic rumen bacteria and fungi in axenic culture and in co-culture with methanogens. Lett. Appl. Microbiol. 9: 195.

Jung, H.G. 1989. Forage lignins and their effects on fiber digestibility. Agron. J. 81: 33.

Kane, P., Machado, P.F. and Cook, R.M. 1989. Effect of the combination of monensin and isoacids on rumen fermentation *in vitro*. J. Dairy Sci. 72: 2767.

Kieliszewski, M.J., Leykam, J.F. and Lamport, D.T.A. 1990. Structure of the threonine-rich extensin from *Zea mays*. Plant Physiol. 92: 316.

Kolattukudy, P.E. 1984. Biochemistry and function of cutin and suberin. Can. J. Botany 62: 2981.

Kudo, H., Cheng, K.-J. and Costerton, J.W. 1987. Electron microscopic study of the methylcellulose-mediated detachment of cellulolytic rumen bacteria from cellulose fibers. Can. J. Microbiol. 33: 267.

Kudo, H., Cheng, K.-J., Hanna, M.R., Howarth, R.W., Goplin, B.P. and Costerton, J.W. 1985. Ruminal digestion of alfalfa strains selected for slow and fast initial rates of digestion. Can. J. Anim. Sci. 65: 157.

Lee, S.F., Forsberg, C.W. and Gibbins, A.M. 1990. Expression of chloramphenicol acetyl transferase (CAT) from a promoter of *Fibrobacter succinogenes* S85. Proc. Can. Soc. Microbiol. GM2.

Li, X.-B., Kieliszewski, M. and Lamport, D.T.A. 1990. A chenopod extensin lacks repetitive tetrahydroxyproline blocks. Plant Physiol. 92: 327-333.

Liebhardt, W.C., Andrews, R.W., Culik, M.N., Harwood, R.R., Janke, R.R., Radke, J.K. and Rieger-Schwartz, S.L. 1990. Crop production during conversion from conventional to low-imput methods. Agron. J. 81: 150.

Lockington, R.A., Attwood, G.T. and Brooker, J.D. 1988. Isolation and characterization of a temperate bacteriophage from the ruminal anaerobe *Selenomonas ruminantium*. Appl. Environ. Microbiol 54: 1575.

Lowe, S.E., Theodorou, M.K. and Trinci, A.P.J. 1987. Growth and fermentation of an anaerobic rumen fungus on various carbon sources and effect of temperature on development. Appl. Environ. Microbiol. 53: 1210.

MacRae, R.J., Hill, S.B., Mehuys, G.R. and Henning, J. 1990. Farm-scale agronomic and economic conversion from conventional to sustainable agriculture. Adv. Agron. 43: 155.

Mann, S.P., Hazlewood, G.P. and Orpin, C.G. 1986. Characterization of a cryptic plasmid (pOM1) in *Butyrivibrio fibrisolvens* by restriction endonuclease analysis and its cloning in *Escherichia coli*. Curr. Microbiol. 13: 17.

Mannarelli, B.M. 1988. Deoxyribonucleic acid relatedness among strains of the species *Butyrivibrio fibrisolvens*. Int. J. Syst. Bacteriol. 38: 340.

Mannarelli, B.M., Evans, S. and Lee, D. 1990. Cloning, sequencing, and expression of a xylanase gene from the anaerobic ruminal bacterium *Butyrivibrio fibrisolvens*. J. Bacteriol. 172: 4247.
Martin, S.A. and Dean, R.G. 1989. Characterization of a plasmid from the ruminal bacterium *Selenomonas ruminantium*. Appl. Environ. Microbiol. 55: 3035.
Mason, V.C., Cook, J.E., Keene, A.S. and Hartley, R.D. 1990. The use of mixtures of ammonium salts with time for upgrading cereal straws. Anim. Feed Sci. Technol. 28: 169.
Matsushita, O., Russell, J.B. and Wilson, D.B. 1990. Cloning and sequencing of a *Bacteroides ruminicola* $B_1$4 endoglucanase gene. J. Bacteriol. 172: 3620.
Matte, A.M., Forsberg, C.W. and Gibbins, A.M. 1988. Pentose metabolizing enzymes of *Bacteroides succinogenes* and *Bacteroides ruminicola*. Proc. Can. Soc. Microbiol. AM2.
McAllister, T.A., Cheng, K.-J., Rode, L.M., and Buchanan-Smith, J.G. 1990a. Use of formaldehyde to regulate digestion of barley starch. Can. J. Anim. Sci. 70: 581.
McAllister, T.A., Rode, L.M., Cheng, K.-J., Forsberg, C.W. and Buchanan-Smith, J.G. 1990b. Digestion of barley, maize, and wheat by selected species of ruminal bacteria. Appl. Environ. Microbiol. 56: 3148-3153.
McDermid, K.P., Forsberg, C.W. and MacKenzie, C.R. 1990a. Purification and properties of an acetylxylan esterase from *Fibrobacter succinogenes* S85. Appl. Environ. Microbiol. 56: 3805-3810.
McDermid, K.P., MacKenzie, C.R. and Forsberg, C.W. 1990b. Esterase activities of *Fibrobacter succinogenes* S85. Appl. Environ. Microbiol. 56: 127.
McGavin, M. and Forsberg, C. 1988. Isolation and characterization of endoglucanases 1 and 2 from *Bacteroides succinogenes*. J. Bacteriol 170: 2914.
McGavin, M. and Forsberg, C.W. 1989. Catalytic and substrate-binding domains of endoglucanase 2 from *Bacteroides succinogenes*. J. Bacteriol. 171: 3310.
McGavin, M.J., Forsberg, C.W., Crosby, B., Bell, A.W., Dignard, D. and Thomas, D.Y. 1989. Structure of the cel-3 gene from *Fibrobacter succinogenes* S85 and characteristic of the encoded gene product, endoglucanase 3. J. Bacteriol. 171: 5587.
Miron, J., Yokoyama, M.T. and Lamed, J. 1989. Bacterial cell surface structures involved in lucerne cell wall degradation by pure cultures of cellulolytic rumen bacteria. Appl. Microbiol. Biotechnol. 32: 218.

Montgomery, L., Flesher, B. and Stahl, D. 1988. Transfer of *Bacteroides succinogenes* Hungate to *Fibrobacter* gen. nov. as *Fibrobacter succinogenes* comb. nov. and description of *Fibrobacter intestinalis* sp. nov. Int. J. Syst. Bacteriol. 38: 430.

Morrison, M., Mackie, R.I. and White, B.A. 1990. Restriction endonucleases from *Ruminococcus albus* 8 and *Ruminococcus flavefaciens* FD-1. Proc. Amer. Soc. Microbiol. H-198.

Mueller-Harvey, I., Hartley, R.D., Harris, P.T. and Curzon, E.H. 1986. Linkage of p-coumaroyl and feruloyl groups to cell-wall polysaccharides of barley straw. Carbohydrate Res. 148: 71.

National Research Council Subcommittee on Feed Composition. 1982. United States-Canadian Tables of Feed Composition. National Academy Press, Washington, DC.

National Research Council. 1989. Alternative Agriculture. National Academy Press, Washington, DC.

Nordin, M. and Campling, R.C. 1976. Digestibility studies with cows given whole and rolled cereal grains. Anim. Prod. 23: 305.

Ohmiya, K., Hoshino, C. and Shimizu, S. 1989a. Cellulose-dependent and penicillin-resistant plasmids isolated from *Ruminococcus albus* Asian-Aust. J. Anim. Sci. 2: 501.

Ohmiya, K., Kajino, T., Kato, A. and Shimizu, S. 1989b. Structure of a *Ruminococcus albus* endo-1,4-Beta-glucanase gene. J. Bacteriol. 171: 6771.

Ohmiya, K., Maeda, K. and Shimizu, S. 1987. Purification and properties of endo-(1–4)–β–glucanase from *Ruminococcus albus.* Carbohydr. Res. 166: 145.

Ohmiya, K., Shimizu, M., Taya, M. and Shimizu, S. 1982. Purification and properties of cellobiosidase from *Ruminococcus albus.* J. Bacteriol. 150: 407.

Ohmiya, K., Shirai, M., Kurachi, Y. and Shimizu, S. 1985. Isolation and properties of β–glucosidase from *Ruminococcus albus.* J. Bacteriol. 161: 432.

Ong, E., Gilkes, N.R., Warren, R.A.J., Miller, R.C. and Miller, D.G. 1989. Enzyme immobilization using the cellulose-binding domain of *Cellulomonas fimi* exoglucanase. BioTechnology 7: 604.

Orpin, C.G. and Joblin, K.N. 1988. The rumen anaerobic fungi. In "The Rumen Microbial Ecosystem," p. 129, Elsevier Science Publishers.

Orpin, C.G., Mathiesen, S.D., Greenwood, Y. and Blix, A.S. 1985. Seasonal changes in the ruminal, microflora of the high-arctic Svalbard Reindeer (*Rangifer tarandus platyrhynchus*). Appl. Environ. Microbiol. 50: 144.

Osborne, J.M. and Dehority, B.A. 1989. Synergism in degradation and utilization of intact forage cellulose, hemicellulose, and pectin by three pure cultures of ruminal bacteria. Appl. Environ. Microbiol. 55: 2247.
Parker, D.S. 1990. Manipulation of the functional activity of the gut by dietary and other means (Antibiotics/Probiotics) in ruminants. J. Nutr. 120: 639.
Porter, K.S., Axtell, J.D., Lechtenberg, V.L. and Colenbrander, V.F. 1978. Phenotype, fiber composition, and *in vitro* dry matter disappearance of chemically induced brown midrib (*bmr*) mutants of sorghum. Crop Sci. 18: 205.
Prins, R.A. and Van Hoven, W. 1977. Carbohydrate fermentation by the rumen ciliate *Isatricha prostoma*. Prostistologica. 13: 549.
Reid, I.D. 1989. Solid-state fermentations for biological delignification. Enzyme Microb. Technol. 11: 786.
Reznikoff, W. and Gold, L. 1986. Maximizing gene expression. Butterworths, Boston.
Romaniec, M.P.M., Davidson, K., White, B.A. and Hazlewood, G.P. 1989. Cloning of *Ruminococcus albus* endo–β1,4–glucanase and xylanase genes. Lett. Appl. Bacteriol. 9: 101.
Rooney, L.W. and Pflugfelder, R.L. 1986. Factors affecting starch digestibility with special emphasis on sorghum and corn. J. Anim. Sci. 63: 1607.
Russell, J.B. and Strobel, H.J. 1989. Effect of ionophores on ruminal fermentation. Appl. Environ. Microbiol. 55: 1.
Russell, J.B. and Wilson, D.B. 1988. Potential opportunities and problems for genetically altered rumen microorganisms. J. Nutr. 118: 271.
Scalbert, A., Monties, B., Lallemand, J.-Y., Guittet, E. and Rolando, C. 1985. Ether linkage between phenolic acids and lignin fractions from wheat straw. Phytochemistry 24: 1359.
Selvendran, R.R. 1983. The chemistry of plant cell walls. In "Dietary Fiber," p. 95, Applied Science Publishers, London.
Sewell, G.W., Utt, E.A., Hespell, R.B., Mackenzie, K.F. and Ingram, L.O. 1989. Identification of the *Butyrivibrio fibrisolvens* xylosidase gene (xylB) coding region and its expression *Escherichia coli*. Appl. Environ. Microbiol. 55: 306.
Shoseyov, O. and Doi, R.H. 1990. Essential 170-kDa subunit for degradation of crystalline cellulose by *Clostridium cellulovorans* cellulase. Proc. Natl. Acad. Sci. USA 87: 2192.

Sipat, A., Taylor, K.A., Lo, R.Y.C. and Forsberg, C.W. 1987. Molecular cloning of a xylanase gene from *Bacteroides succinogenes* and its expression in *Escherichia coli*. Appl. Environ. Microbiol. 53: 477.
Teather, R.M. 1982. Isolation of plasmid DNA from *Butyrivibrio fibrisolvens*. Appl. Environ. Microbiol. 43: 298.
Teather, R.M. 1985. Application of gene manipulation to rumen microflora. Can. J. Anim. Sci. 65: 563.
Teather, R.M. and Erfle, J.D. 1990. DNA sequence of a *Fibrobacter succinogenes* mixed-linkage β–glucanase (1,3–1,4–β–D-glucan 4-glucanohydrolase) gene. J. Bacteriol. 172: 3837.
Thomson, A.M. and Flint, H.J. 1989. Electroporation induced transformation of *Bacteroides ruminicola* and *Bacteroides uniformis* by plasmid DNA. FEMS Microbiol. Lett. 61: 101.
Ushida, K. and Jouany, J.P. 1990. Effect of defaunation on fiber digestion in sheep given two isonitrogenous diets. Anim. Feed Sci. Technol. 29: 153.
Van Nevel, C.J. and Demeyer, D.I. 1988. Manipulation of rumen fermentation. In "The Rumen Microbial Ecosystem," p.387, Elsevier Applied Science, London.
Van Soest, P.J. 1982. Nutritional Ecology of the Ruminant. Durham and Downey Inc., Portland, OR.
Varel, V.H. and Jung, H.G. 1986. Influence of forage phenolics on ruminal fibrolytic bacteria and *in vitro* fiber digestion. Appl. Environ. Microbiol. 52: 275.
Waite, R. and Gorrod, A.R.N. 1959. The structural carbohydrates of grasses. J. Sci. Food Agr. 10: 308.
Wallace, R.J. and Cotta, M.A. 1988. Metabolism of ntirogen-containing compounds. In "The Rumen Microbial Ecosystem," p. 217, Academic Press. London and New York.
Wallace, R.J., and Joblin, K.N. 1985. Proteolytic activity of a rumen anaerobic fungus. FEMS Microbiol. Lett. 29: 19.
Wang, W. and Thomson, J.A. 1990. Nucleotide sequence of the *celA* gene encoding a cellodextrinase of *Ruminococcus flavefaciens* FD-1. Mol. Gen. Genet. 222: 265.
Ware, G.E., Bauchop, T. and Gregg, K. 1989. The isolation and comparison of cellulase genes from two strains of *Ruminococcus albus*. J. Gen. Microbiol. 135: 921.
Wheeler, J.L. and Corbett, J.L. 1989. Criteria for breeding forages of improved feeding value: results of a Delphi survey. Grass Forage Sci. 44: 77.

Whitehead, T.R. and Hespell, R.B. 1989. Cloning and expression in *Escherichia coli* of a xylanase gene from *Bacteroides ruminicola* 23. Appl. Environ. Microbiol. 55: 893.

Whitehead, T.R. and Hespell, R.B. 1990. Heterologous expression of the *Bacteroides ruminicola* xylanase gene in *Bacteroides fragilis* and *Bacteroides uniformis*. FEMS Microbiol. Lett. 66: 61.

Wilkie, K.C.B. 1979. The hemicelluloses of grasses and cereals. Adv. Carbohydr. Chem. Biochem. 36: 215.

Williams, A.G. and Coleman, G.S. 1988. The rumen protozoa. In "The Rumen Microbial Ecosystem," p. 77, Academic Press, London and New York.

Williams, A.G. and Orpin, C.G. 1987a. Polysaccharide degrading enzymes formed by three species of anaerobic rumen fungi grown on a range of carbohydrate substrates. Can. J. Microbiol. 33: 418.

Williams, A.G. and Orpin, C.G. 1987b. Glycoside hydrolase enzymes present in the zoospore and vegetative growth stages of the rumen fungi *Neocallimastix patriciarum, Piromonas communis,* and an unidentified isolate, grown on a range of carbohydrates. Can. J. Microbiol. 33: 427-434.

Wood, T.M., Wilson, C.A., McCrae, S.I. and Joblin K.N. 1986. A highly active extracellular cellulase from the anaerobic rumen fungus *Neocallimastix frontalis*. FEMS Microbiol. Lett. 34: 37.

Woods, J.R., Hudman, J.F. and Gregg, K. 1989. Isolation of an endoglucanase gene from *Bacteroides ruminicola* subsp. *brevis* J. Gen. Microbiol. 135: 2543.

Wu, J.H.D., Orme-Johnson, W.H. and Demain, A.L. 1988. Two components of an extracellular protein aggregate of *Clostridium thermocellum* together degrade crystalline cellulose. Biochemistry 27: 1703.

Genetic and Molecular Genetic Regulation of Soluble and Insoluble Carbohydrate Composition in Tomato

Alan B. Bennett, Ellen M. Klann, Coralie C. Lashbrook, Serge Yelle
Mann Laboratory
Department of Vegetable Crops
University of California, Davis, CA 95616
Roger T. Chetelat, Joseph W. De Verna
Campbell's Institute for Research and Technology
Davis, CA 95616
Robert L. Fischer
Department of Plant Biology
University of California, Berkeley, CA 94720

Fruit quality is determined by the amount and composition of soluble and insoluble solids. In tomato fruit the soluble solids are comprised primarily of mono and disaccharides, whereas the insoluble solids are comprised primarily of cell wall polysaccharides. To assess the feasibility of improving tomato fruit quality by manipulating monogenic traits, genetic and molecular genetic manipulation of soluble and insoluble carbohydrate composition has been carried out. A trait reversing the monosaccharide to disaccharide ratio of soluble carbohydrate in tomato fruit has been transferred from *Lycopersicon chmielewskii* to *Lycopersicon esculentum* and shown to be governed by a single recessive gene. Biochemical characterization of this trait has led to the tentative cloning of a gene that may control this trait, thus providing the means to genetically engineer one aspect of soluble sugar composition. Insoluble polysaccharides, including pectins and hemicelluloses, are degraded during tomato fruit ripening. To manipulate the extent and timing of this change in cell wall polymer structure, we have cloned genes encoding pectin and hemicellulose-degrading enzymes and are using these genes to modify their expression in transgenic plants. Together, these experiments are providing the basis to assess the feasibility of achieving incremental enhancement of complex traits that control components of fruit quality by identifying specific biochemical targets for genetic manipulation.

Introduction

Tomato is a popular vegetable for genetic modification because there is a wealth of well-characterized germplasm and it is well-suited for molecular genetic manipulation. In this context, tomato has been intensively bred to improve plant performance and fruit quality. Tomato fruit quality is a term that encompasses fruit color, vitamin context, texture, and flavor. Attributes that primarily influence tomato processing characteristics such as viscosity and yield include the soluble solids concentration and the structure of insoluble solids.

Traits associated with tomato fruit quality are often polygenic and inherited quantitatively (DeVerna and Paterson, 1990). For instance, fruit soluble solids concentration has been reported to be determined by three genes (Ibaria and Lambreth, 1969) and restriction fragment length polymorphism (RFLP) analysis has defined three chromosomal segments associated with a high soluble solid derivative of *Lycopersicon chmielewskii* (Tanksley and Hewitt, 1988). There are, however, examples of monogenic traits associated with tomato fruit quality. This is especially true in fruit pigmentation where og^c, *hp* and *dg* each enhance lycopene content. Two of these genes, *hp* and *dg,* also confer pleiotropic effects, including reduced plant vigor, presumably by altering carotenoid composition in vegetative tissues (Jarret *et al.,* 1984).

In spite of the polygenic inheritance of many traits associated with fruit quality, it should be possible to identify specific biochemical targets for genetic manipulation that may confer incremental enhancement of a complex trait. Such a strategy assumes that single biochemical steps in a complex pathway are either rate-controlling or otherwise critical for final production formation. If these key biochemial steps can be identified and specifically modified by direct biochemical selection or by genetic engineering it may be possible to modify complex traits incrementally. To test the validity of this strategy we have utilized wild tomato relatives and recombinant DNA techniques to modify specific biochemical steps contributing to soluble sugar composition and to the structure of insoluble cell wall polymers.

Metabolism of Soluble Carbohydrate

Soluble solids of tomato fruit are primarily comprised of sugars, organic acids, and salts (Figure 1). Collectively, the soluble solids content of the fruit is a major determinant of fruit quality, especially for processing tomatoes. Although soluble solids content represents a commercially valuable trait, progress in enhancing this trait by classical genetic approaches has been slow. Indeed, because of a strong negative correlation between yield and soluble solids content (Stevens, 1986; Stevens and Rudich, 1978), soluble solids levels have tended to decrease in association with the development of high yielding tomato varieties.

Several strategies to increase soluble solids levels have been explored, including the transfer of high soluble solids from a wild *Lycopersicon* species (Hewitt and Garvey, 1987; Rick, 1974). Recent efforts have attempted to use restriction fragment length polymorphism DNA markers to facilitate the introgression of high soluble solids-conferring genes from *L. chmielewskii* to *L. esculentum* (Osborn *et al.*, 1987; Paterson *et al.*, 1988; Tanksley and Hewitt, 1988). However, in these efforts, selections for increased soluble solids were made primarily in backcross F_1 generations and so only dominant genes that contributed to high soluble solids content were selected. In contrast, our results suggest that recessive genes may control specific biochemical traits important in soluble solids accumulation (see below).

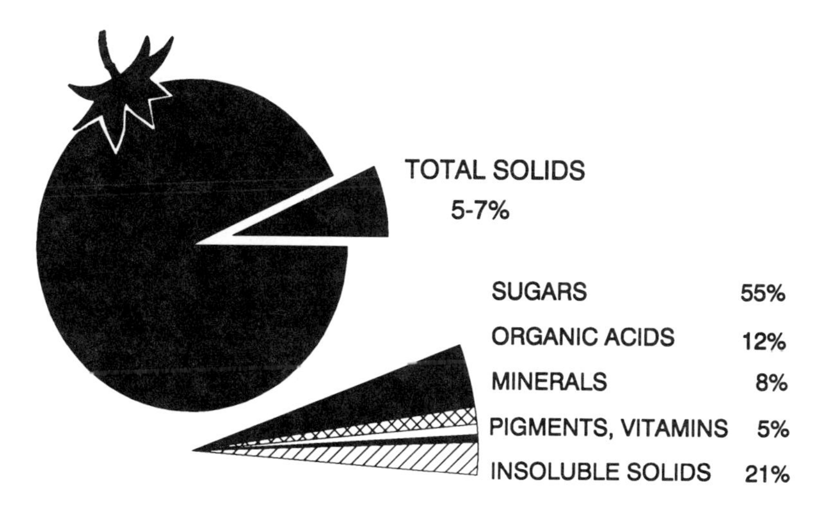

Figure 1. Approximate composition of tomato fruit (data from Davies and Hobson, 1981).

As described above, an alternative approach to the enhancement of tomato fruit soluble solids is to evaluate components of carbohydrate metabolism in developing fruit tissue in order to identify biochemical steps whose modification may lead to increased soluble carbohydrate content in the fruit. Once identified, these biochemical processes could then be modified by classical genetic means, using direct biochemical selection, or by molecular genetic strategies. Several studies have described sink metabolism in *L. esculentum (Damon et al.*, 1988; Ehret and Ho, 1986; Manning and Maw, 1975; Nakagawa *et al.*, 1971; Robinson *et al.*, 1988; Walker and Ho, 1977). We expanded upon this approach several years ago by examining carbohydrate metabolism in a wild *Lycopersicon species, L. chmielewskii*, that is characterized by high (> 10 percent) soluble solids content (Yelle *et al.*, 1988). It was reasoned that unique biochemical traits identified in this species may provide insights into mechanisms conferring high soluble solids content to tomato fruit and provide a basis for transferring the appropriate biochemical trait(s) to *L. esculentum*.

Fruit of *L. chmielewskii* were found to accumulate sucrose as the predominant soluble sugar (Yelle *et al.*, 1988), unlike *L. esculentum* fruit which accumulate hexose (Davies and Hobson, 1981). It was reasoned that sucrose, as opposed to hexose, accumulation may contribute to enhanced soluble solids levels in several ways. First, osmotic considerations suggest that sucrose-accumulating fruit can accumulate twice as much soluble carbohydrate as hexose-accumulating fruit while maintaining an equivalent osmotic potential. In this regard it is interesting to note that while *L. chmielewskii* fruit accumulated twice as much total soluble carbohydrate when expressed on a glucose equivalent basis, the osmotic concentration of soluble carbohydrate was approximately the same in *L. chmielewskii* as in *L. esculentum* (Yelle *et al.*, 1988). To the extent that soluble sugar uptake and accumulation is regulated by turgor (Wyse *et al.*, 1986) or by osmotic potential (Ehret and Ho, 1986) the trait of sucrose accumulation might favor elevated soluble sugar accumulation. Secondly, sucrose is metabolically less active than hexose (Pontis, 1978) and may be less accessible for loss through respiration, contributing to higher levels of accumulation. It is significant to note that storage organs accumulating very high levels of sugars do so by accumulating sucrose (Giaquinta, 1979; Hatch *et al.*, 1963; Kato and Kubota, 1978; Schaffer *et al.*, 1987). For instance, in sweet melon sugar accumulation occurs in two stages with only low levels of hexose accumulating early in development followed by high levels of sucrose accumulating in later stages of development (Schaffer *et al.*, 1987). These considerations suggest that sucrose accumulation may contribute to the ability of the wild tomato species to accumulate high levels of total soluble solids and introgression of this trait into *L. esculentum* supports this contention (see below).

Genetic analysis of sucrose accumulation. To analyze the genetic basis for sucrose accumulation, a cross between *L. esculentum* and *L. chmielewskii* was made and the progeny analyzed. No sucrose-accumulating progeny were identified in the F_1 hybrids, suggesting that recessive genes control this trait. Among the segregating population of F_2 progeny, several sucrose-accumulating individuals were identified and one was self-pollinated and crossed to *L. esculentum* to produce F_3 and BC_1F_1 seed, respectively. All plants analyzed in the F_3 population accumulated sucrose rather than hexose, and the total sugar content was 57 percent higher than in *L. esculentum*. The fact that true-breeding, sucrose-accumulating lines could be obtained indicated that sucrose accumulation could be transferred from *L. chmielewskii* to *L. esculentum* as a stable trait.

As expected from analysis of the F_1 population, all fruit of the BC_1F_1 population accumulated hexose. The BC_1F_1 plants were self-pollinated to produce BC_1F_2 seed. Of the 62 BC_1F_2 plants that were grown, only 56 were analyzed since six were sterile and did not set fruit. Of the 56 plants analyzed, 46 were hexose accumulators and ten were sucrose accumulators (Table 1). Thus, the ratio of sucrose- to hexose-accumulating plants was approximately 1:6 in the BC_1F_2 population. In mature fruit (40 d after anthesis) 32 percent more sugar was present in sucrose-accumulating than in hexose-accumulating fruit of the same BC_1F_2 population (Table 1). The increased level of soluble sugar in mature fruit resulted in a comparable (30 percent) increase in total soluble solids of sucrose-accumulating fruit relative to hexose-accumulating fruit in the same BC_1F_2 population. This result is consistent with our expectation that the trait of sucrose accumulation may enhance total soluble solids levels in tomato fruit.

The BC_1F_2 population was further studied by restriction fragment length polymorphism analysis and the trait of sucrose accumulation was found to be associated with a single region on chromosome 3. This chromosomal region has been shown to carry a recessive allele associated with sterility in crosses of *L. chmielewskii* with *L. esculentum* (A. Paterson, personal communication) and as mentioned above, six plants in the BC_1F_2 population failed to set fruit. Interestingly, all six of these plants were homozygous for the RFLP marker linked to the trait of sucrose accumulation suggesting that, if fruit had been present on these plants, they would have been sucrose accumulators. If this presumption is correct the true ratio of sucrose- to hexose-accumulators in the BC_1F_2 population was approximately 1:4 (16:62). We therefore conclude that it is likely that the trait of sucrose accumulation derived from *L. chmielewskii* is controlled by a single recessive gene.

Table 1. Level of hexose and sucrose in tomato fruits of the BC_1F_2 population.

Population	Age (day)	Hexose (μmol glc/g fr wt)	Sucrose (μmol glc/g fr wt)	Sucrose Hexose ratio	Total Sugar (μmol glc/g fr wt)
Hexose Accumulator	20	70.9 ± 3.2	7.2 ± 1.4	0.11 ± 0.02	78.1 ± 3.9
(46/56)	40	176.3 ± 6.8	5.1 ± 1.8	0.05 ± 0.01	181.4 ± 6.5*
Sucrose Accumulator	20	36.6 ± 5.6	48.9 ± 4.51	1.34 ± 0.23	85.5 ± 7.36
(10/56)	40	71.9 ± 12.6	165.0 ± 18.9	4.22 ± 1.26	236.5 ± 18.4*

* Values for total sugar are different at the 99.9% confidence level. (Data from Yelle *et al.*, 1990a. Reproduced with permission.)

Biochemical analysis of sucrose accumulation. In order to assess the role of specific metabolic processes contributing to the *L. chmielewskii*-derived trait of sucrose accumulation, invertase and sucrose synthase activity were assayed in fruit of the BC_1F_2 population (Table 2). Fruit of the 46 hexose accumulators showed a high level of both invertase and sucrose synthase, while fruit of the ten sucrose accumulators had high sucrose synthase but low levels of invertase. These results indicate that low invertase levels are closely associated with sucrose-accumulation in this population. Invertase was purified from *L. esculentum* and found to resolve into three highly similar isoforms when analyzed by chromatofocusing (Figure 2). Invertase purified from fruit of *L. chmielewskii* or of a sucrose-accumulating BC_1F_2 plant was far less abundant but was comprised of identical isoforms (data not shown). An antibody raised to carrot invertase

(Lauriere *et al.*, 1988; courtesy of Dr. M. Chrispeels, UC San Diego) was used to probe immunoblots of protein from *L. esculentum, L. chmielewskii* and sucrose- and hexose-accumulators of the BC_1F_2 population (Figure 3). In all cases low invertase activity was associated with a low level of immunologically detectable invertase protein, indicating that the low invertase in sucrose-accumulating fruit is caused by lack of enzyme rather than by inactivation of the enzyme, perhaps by an invertase inhibitor proposed to be present in senescent tomato fruit (Nakagawa *et al.*, 1980; Nakagawa *et al.*, 1971).

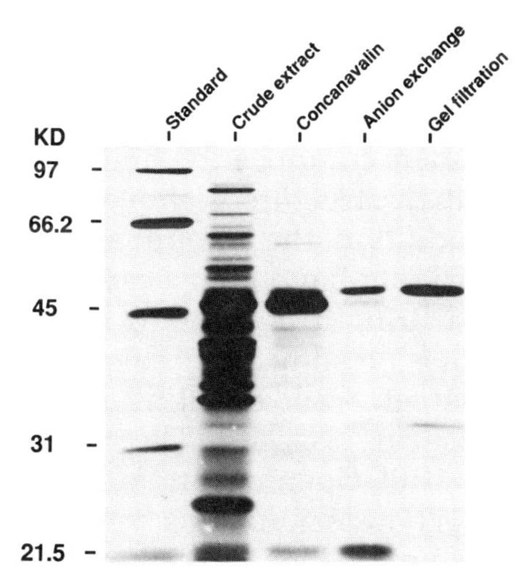

Figure 2. **SDS polyacrylamide gel electrophoresis of tomato fruit invertase at various stages of purification. The amount of protein in each sample lane was 30 μg, crude extract; 15 μg, Concanavalin; 5 μg, anion exchange; and 3 μg, gel filtration. The gel was stained with Coomassie blue (Adapted from Yelle *et al.*, 1990b. Reproduced with permission.)**

Hexose | Sucrose
L. esc | BC1F2 | L. chm | BC1F2
20 40 20 40 20 40 20 40 std

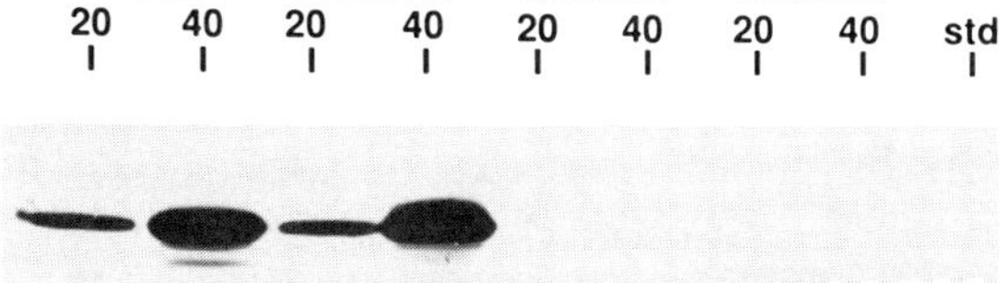

Figure 3. **Immunological detection of invertase in *L. esculentum* or *L. chmielewskii* and in hexose- or sucrose-accumulating BC_1F_2 fruit. Protein from hexose-accumulating fruit (*L. esculentum* or BC_1F_2) or from sucrose-accumulating fruit (*L. chmielewskii* or BC_1F_2) at either 20 or 40 d after anthesis was analyzed. 100 μg of protein was loaded per lane, blotted to PVDF membrane and invertase detected with antibodies to carrot invertase (adapted from Yelle *et al.*, 1990b. Reproduced with permission.)**

The results described above suggest that the single gene controlling sucrose accumulation may encode tomato fruit invertase or a factor regulating expression of the invertase gene. To test this hypothesis we have used sequence information from the invertase protein and have tentatively identified the corresponding cDNA clone from *L. esculentum*. This molecular clone can now be used to determine the chromosomal location of the invertase gene in relation to the region of chromosome three that we have shown to be associated with the trait of sucrose accumulation. In addition, the role of invertase in controlling sucrose accumulation can be directly tested in transgenic plants in which invertase expression is modified.

Degradation of Cell Wall Polymers

Cell wall polymers comprise the bulk of the insoluble solids of tomato fruit and their structure is thought to be a determinant of fruit firmness. In addition, the amount and structure of the insoluble solids influence water-binding capacity and viscosity of processed tomato products. During fruit ripening several cell wall polymers are degraded and it is presumed that this degradation compromises the rigidity of the cell wall, resulting in fruit softening and an associated decrease in viscosity. The major cell wall polymers degraded during fruit ripening are pectins and hemicelluloses (Huber, 1983b). It is not clear whether the degradation of either one or both cell wall components is required for fruit softening. One approach to assess the role of each cell wall polymer is to selectively alter the expression of genes encoding pectinases and hemicellulases in ripening fruit and determine the effect on cell wall structure and on fruit rheological properties.

Polygalacturonase. Pectin degradation has been considered to be of paramount importance in determining textural changes that occur during fruit ripening (Huber, 1983a) and recent research has focused on the implementation of molecular genetic strategies to alter pectin degradation by modifying expression of the gene encoding polygalacturonase (Giovannoni *et al.*, 1989; Sheehy *et al.*, 1988; Smith *et al.*, 1988). One series of experiments introduced a chimeric polygalacturonase gene into a ripening-impaired tomato mutant (rin) that normally fails to express polygalacturonase (Giovannoni *et al.*, 1989). Upon induction of chimeric gene expression in transgenic rin fruit, pectins were degraded, but this was not sufficient to induce fruit softening (Giovannoni *et al.*, 1989; DellaPenna *et al.* 1990). Other experiments utilized antisense polygalacturonase genes to reduce polygalacturonase enzyme levels in ripening tomato fruit (Sheehy *et al.*, 1988; Smith *et al.*, 1988). In these experiments reductions in polygalacturonase of up to 99 percent failed to reduce fruit softening (Smith *et al.*, 1990), however a significant increase in viscosity of processed fruit carrying the antisense polygalacturonase gene was noted (Kramer *et al.*, 1990). Taken together, these results indicate that while polygalacturonase plays an important role in determining rheological properties of the ripe tomato fruit cell wall, it is not the sole determinant of fruit softening and of the textural changes that accompany fruit ripening.

Endo-β-1,4-glucanase. A class of enzymes, commonly referred to as cellulases, that cleave β-1,4-glucan linkages are widely present in plant tissues and have been proposed to play major roles in a number of processes, including cell growth, abscission and fruit softening (Hayashi *et al.*, 1984; Tucker *et al.*, 1988; Pesis *et al.*, 1978; Christofferson *et al.*, 1984). It is unfortunate that these enzymes have been referred to as cellulases since it is well documented that they do not degrade crystalline cellulose as it occurs in situ (Hatfield and Nevins, 1986). Although the endogenous substrate of endo-β-1,4-glucanases is not established, it has been shown that pea stem and avocado fruit endo-β-1,4-glucanases can degrade xyloglucan, suggesting that this may be the substrate in situ (Hayashi *et al.*, 1984; Hatfield and Nevins, 1986) and that these enzymes may therefore act, at least in part, as hemicellulases. Given the complexity of cell wall polymers and the multiplicity of endo-β-1,4-glucanases in plants, it may be reasonable to speculate that this class of enzyme acts on more than one substrate in the cell wall. To determine the extent to which endo-β-1,4-glucanases participate in changes in cell wall structure during tomato fruit ripening, we have isolated cDNA clones encoding tomato fruit endo-β-1,4-glucanases and have initiated analysis of transgenic plants in which their expression is altered.

Cloning tomato endo-β-1,4-glucanase. Plant endo-β-1,4-glucanases have been cloned from avocado and bean (Christoffersen *et al.*, 1984; Tucker *et al.*, 1987; Tucker *et al.*, 1988). Unfortunately, these plants are not suitable for gene transfer experiments so it has not been feasible to directly assess the function of the endo-β-1,4-glucanases using, for instance, antisense gene constructs to inhibit expression in these plants. In order to clone the corresponding cDNAs from tomato, a plant suitable for gene transfer experiments, we examined the avocado and bean endo-β-1,4-glucanase sequences and identified two highly conserved peptide sequence domains (Figure 4). These sequences were used to design corresponding degenerate oligonucleotide probes and a number of cDNA clones were isolated from a ripe tomato fruit cDNA library. One of the tomato endo-β-1,4-glucanase cDNA clones (pTCEL1) was full length and was sequenced to determine its overall similarity to the bean and avocado sequences. The sequence was approximately 50 percent identical to bean and avocado endo-β-1,4-glucanase over the entire sequence and the conserved domains used to construct the oligonucleotide probes were completely conserved in the tomato fruit endo-β-1,4-glucanase cDNA as well (Figure 4).

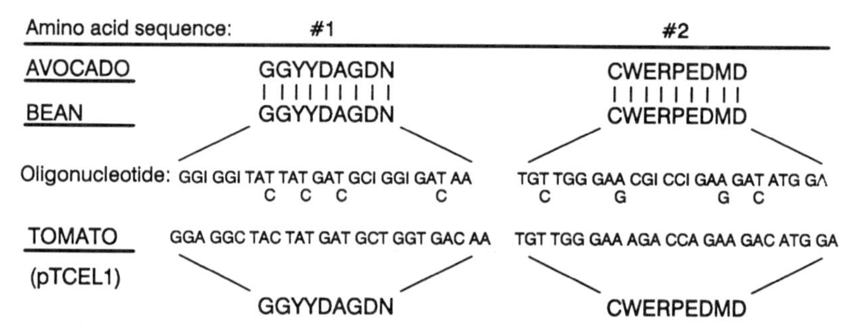

Figure 4. Conserved amino acid domains in avocado and bean endo-β-1,4-glucanase and corresonding oligonucleotides used to isolate tomato fruit endo-β-1,4-glucanase cDNAs. The conserved sequences are labeled as amino acid sequences 1 and 2, with 1 corresponding to the more N-terminal domain. Each sequence was used to design the corresponding degenerate oligonucleotide that in turn was used to isolate a tomato fruit endo-β-1,4-glucanase cDNA (pTCEL1). The sequence and deduced amino acid sequence corresponding to the conserved endo-β-1,4-glucanase sequences are also shown.

The pattern of pTCEL1 mRNA accumulation was determined throughout tomato fruit ripening and found to parallel the timing of expression of polygalacturonase (Figure 5). However, the maximal level of mRNA accumulation was approximately 100-fold less for endo-β-1,4-glucanase mRNA than for polygalacturonase mRNA. This temporal pattern of mRNA accumulation is similar to the progression of fruit softening, consistent with endo-β-1,4-glucanase contributing to this process.

Recent results, however, indicate that multiple mRNAs encoding endo-β-1,4-glucanase, in addition to that corresponding to pTCEL1, accumulate in ripe tomato fruit. The evidence supporting this con-

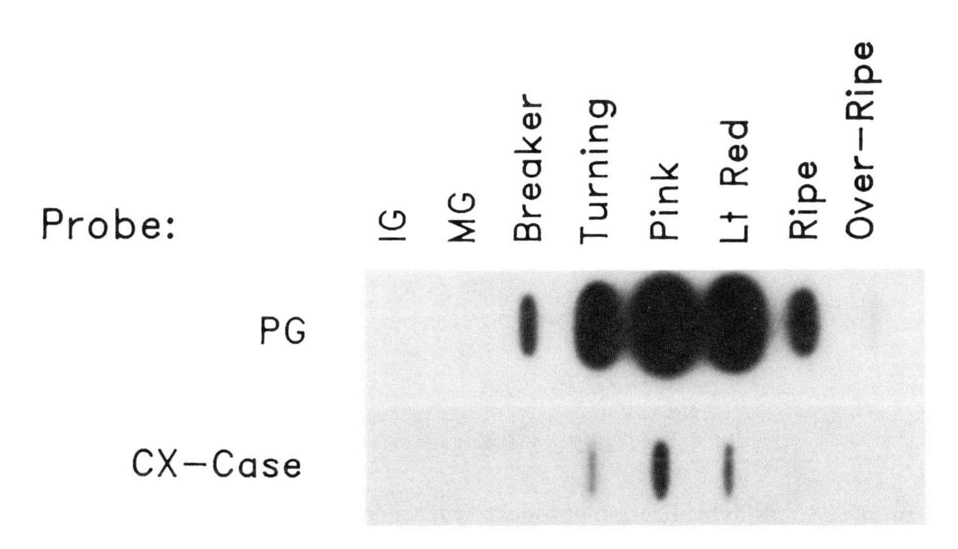

Figure 5. Endo-β-1,4-glucanase mRNA accumulation in ripening tomato fruit. Two μg of poly (A)+ mRNA were blotted to nitrocellulose and hybridized with ^{32}P-labeled endo-β-1,4,-glucasanse (upper panel) or polygalacturonase (lower panel) cDNA. Quantitative analysis of hybridization indicated that polygalacturonase mRNA is approximately 100-fold more abundant than endo-β-1,4-glucanase mRNA at the maximal level of induction for both mRNAs.

tention is that the cDNAs isolated from the ripe tomato fruit library using conserved oligonucleotides can be differentiated into at least three distinct classes based on cross-hybridization (Figure 6). All of these cDNAs have not been sequenced but their failure to cross-hybridize suggests that they are quite divergent in their overall sequence, in spite of the fact that all hybridize to one of the conserved oligonucleotides (Figure 6). These results are somewhat different from the situation in avocado, where it appears that a single endo-β-1,4-glucanase gene is expressed in ripening fruit (Cass *et al.*, 1990). Nevertheless, it appears that in tomato, multiple endo-β-1,4-glucanase genes are expressed in ripening fruit. The requirement for expression of multiple related genes in tomato is provocative. It is interesting to speculate that each gene product may have a unique substrate specificity, working in concert, to bring about the major alterations in cell wall structure that are associated with tomato fruit ripening.

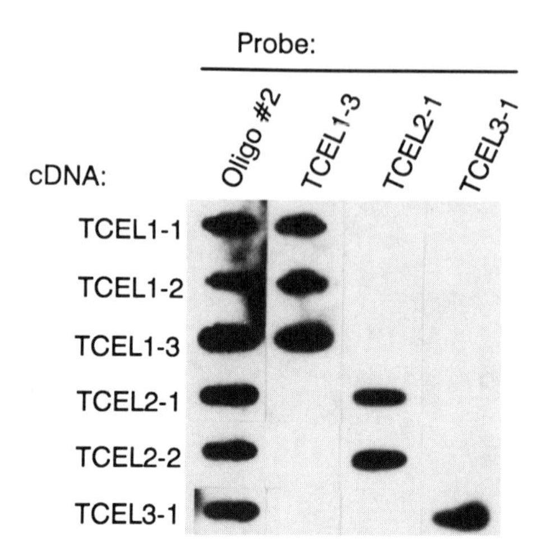

Figure 6. Cross-hybridization of tomato fruit endo-β-1,4-glucanase cDNAs. Each cDNA was isolated by hybridizationto oligonucleotide probes corresponding to conserved amino acid sequence domains. Cross-hybridization indicated that three cDNAs corresponded to TCEL1, two to TCEL2 and a third unique cDNA corresponded to TCEL3. TCEL1, TCEL2, and TCEL3 all hybridized to the oligonucleotide derived from conserved amino acid sequence 2.

Function of endo-β-1,4-glucanase. To determine the function of endo-β-1,4-glucanases in tomato we have initiated experiments using antisense gene constructs to depress the expression of pTCEL1 in ripening transgenic tomato fruit. To target expression of the antisense gene to ripening fruit, a regulatory element from a ripening-induced gene, E8 (Deikmen and Fischer, 1988), was used in the construction of the chimeric antisense gene. Because several reports suggest that endo-β-1,4-glucanases act in concert with polygalacturonase (Lewis *et al.*, 1975; Ben-Arie *et al.*, 1979), chimeric antisense polygalacturonase genes have also been incorporated into transgenic tomato plants, both alone and in conjunction with chimeric antisense endo-β-1,4-glucanase genes. The plants carrying these antisense genes develop normally, flower and set fruit that appear to develop normally. Analysis of these fruit should provide the basis for determining the function of the endo-β-1,4-glucanase gene encoded by pTCEL1, both alone and in relation to polygalacturonase action. As described above, however, there appear to be multiple endo-β-1,4-glucanase genes expressed in ripening tomato, each of which may have unique functions or indeed unique substrates in the cell wall. Full assessment of the function(s) of the endo-β-1,4-glucanases in tomato fruit ripening will most likely require detailed analysis of an extensive population of transgenic plants, each with the expression of one or more cell wall hydrolase genes modified by chimeric sense or antisense genes. This effort, while time-consuming, is well within the state of available technology and may ultimately unravel some of the complexity of cell wall structure, by specifically modifying the enzymic determinants of cell wall degradative processes.

Conclusions

In general, the prospects for improvement of fruit quality by manipulation of single genes are promising. As mentioned above, some important traits such as fruit color are controlled by single genes. In addition, complex polygenic traits such as soluble solids content may be modified incrementally by addressing component biochemical processes. The means for accomplishing such modification, using either exotic germplasm or molecular genetic techniques, are well-established for some species. However, the biochemical information necessary to identify appropriate targets for genetic manipulation is often limiting. Fortunately, in addition to providing the means to ultimately manipulate specific biochemical traits, molecular genetic techniques and the analysis of transgenic plants are providing the means to critically assess the function(s) of single biochemical steps in complex physiological processes.

References

Ben-Arie, R., Kislev, N. and Frenkel, C. 1979. Ultrastructural changes in the cell walls of ripening apple and pear fruit. Plant Physiol. 64: 197-202.

Cass, L.G., Kirven, K.A. and Christoffersen, R.E. 1990. Isolation and characterization of a cellulase gene family member expressed during avocado fruit ripening. Molec. Gen. Genet. 223: 76-86.

Christofferson, R.E., Tucker, M.L. and Laties, G.G. 1984. Cellulase gene expression in ripening avocado fruit: the accumulation of cellulase mRNA and protein as demonstrated by cDNA hybridization and immunodetection. Plant Molec. Biol. 3: 385-391.

Damon, S., Hewitt J., Neider M. and Bennett, A.B. 1988. Sink metabolism in tomato fruit. II. Phloem unloading and sugar uptake. Plant Physiol. 87: 731-736.

Davies, J.N. and Hobson, G. 1981. The constituents of tomato fruit—the influence of environment, nutrition and genotype. CRC Crit. Rev. Food Sci. Nutr. 15: 205-280.

Deikman, J. and Fischer, R. 1988. Interaction of a DNA binding factor with the 5′ flanking region of an ethylene-responsive fruit ripening gene from tomato. EMBO J. 7: 3315-3320.

DellaPenna, D., Lashbrook, C., Toenjes, K., Giovannoni, J.J., Fischer, R.L. and Bennett, A.B. 1990. Polygalacturonase isozymes and pectin depolymerization in transgenic rin tomato fruit. Plant Physiol. (In press).

DeVerna, J.W. and Paterson, A.H. 1990. Genetics of *Lycopersicon*. In "Genetic Improvement of Tomato," Theor. Appl. Genet. Monograph Series, Springer-Verlag, Berlin (In press).

Ehret, D. and Ho, L.C. 1986. Effects of salinity on dry matter partitioning and fruit growth in tomatoes grown in nutrient film culture. J. Am. Soc. Hort. Sci. 61: 361-367

Giaquinta, R.T. 1979. Sucrose translocation and storage in the sugar beet. Plant Physiol. 63: 828-832

Giovannoni, J., DellaPenna, D., Bennett, A.B. and Fischer, R.L. 1989. Expression of a chimeric polygalacturonase gene in transgenic *rin* (ripening inhibitor) tomato fruit results in polyuronide degradation but not fruit softening. Plant Cell 1: 53-63.

Hatch, M.D., Sacher, J.A. and Glasziou K.T. 1963. Sugar accumulation cycle in sugar cane. II. Relationship of invertase activity to sugar content and growth rate in storage tissues of plant grown under controlled environments. Plant Physiol. 38: 344-348

Hatfield, R. and Nevins, D. 1986. Characterization of the hydrolytic activity of avocado cellulase. Plant and Cell Physiol. 27: 541-552.

Hayashi, T., Wong, Y.S. and MacLachlan, G.A. 1984. Pea xyloglucan and cellulose. II. Hydrolysis by pea endo-1,4-β glucanases. Plant Physiol. 75: 605-610.

Hewitt, J.D. and Garvey, T.C. 1987. Wild sources of high soluble solids in tomato. In "Tomato Biotechnology," p. 45, Alan R. Liss, Inc., New York.

Ho, L.C. 1988. Metabolism and compartmentation of imported sugars in sink organs in relation to sink strength. Ann. Rev. Plant Physiol. & Plant Mol. Biol. 39: 355-378.

Huber, D. 1983a. The role of cell wall hydrolases in fruit softening. Hort. Rev. 5: 169-219.

Huber, D. 1983b. Polyuronide degradation and hemicellulose modifications in ripening tomato fruit. J. Am. Soc. Hort. Sci. 108: 405-409.

Ibaria, E.A. and Lambeth, V.N. 1969. Inheritance of soluble solids in a large/small-fruited tomato cross. J. Amer. Soc. Hort. Sci. 94: 496-498.

Jarret, R.L., Sayama, H. and Tigchelaar, E.C. 1984. Pleiotropic effects associated with the chlorophyll intensifier mutations *high pigment* and *dark green* in tomato. J. Amer. Soc. Hort. Sci. 109: 873-878.

Kato, T. and Kubota, M. 1978. Properties of invertase in sugar storage tissues of citrus fruit and changes in their activities during maturation. Plant Physiol. 42: 67-72

Kramer, M., Sanders, R.A., Sheehy, R.E., Melis, M., Kuehn, M. and Hiatt, W.R. 1990. Field evaluation of tomatoes with reduced polygalacturonase by antisense RNA. In "Horticultural Biotechnology," p. 347, Wiley-Liss, Inc., New York.

Lauriere, C., Lauriere, M., Sturm, A., Faye, L. and Chrispeels, M.J. 1988. Characterization of β-fructosidase, an extracellular glycoprotein of carrot cells. Biochimie 70: 1483-1491.

Lewis, L.N., Linkins, A.E., O'Sullivan, S.A. and Reid, P.D. 1975. Two forms of cellulase in bean plants. In "Proc. Eighth Int. Conf. Plant Growth Substances," pp. 708-718, Hirokawa Publishing Co. Inc., Tokyo.

Manning, K. and Maw, G.A. 1975. Distribution of acid invertase in the tomato plant. Phytochemistry 14: 1965-1969

Nakagawa, H., Iki, K., Hirata, M., Ishigami, S. and Ogura, N. 1980. Inactive fructofuranosidase molecules in senescent tomato fruit. Phytochemistry 19: 195-197.

Nakagawa, H., Kawasaki, Y., Ogura, N. and Takehana, H. 1971. Purification and some properties of two types of fructofuranosidase from tomato fruit. Agric. Biol. Chem. 36: 18-26.

Osborn, T.C., Alexander, D.C. and Fobes, J.S. 1987. Identification of restriction fragment length polymorphism linked to genes controlling soluble solids content in tomato fruit. Theor. Appl. Genet. 73: 350-356.

Paterson, A.H., Lander, E.S., Hewitt, J.D., Peterson, S., Lincoln, S. and Tanksley, S.D. 1988. Resolution of quantitative traits into Mendelian factors by using a complete linkage map of restriction fragment length polymorphisms. Nature 355: 721-726.

Pesis, E., Fuchs, Y. and Zauberman, G. 1978. Cellulase activity and fruit softening in avocado. Plant Physiol. 61: 416-419.

Pontis, H.G. 1978. On the scent of the riddle of sucrose. Trends Biochem. Sci. 3: 137-139

Rick, C.M. 1974. High soluble-solids content in large fruited tomato lines derived from a wild-green fruited species. Hilgardia 42: 493-510

Robinson, N.L., Hewitt, J.D. and Bennett, A.B. 1988. Sink metabolism in tomato fruit. I. Developmental changes in carbohydrate metabolizing enzymes. Plant Physiol. 87: 727-730.

Schaffer, A.A., Aloni, B. and Fogelman, E. 1987. Sucrose metabolism and accumulation in developing fruit of sweet and non-sweet genotypes of *Cucumis*. Phytochemistry 26: 1883-1887.

Sheehy, R.E., Kramer, M. and Hiatt, W.R. 1988. Reduction of polygalacturonase activity in tomato fruit by antisense RNA. Proc. Natl. Acad. Sci. USA 85: 8805-8809.

Smith, C.J.S., Watson, C.F., Morris, P.C., Bird, C.R., Seymour, G.B., Gray, J.E., Arnold, C., Tucker, G.A., Schuch, W., Harding, S. and Grierson, D. 1990. Inheritance and effect on ripening of antisense polygalacturonase genes in transgenic tomatoes. Plant Molec. Biol. 14: 369-379.

Smith, C.J.S., Watson, C.F., Ray, J., Bird, C.R., Morris, P.C., Schuch, W. and Grierson, D. 1988. Antisense RNA inhibition of polygalacturonase gene expression in transgenic tomatoes. Nature 334: 724-726.

Stevens, M.A. 1986. Inheritance of tomato fruit quality components. Plant Breeding Rev. 4: 274-310.

Stevens, M.A. and Rudich, J. 1978. Genetic potential for overcoming physiological limitations on adaptability, yield and quality in the tomato. Hort. Sci. 13: 673-678.

Tanksley, S.D. and Hewitt, J. 1988. Use of molecular markers in breeding for soluble solids content in tomato—a re-examination. Theor. Appl. Genet. 75: 811-823.

Tucker, M.L., Durbin, M.L., Clegg, M.T. and Lewis, L.N. 1987. Avocado cellulase: nucleotide sequence of a putative full length cDNA clone and evidence for a small gene family. Plant Molec. Biol. 9: 197-203.

Tucker, M.L., Sexton, R., del Campillo, E. and Lewis, L.N. 1988. Bean abscission cellulase: characterization of a cDNA clone and regulation of gene expression by ethylene and auxin. Plant Physiol. 88: 1257-1262.

Walker, A.J. and Ho, L.C. 1977. Carbon translocation in the tomato: carbon import and growth. Ann. Bot. 43: 813-823

Wyse, R.E., Zamskie, E. and Tomos, A.D. 1986. Turgor regulation of sucrose transport in sugar beet taproot tissue. Plant Physiol. 81: 478-481.

Yelle, S., Chetelat, R.T., DeVerna, J.W. and Bennett, A.B. 1990a. Sink metabolism in tomato fruit. IV. Genetic and biochemical analysis of sucrose accumulation. Plant Physiol. (Submitted).

Yelle, S., Dorais, M. and Bennett, A.B. 1990b. Sink metabolism in tomato fruit. VI. Purification and characterization of invertase in hexose and sucrose-accumulating fruit. Plant Physiol. (Submitted).

Yelle, S., Hewitt, J.D., Robinson, N.L., Damon, S. and Bennett, A.B. 1988. Sink metabolism in tomato fruit. III. Analysis of carbohydrate assimilation in a wild species. Plant Physiol. 87: 737-740.

Causes for Concern and Opportunities for Enhanced Nutrition in the Modification of Dietary Carbohydrate Composition

---

Sheldon Reiser, Ph.D
Carbohydrate Nutrition Laboratory
Beltsville Human Nutrition Research Center
Agricultural Research Service
United States Department of Agriculture
Beltsville, MD 20705

Recent recommendations for modification of the diet presently consumed in the United States have called for an increase in carbohydrate intake from present levels of about 47 percent of total kilocalories to levels between 55-60 percent of total kilocalories. Depending on the way in which these recommendations are implemented, increased intake of carbohydrate may either raise serious concern regarding the resulting effects on metabolic risk factors associated with chronic degenerative diseases or offer an opportunity for lowering the levels of these metabolic risk factors. This chapter will focus on two aspects of carbohydrate nutrition in humans that are related to biotechnological modification of carbohydrate composition: 1) the metabolic effects of the increased intake of fructose through the advent of high-fructose corn sweeteners, and 2) the metabolic effects of the intake of starch in the form of amylose as compared to amylopectin. These effects will be discussed in relation to risk factors associated with heart disease and diabetes.

Introduction

Information obtained by the United States Department of Agriculture (USDA) based on a nationwide food consumption survey showed that in 1985 carbohydrate, fat and protein provided about

47 percent, 37 percent, and 16 percent, respectively, of the total kilocalories consumed in this country (USDA 1985a, 1985b). Recommendations for modification of the present diet consumed in the United States have been suggested by a number of sources including the American Heart Association (American Heart Association, 1986) and the American Diabetes Association (American Diabetes Association, 1987). These recommendations have consistently included an increase in the intake of carbohydrate to 55-60 percent of total kilocalories, primarily at the expense of saturated fat.

A major objective of studies carried out in the area of carbohydrate nutrition is to determine the effects of the type and amount of the various components comprising dietary carbohydrate on the levels of metabolic risk factors associated with diseases such as heart disease and diabetes, thereby establishing a basis for dietary recommendations pertaining to carbohydrate intake. In light of the recommendations for increased intake of carbohydrates, this objective takes on added importance. Metabolic risk factors associated with heart disease include elevated levels of blood cholesterol associated with low density lipoprotein (LDL) and very low density lipoprotein (VLDL) (Witzum and Schonfeld, 1979, Shekelle and Stamler, 1989), fasting triglycerides (Aberg *et al.*, 1985; Austin *et al.*, 1990), insulin (Stout, 1985; Flodin, 1986), uric acid (Jacobs, 1977; Fox *et al.*, 1985) and lowered levels of cholesterol associated with high density lipoprotein (HDL) (Miller *et al.*, 1977). Although elevated fasting and response levels of blood glucose are the accepted criteria for determining the presence of diabetes, an elevated insulin response to a glycemic stress indicative of insulin insensitivity is believed to be one of the earliest signs of noninsulin-dependent diabetes (Jackson *el al.*, 1972).

It is evident that progress in biotechnology now presents us with the opportunity to modify the composition of the carbohydrate components of our diet. Depending on how this technology is utilized, the products formed may either raise serious areas of concern regarding the resultant effects on human health or offer an opportunity for improved nutrition. This chapter will focus on two aspects of carbohydrate nutrition in humans that are related to biotechnological modification of carbohydrate composition: 1) the metabolic effects of the increased intake of fructose through the advent of high-fructose corn sweeteners (HFCS), and 2) the metabolic effects of the intake of starch in the form of amylose as compared to amylopectin. These effects will be discussed in relation to levels of metabolic risk factors associated with heart disease and diabetes.

Fructose

Table 1 presents the per capita consumption of total caloric sweeteners in the United States between 1975 and 1989. These data were compiled by the U.S. Department of Agriculture and are based on the disappearance of foods and calculated on the basis of a daily intake of 3,400 kilocalories (Buzzanell and Barry, 1989). The major sources of dietary fructose are from sucrose and HFCS. With the introduction of HFCS in 1967, their manufacture and use have registered impressive growth. The contribution of HFCS to the total intake of caloric sweeteners has increased ten-fold between 1975 and 1989, mainly at the expense of sucrose (Table 1). This increase in HFCS consumption has been largely responsible for the 14 percent increase in total caloric sweetener intake during this time period. Fructose from all dietary sources now accounts for about 70g/day or about 8 percent of total caloric intake.

Table 1. Per capita consumption of caloric sweeteners in the United States

Year	Refined Sucrose		HFCS*		Total Caloric Sweeteners	
	g/day	% Kcal+	g/day	% Kcal+	g/day	% Kcal+
1975	111	13.1	6	0.7	147	17.3
1980	104	12.2	23	2.6	154	18.2
1985	79	9.3	55	6.5	165	19.5
1989 (Preliminary)	77	9.1	61	7.2	167	19.7

*High Fructose Corn Sweeteners
+Based on 3400 Kcal/day
Adapted from Buzzanell and Barry (1989)

HFCS produced from the isomerization of dextrose obtained from hydrolyzed corn starch typically contains about 42 percent fructose, 52 percent unconverted dextrose and 6 percent oligosaccharides (Coker and Venkatasubramanian, 1985). However, technological advances now permit the production of enriched HFCS containing 90 percent fructose. It is therefore evident that the biotechnology now exists to significantly increase the level of free fructose in the food supply. This section will deal with the effects observed after the feeding of fructose or sucrose to humans on the levels of blood lipids (especially triglycerides), insulin, uric acid, lactic acid and on copper status. The inclusion of sucrose in this discussion is appropriate because many of the undesirable metabolic effects of sucrose appear to be due to the fructose rather than the glucose moiety.

Blood Lipids. No other metabolic effect of fructose or sucrose intake has attracted as much attention as their effect on the levels of endogenous triglycerides. Elevated levels of triglycerides are characteristic of type IV hyperlipoproteinemia. It is estimated that 9–17 percent of the adult population of the United States is hypertriglyceridemic (triglyceride levels above 150 m/dl) (Wood *et al.*, 1972; Brown and Daudiss, 1973). Members of the population showing this genetic predisposition toward elevated triglycerides have been called carbohydrate-sensitive (Reiser *et al.*, 1979) because their lipemia is carbohydrate-induced. There appears to be a strong positive association between hypertriglyceridemia and hyperinsulinemia (Reiser *et al.*, 1981; Steiner, 1986). It has been proposed that the primary metabolic defects leading to endogenous hypertriglyceridemia in humans are insulin resistance followed by compensatory hyperinsulinism (Olefsky *et al.*, 1974).

In studies with subjects exhibiting hypertriglyceridemia and/or hyperinsulinemia, sucrose at levels approximating those consumed in this country and in the context of the diet presently consumed in the United States has usually produced significant increases in blood triglycerides (Little *et al.*, 1970; Nikkila, 1974; Reiser *et al.*, 1981; Liu *et al.*, 1984; Coulston *et al.*, 1985; Emanuele *et al.*, 1986). Table 2 presents the results from five studies in which fructose was fed to subjects exhibiting these same metabolic characteristics at levels no higher than 21 percent of the total kilocalories. Increases in trigylycerides due to fructose ranged from 10–46 percent. Differences in the magnitude of the increase in triglycerides due to fructose could be due to differences in the polyunsaturated/saturated fat ratio (Mann *et al.*, 1973; Porikos and Van Itallie, 1983) and in the levels of dietary fiber (Albrink and Ullrich, 1986) and cholesterol (Birchwood *et al.*, 1970) or a combination of these factors in the various diets used. These results support the conclusion that it is the fructose moiety of sucrose that is responsible for the increased levels of trigylycerides observed after sucrose feeding. Studies utilizing human subjects with triglyceride and insulin levels within the normal range indicate that fructose levels above 21 percent of kilocalories would be required to produce a significant increase in blood triglycerides (Reiser *et al.*, 1989a).

Table 2. Increases in fasting blood triglycerides in hypertriglyceridemic and/or hyperinsulinemic humans after consumption of fructose at levels approximating those presently consumed in the United States

Dietary Conditions	Subjects n	Blood Triglycerides (mg/dl) with fructose	without fructose	Increase %	Reference
14% of starch Kcal replaced by fructose for 14 days	5	201	177	14	Nikkila, 1974
7.5% of wheat starch Kcal replaced by fructose for 5 weeks	12	132	107	23	Hallfrish *et al.*, 1983a
13% of sucrose Kcal replaced by fructose for 14 days	5	363*	320*	13	Crapo *et al.*, 1986
21% of starch Kcal replaced by fructose for 8 days	12	234	212	10	Bantle *et al.*, 1986
20% of high-amylose cornstarch Kcal replaced by fructose for 5 weeks	10	213	146	46	Reiser *et al.*, 1989a

*Estimated from bar graph

In addition to its association with elevated insulin levels, high levels of triglycerides are also associated with a number of other undesirable metabolic conditions. Numerous studies have shown a significant negative relationship between the levels of VLDL triglycerides and HDL cholesterol (Schaffer *et al.*, 1978; Moberg and Wallentin, 1981; Moorjani *et al.*, 1986; Zavaroni *et al.*, 1989; Austin *et al.*, 1990). In one of these studies (Zavaroni *et al.*, 1989), increased levels of triglycerides were also associated with increased systolic and diastolic blood pressure. Three studies have shown a direct linkage between fructose-induced increases in triglycerides and decreases in HDL-cholesterol (Macdonald, 1978; Reiser *et al.*, 1981; Albrink and Ullrich, 1982). In view of the inverse relationship between the concentration of HDL cholesterol and the risk of heart disease, the effect of fructose on this aspect of lipid metabolism is undesirable.

Another potentially harmful effect of hypertriglyceridemia appears to be the overproduction of the superoxide free radical. It was recently shown that the release of the superoxide free radical from stimulated mononuclear white blood cell was about 2–5 times greater from both hypertriglyceridemic nondiabetic and diabetic subjects than their nonhypertriglyceridemic controls (Hiramatsu and Arimori, 1988). The superoxide free radical carries out necessary physiological functions such as killing bacteria and inactivating viruses. However, overproduction of free radicals can also induce lipid peroxidation, DNA depolymerization, enzyyme inactivation and erythrocyte hemolysis. Increased activity of free radicals has also been associated with the aging process.

In two studies, humans with elevated levels of triglycerides and/or insulin showed small increases in total cholesterol when fed diets containing fructose as compared to starch (Hallfrisch *et al.*, 1983a; Reiser *et al.*, 1989a). These increases may in part be attributable to the cholesterol present in VLDL produced in response to the elevated triglycerides and their resultant breakdown products.

It is reasonable to postulate that reduction in the concentration of triglycerides that may occur with the decreased intake of fructose containing carbohydrates would also have beneficial effects on the levels of other risk factor metabolites associated with high triglyceride levels.

Insulin. The effect of fructose on insulin levels is of particular interest in view of the strong relationship between insulin and triglyceride levels and since fructose has been suggested as being a sugar that can be safely consumed by individuals with impaired glucose tolerance. This suggestion emanates from studies showing that fructose when consumed alone by humans elicits neither a significant insulin or glucose response (Macdonald *et al.*, 1978; Bohannon *et al.*, 1980; Crapo *et al.*, 1980a) nor is insulin required for its metabolism. However, fructose is rarely consumed in the absence of a glucose-containing component (e.g., HFCS, sucrose). In studies where blood glucose levels were elevated to postprandial levels either by infusion (Lawrence *et al.*, 1980) or consumption of glucose-containing drinks (Reiser *et al.*, 1987), systemic levels of at least 10 mg/dl fructose were shown to significantly increase insulin levels 60–288 percent while not significantly increasing blood glucose levels. Under these conditions, blood insulin levels were positively correlated with blood fructose levels but not blood glucose levels. Mechanisms proposed to explain this insulinogenic effect of fructose were the involvement of a specific fructose receptor on the pancreatic beta cell following glucose priming (Lawrence *et al.*, 1980) and an increase in the secretion of the enteric hormone gastric inhibitory polypeptide (Reiser *et al.*, 1987) which enhances pancreatic insulin secretion when blood glucose levels are elevated (Dupre *et al.*, 1973).

The extended feeding of fructose may also adversely affect insulin status. Both normal and hyperinsulinemic men were shown to have significantly higher insulin responses to a glycemic stress after consuming diets containing 15 percent of kiloalories as fructose at the expense of wheat starch for 5 weeks (Hallfrish *et al.*, 1983b). The increases were of greater magnitude in the hyperinsulinemic men. After consuming a liquid formula diet containing 17 percent of the kilocalories as either fructose or dextromaltose for 2 weeks, 6 hypertriglyceridemic men were reported to have a 28 percent greater insulin response after the diet containing fructose (Turner *et al.*, 1979). Fructose feeding may increase insulin levels by reducing the affinity of the insulin receptor. Humans free of overt disease fed their normal diet plus 1,000 kilocalories of fructose per day for one week showed a significant reduction in both insulin binding to isolated monocytes and insulin sensitivity (Beck-Nielsen *et al.*, 1980). In contrast, the feeding of an equivalent amount of glucose under the same experimental conditions affected neither insulin binding nor sensitivity. Under more relevant dietary conditions, hyperinsulinemic men showed a decrease in the binding of insulin to erythrocytes at concentrations encompassing the range of postprandial insulin response after consuming diets containing 20 percent fructose as compared to high-amylose cornstarch for 5 weeks (Table 3). Under the same conditions, nonhyperinsulinemic men showed no decrease in insulin binding. The fructose-induced decrease of insulin binding may be mediated through an increase in triglyceride levels since it has been reported that monocytes and erythrocytes from hypertriglyceridemic men as compared to normotriglyceridemic men bound significantly less insulin (Bieger *et al.*, 1984).

Table 3. Insulin binding to erythrocytes from 10 hyperinsulinemic and 11 nonhyperinsulinemic men after consuming diets containing 20 percent of kilocalories as fructose or high-amylose cornstarch for five weeks

Diet	Subjects	287 pmol insulin/L	2870 pmol insulin/L
		fmol insulin bound/1.7×10^9 erythrocytes	
Fructose	Hyperinsulinemic	0.104	0.337*
Starch	Hyperinsulinemic	0.158	1.119
Fructose	Nonhyperinsulinemic	0.200	0.743
Starch	Nonhyperinsulinemic	0.149	0.713

*Significantly lower ($p < 0.05$) than corresponding starch value as determined by Duncan's Multiple-Range Test.
Adapted from Reiser *et al.* (1989b).

Uric Acid. Uric acid is the end product of the catabolism of purine nucleotides in humans and higher primates. The formation of uric acid from xanthine, a reaction catalyzed by the enzyme xanthine oxidase, has been shown to liberate the superoxide free radical (Joannidis *et al.*, 1990). The ability of a carbohydrate to increase blood uric acid levels following short-term oral intake appears to be dependent on the presence of the fructose moiety (Macdonald *et al.*, 1978; Reiser *et al.*, 1984). In an extended feeding study, 21 men consumed diets containing 20 percent of the kilocalories from fructose as compared to high-amylose constarch for five weeks each in a crossover design (Reiser *et al.*, 1989a). Figure 1 presents the average uric acid levels before and 30-180 minutes after the men consumed their respective diets at breakfast, lunch and dinner. Uric acid levels were significantly higher ($p < 0.05$) when the men consumed the fructose than when they consumed the starch at all time points. In addition, there were significant increases over zero time uric acid levels after the subjects consumed the fructose-containing diet but not the starch-containing diet. These results indicate that fructose as compared to starch feeding produces both chronic and acute increases in uric acid levels.

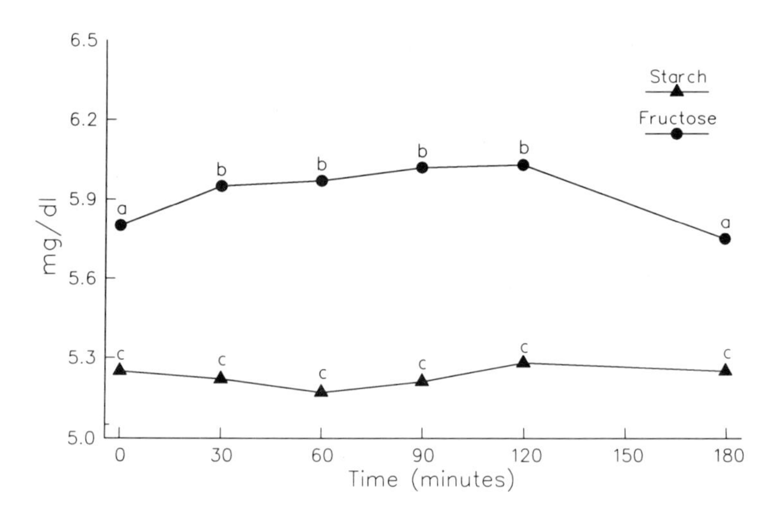

Figure 1. Average uric acid levels before and 30, 60, 90, 120 and 180 minutes after 21 men consumed breakfast, lunch and dinner. Each value represents the mean from 63 determinations (21 subjects, 3 meals). Analysis of variance for diet ($p<0.03$), for time ($p<0.05$) and for interaction between diet and time ($p<0.03$). Each value not sharing a common superscript letter is significantly different ($p<0.05$) according to Duncan's multiple range test. Adapted from Reiser *et al.* (1989a).

The increased levels of uric acid observed after the intake of sucrose or fructose may be explained on the basis of fructose metabolism in the liver (Woods *et al.*,1970). The phosphorylation of fructose has been shown to produce a rapid decrease of both hepatic ATP and inorganic phosphate. In this regard the infusion of fructose, but not glucose, into human liver produced a significant decrease of ATP content (Bode *et al.*, 1971). Since ATP inhibits 5′ nucleotidase and inorganic phosphate inhibits AMP deaminase, the AMP formed in the liver as a result of the phosphorylation of fructose is readily converted to IMP and then to inosine. Inosine is then oxidized to uric acid which appears in the blood. The increase in uric acid caused by fructose-containing carbohydrates is therefore not only a risk factor associated with heart disease, but also suggests a short-term and adaptive change in metabolism consistent with decreased hepatic levels of ATP which could potentially affect the myriad of metabolic processes dependent on ATP.

Lactic Acid. Lactic acid is the end product of anaerobic glycolysis that occurs in most mammalian tissues. It is also a normal component of human blood. High levels of blood lactic acid in humans produce the extremely toxic condition, lactic acidosis. Many studies using various preparations of rat liver have repeatedly demonstrated that lactate formation occurs much more rapidly from fructose than from glucose (Reiser, 1987). Fructose infusions in humans have resulted in dangerous increases in blood lactic acid (Reiser, 1987), especially in patients with preexisting acidotic conditions such as anoxia, diabetes, postoperative stress or uremia (Van den Berghe, 1986).

A significant increase in blood lactic acid in response to an oral load of fructose, but not glucose, has been reported in two studies (Cook, 1971; Macdonald *et al.*, 1978). The increase in lactic acid observed after an oral load of sucrose (Macdonald *et al.*, 1978) must therefore be due to the fructose moiety. Feeding fructose-containing carbohydrates for more extended time periods also appears to increase blood lactic acid. Both type I and type II diabetics showed higher peak postrandial levels of lactic acid after being fed diets containing 21 percent of kilocalories from fructose and sucrose as compared to starch for eight days (Bantle *et al.*, 1986). Lactate increased in another study in which self-selected diets of type II diabetics were supplemented with fructose (Anderson *et al.*, 1989). These lactate levels were much lower and less dangerous than those reached after fructose infusion, but would not be considered a desirable metabolic occurrence.

Copper Status. In most studies in which the metabolic effects of copper deficiency in experimental animals have been reported, sucrose has usually been the source of the dietary carbohydrate. The rationale for the use of sucrose is that it is an extremely pure dietary carbohydrate and would not be expected to provide any impurities containing copper. However, a number of studies have shown that the severity of the signs of copper deficiency in rats (Fields *et al.*, 1983; Reiser *et al.*, 1983; Fields *et al.*, 1984; Johnson, 1986; Babu *et al.*, 1987) and pigs (Scholfied *et al.*, 1990) were exacerbated when the dietary carbohydrate was either sucrose or fructose as compared to either starch or glucose. From these results it can be inferred that the fructose moiety of sucrose is responsible for the adverse effects of this sugar on copper status.

Diets consumed in the United States contain a relatively high level of fructose-containing carbohydrates and provide well below the 1.5 mg of copper per day considered to be adequate (Holden *et al.*, 1979; Klevay and Viestenz, 1981). A human study was therefore initiated to determine whether the same type of interaction between copper status and dietary fructose found in experimental animals also exisits in humans (Reiser *et al.*, 1985). Table 4 presents the activity and level of the enzyme superoxide dismutase (SOD) in erythrocytes and hemoglobin from 20 men after they consumed a diet low in copper (1 mg per day) and containing 20 percent of kilocalories from either fructose or cornstarch for 11 weeks. SOD is a copper-metallo enzyme found in tissues such as liver and erythrocytes which appears to be a very sensitive indicator of copper status in humans (Klevay *et al.*, 1984). SOD activities were found to be significantly

Table 4. SOD activity and level after 20 men were fed a low copper diet (1 mg per day) containing 20 percent of kilocalories from either fructose or constarch for 11 weeks

Diet	SOD	
	U/10^9 erythrocytes	μg/g hemoglobin
Fructose	31.6 ± 1.4^a (9)	436 ± 30^a (9)
Starch	38.6 ± 1.5^b (11)	513 ± 32^b (11)

Each value represents the mean ± SEM from the number of men shown in parentheses. Values in a column not sharing a common superscript letter are significantly different ($p < 0.05$) according to an unpaired difference t test.

lower after men consumed the fructose diet than after the starch diet. Repletion of the diets with 3 mg copper per day for three weeks significantly increased SOD only in the men previously fed fructose (Reiser *et al.*, 1985), confirming the existence of a relative copper deficiency in the fructose-fed men.

The function of SOD appears to be to scavenge the intermediates of oxygen reduction such as the superoxide free radical. In this capacity, SOD is involved with the prevention of the undesirable effects of excess free radicals described earlier. From the material presented it appears that there are at least three processes that can link fructose metabolism to the formation and processing of free radicals: 1) increased release of free radicals due to hypertriglyceridemia which can be produced by fructose, 2) increased uric acid formation and 3) decreased processing of free radicals due to reduced SOD levels. A combination of these metabolic events would link fructose intake with increased levels of free radicals.

Conclusion: It appears that there are still a number of potentially serious metabolic problems associated with the intake of fructose that should be addressed and resolved before further efforts are made toward increasing the level of fructose in our food supply.

Amylose and Amylopectin

Of the total 47 percent of kilocalories supplied by utilizable dietary carbohydrate, about 22 percent is provided by starch found in foods such as cereals and vegetables (Woteki *et al.*, 1982) and is classified as complex carbohydrate. Starch occurs in two molecular forms, amylose and amylopectin. Amylose consists of long, unbranced chains in which D-glucose units are joined in α(1–4) linkages. The chains vary in molecular weight from a few thousand to 500,000. In contrast, amylopectin is a highly branched polysaccharide. The basic structure of amylopectin, like amylose, is composed of glucose units bound in α(1–4) linkages. The branching points consist of α(1–6) linkages. The branching is catalyzed by a 1, 4 α-glucan branching enzyme present in most plant tissues which transfers a terminal fragment of from 24 to 30 glucose residues to the 6-hydroxy group of a glucose residue on the same or another amylopectin chain. The molecular weight of amylopectin may be as high as 100 million.

The source of the starch used in conducting studies on the metabolic effects of carbohydrates has usually been treated casually with the inference that starches from various food sources all produce essentially the same metabolic effects. However, on the basis of some recent studies, it is now apparent that complex carbohydrates from various food sources can differentially affect levels of metabolic risk factors associated with diseases. This section will describe some of these studies with special emphasis on differences in metabolic effects that can be ascribed to differences in the comparative levels of amylose and amylopectin.

One of the first indications that complex carbohydrates from various food sources could differentially influence metabolic parameters came from a series of studies conducted by Phyllis Carpo and her co-workers. In these studies the glucose and insulin responses of subjects free of overt disease (Crapo *et al.*, 1977), or subjects with impaired glucose tolerance (Crapo *et al.*, 1980b), and subjects with noninsulin-dependent diabetes (Crapo *et al.*, 1981) were determined after the intake of the equivalent of 50 grams of dextrose from potato, bread, corn and rice and compared with responses after the intake of 50 grams of dextrose alone. Similar results were obtained with each subject group. In general, potatoes and dextrose produced similar glucose and insulin responses while corn and rice produced the lower responses with bread giving an intermediate response. It may be concluded that the intake of different foods providing complex carbohydrate could produce a wide range of glucose and insulin responses and that metabolic benefits could be achieved by the intake of starches producing the lower glucose and insulin responses.

In an attempt to quantitate the effect of different foods providing dietary carbohydrate on blood glucose levels, the concept of the glycemic index was introduced (Jenkins *et al.*, 1981). The index is defined as the area under the blood glucose response curve between zero and two hours after the intake of 50 grams of carbohydrate from a given food expressed as the percentage of the area after the intake of 50 grams of glucose alone. It is reasonable to assume that the glycemic index will also reflect the magnitude of the insulin responses produced by the carbohydrate source tested. A large variation was found in the glycemic index produced by different souces of complex carbohydrate. As a group, cereals had a nearly two-fold higher index (60) than legumes (31). Although many components present in these foods could contribute to the observed differences in glucose response, the type of starch could be an important factor. The amylose content of starches from potato, wheat, corn and barley was reported to contribute 20, 25, 26 and 27 percent, respectively, of the total starch (Young, 1984). In contrast, the amylose content of legumes ranges between 34–40 percent of the total starch (Wursch *et al.*, 1986). It therefore appears that a negative relationship may exist between the level of the amylose content of the starch in a food and the magnitude of the glucose and insulin responses produced.

In a study designed to more directly assess the effect of the amylose content of foods on glucose and insulin responses, humans were fed three varieties of rice calculated to yield 50 grams of glucose upon hydrolysis in a crossover design (Goddard *et al.*, 1984). The rices contained either zero, 14–17 percent or 23–25 percent of carbohydrate in the form of amylose. In general, peak glucose and insulin response 30 and 60 minutes after the meals were lowest with the rice containing 23–25 percent amylose, intermediate with the rice containing 14–17 percent amylose and highest with the amylose-free rice. While these results are consistent with the contention that higher amylose levels in starch are associated with improved glucose tolerance, it is still possible that differences in micro- and/or macronutrients (e.g., fiber) in the rices could have been responsible for the observed effects.

In order to more definitively determine whether the structure of the starch can influence metabolic parameters, studies using purified cornstarch with different levels of amylose and amylopectin were initiated in this laboratory. These studies were made possible by the availability of a cornstarch (Amylomaize VII, American Maize-Producing Company, Hammond, Indiana) produced from a special type of hybrid corn that has an amylose content of 70 percent as compared to the 25–30 percent found in the usual forms of corn. In a response study (Behall *et al.*, 1988) 12 women and 13 men in general good health were given meals containing cornstarch with 70 percent of the starch in the form of either amylose (Amylomaize VII) or amylopectin in a crossover design. The meals consisted of one gram of the respective starches and 0.33 gram of fat per kilogram body weight. The glucose response to the starch meals is shown in Figure 2. The high-amylose meal produced a significantly lower peak glucose response at 30 minutes than did the amylopectin meal. In addition the amylose meal, but not the amylopectin meal, prevented the rebound fall of blood glucose below baseline levels

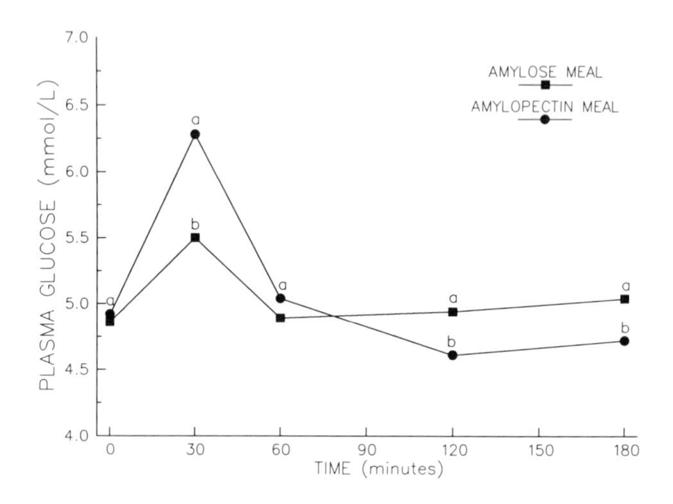

Figure 2. Plasma glucose responses to a meal containing 1 gram of either amylose or amylopectin per kilogram body weight. Each time point represents the mean ± SEM from 25 subjects. Time points with different letters are significantly different ($p<0.05$) according to ANOVA. Adapted from Behall *et al.* (1988).

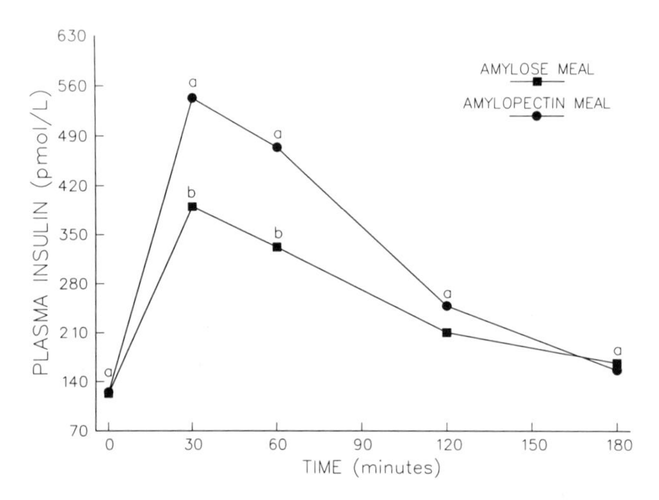

Figure 3. Plasma insulin responses to a meal containing 1 gram of either amylose or amylopectin per kilogram body weight. Each time point represents the mean ± SEM from 25 subjects. Time points with different letters are significantly different ($p<0.05$) according to ANOVA. Adapted from Behall *et al.* (1988).

120 and 180 minutes after the meals. The corresponding insulin responses to the starch meals are presented in Figure 3. The high-amylose meal produced a significantly lower insulin response both at 30 and 60 minutes as compared to the high-amylopectin meal.

A second study involved the extended feeding of diets containing cornstarch high in amylose as compared to amylopetcin (Behall *et al.*,1989a). The purpose of this study was two-fold. It was important to determine whether adaptation to amylose would occur after its long term feeding that could either reduce or eliminate the beneficial effects observed on parameters of glucose tolerance in the response study. In addition, since increased levels of insulin are associated with increased levels of triglycerides, it was of interest to determine the dietary effects of the two starches on blood lipids. Twelve men consumed a diet containing 34 percent of their kilocalories as either 70 percent amylose or amylopectin for five weeks in a crossover design. In agreement with the response study, glucose and insulin responses were significantly lower 30 and 60 minutes after a meal containing amylose as compared to amylopectin was consumed after an overnight fast during the fifth week of the study. Table 5 shows the fasting levels of triglycerides, total cholesterol and HDL cholesterol after the starches were fed for five weeks. Triglycerides and total cholestrol levels were significantly lower after the diet high in amylose as compared to amylopectin was consumed. HDL cholesterol levels were not significantly different, indicating that amylose reduced the cholesterol associated with LDL and/or VLDL. The formation of resistant starch from the amylose after cooking and cooling did not appear to be responsible for the metabolic effects observed (Behall *et al.*, 1989a, 1989b).

Table 5. Mean fasting levels of triglycerides, total cholesterol and HDL cholesterol five weeks after 12 men consumed diets containing either amylose or amylopectin in a crossover design

	Triglycerides	Total Cholesterol	HDL Cholesterol
		mmol/L	
Amylose Diet	0.87 ± 0.04^a	4.72 ± 0.10^a	1.18 ± 0.03^a
Amylopectin Diet	1.07 ± 0.06^b	5.03 ± 0.12^b	1.22 ± 0.04^a

Each value represents the mean ± SEM. Values in a column with different superscript letters are significantly different ($p<0.05$) according to ANOVA. Adapted from Behall *et al.* (1989a).

Conclusions. The decreases in glucose and insulin responses due to amylose as compared to amylopectin both after the acute meal response and 10-week feeding study were in strong agreement. These results indicate that the long term intake of cereals and vegetables containing high levels of amylose may be of benefit in improving glucose tolerance not only to noninsulin-dependent diabetics or carbohydrate-sensitive individuals, but also to subjects with desirable metabolic parameters, such as these used in our studies, by decreasing insulin and glucose response levels. In addition, the reduced levels of blood triglycerides and cholesterol not associated with HDL observed after the intake of amylose may decrease the risk of heart disease. Efforts should therefore be made to genetically modify cereals so as to reduce or eliminate the activity of the branching enzyme and thereby increase their content of amylose relative to amylopectin.

References

Aberg, H., Lithell, H.., Selinus, I. and Hedstrand, H. 1985. Serum triglycerides are a risk factor for myocardial infarction but not for angina pectoris. Results from a 10-year follow-up of Uppsala primary preventive study. Atherosclerosis 54: 89.

Albrink, M.J. and Ullrich, I.H. 1982. Effects of dietary fiber on sucrose-induced lipemia in humans. In "Metabolic Effects of Utilizable Dietary Carbohydrates," p. 221. Marcel Dekker, New York.

Albrink, M.J. and Ullrich, I.H. 1986. Interaction of dietary sucrose and fiber of serum lipids in healthy young men fed high carbohydrate diets. Am. J. Clin. Nutr. 43: 419.

American Diabetes Association. 1987. Nutritional recommendations and principles for individuals with diabetes mellitus. Diabetes Care 10: 126.

American Heart Association. 1986. Dietary guidelines for healthy American adults. A statement for physicians and health professionals by the Nutrition Committee, American Heart Association. Circulation 74: 1465A.

Anderson, J.W., Story, L.J., Zettwoch, N.C., Gustafson, N.J. and Jefferson, B.S. 1989. Metabolic effects of fructose supplementation in diabetic individuals. Diabetes Care 12: 337.

Austin, M.A., King, M-C., Vranizan, K.M. and Krauss, R.M. 1990. Atherogenic lipoprotein phenotype. A proposed genetic marker for coronary heart disease risk. Circulation 82: 495.

Babu, U., Failla, M.L. and Caperna, T.J. 1987. Fructose attenuates humoral immunity in rats with moderate copper deficiency. Fed. Proc. 46: 568. (Abstract)

Bantle, J.P., Laine, D.C. and Thomas, J.W. 1986. Metabolic effects of dietary fructose and sucrose in types I and II diabetic subjects. JAMA 256: 241.

Beck-Nielson, H., Pedersen, O. and Linkskov, H.O. 1980. Impaired cellular insulin binding and insulin sensitivity induced by high-fructose feeding in normal subjects. Am. J. Clin. Nutr. 33: 273.

Behall, K.M., Scholfield, D.J. and Canary, J.J. 1988. Effect of starch structure on glucose and insulin responses in adult subjects. Am. J. Clin. Nutr. 47: 428.

Behall, K.M., Scholfield, D.J., Yuhaniak, I. and Canary, J.J. 1989a. Diets containing high amylose vs amylopectin starch: effects on metabolic variables in human subjects. Am. J. Clin. Nutr. 49: 337.

Behall, K.M. and Scholfield, D.J. 1989b. Reply to JMM van Amelsvoort and JA Weststrate. Am. J. Clin. Nutr. 50: 2472.

Bieger, W.P., Michel, G., Barwich, D., Biehl, K. and Wirth, A. 1984. Diminished insulin receptors on monocytes and erythrocytes in hypertriglyceridemia. Metabolism 33: 982.

Birchwood, B.L., Little, J.A., Antar, M.A., Lucas, C., Buckley, G.C., Csima, A. and Kallus, A. 1970. Interrelationship between the kinds of dietary carbohydrate and fat in hyperlipoproteinemic patients. Part 2. Sucrose and starch with mixed saturated and polyunsaturated fats. Atherosclerosis 11: 183.

Bode, C., Schumacher, H., Goebell, H., Zelder, O. and Pelzel, H. 1971. Fructose-induced depletion of liver adenine nucleotides in man. Hormone Metab. Res. 3: 289.

Bohannon, N.V., Karam, J.H. and Forsham, P.H. 1980. Endocrine responses to sugar injestion in man. J. Am. Diet. Assoc. 76: 555.

Brown, D.F. and Daudiss, K. 1973. Hyperlipoproteinemia prevalence in a free-living population in Albany, New York. Circulation 47: 558.

Buzzanell, P. and Barry, R. 1989. Economic Research Service, USDA. Sugar and sweeteners. Situation and outlook report yearbook, p. 81, U.S. Government Printing Office, Washington, D.C.

Coker, L.E. and Venkatasubramanian, K. 1985. High fructose corn syrup. In "Biotechnology Applications and Research," p. 165. Technomic Publishing AG, Lancaster, PA.

Cook, G.C. 1971. Absorption and metabolism of D (-) fructose in man. Am. J. Clin. Nutr. 24: 1302.

Coulston, A.M., Hollenbeck, C.B., Donner, C.C., Williams, R., Chiou, Y.-A.M. and Reaven, G.M. 1985. Metabolic effects of added dietary sucrose on individuals with noninsulin-dependent diabetes mellitus (NIDDM). Metabolism 34: 962.

Crapo, P.A., Reaven, G. and Olefsky, J. 1977. Postprandial plasma-glucose and -insulin responses to different complex carbohydrates. Diabetes 26: 1178.

Crapo, P.A., Kolterman, O.G. and Olefsky, J.M. 1980a. Effects of oral fructose in normal, diabetic, and impaired glucose tolerance subjects. Diabetes Care 3: 575.

Crapo, P.A., Kolterman, O.G., Waldeck, N., Reaven, G. and Olefsky, J.M. 1980b. Postprandial hormonal responses to different types of complex carbohydrate in individuals with impaired glucose tolerance. Am. J. Clin. Nutr. 33: 1723.

Crapo, P.A., Insel, J., Sperling, M. and Kolterman, O.G. 1981. Comparison of serum glucose, insulin and glucagon responses to different types of complex carbohydrate in noninsulin-dependent diabetic patients. Am. J. Clin. Nutr. 34: 184.

Crapo, P.A., Kolterman, O.G. and Henry, R.R. 1986. Metabolic consequence of two-week fructose feeding in diabetic subjects. Diabetes Care 9: 111.

Dupre, J., Ross, S.A., Watson, D. and Brown, J.C. 1973. Stimulation of insulin secretion by gastric inhibitory polypeptide in men. J. Clin. Endocrinol. Metab. 37: 826.

Emanuele, M.A., Abraira, C., Jellish, W.S. and Debartolo, M. 1986. A crossover trial of high and low sucrose-carbohydrate diets in type II diabetics with hypertriglyceridemia. J. Am. Coll. Nutr. 5: 429.

Fields, M., Ferretti, R.J., Smith, Jr., J.C. and Reiser, S. 1983. Effect of copper deficiency on metabolism and mortality in rats fed sucrose or starch diets. J. Nutr. 113: 1135.

Fields, M., Ferretti, R.J., Smith, Jr., J.C. and Reiser, S. 1984. The interaction of type of dietary carbohydrates with copper deficiency. Am. J. Clin. Nutr. 39: 289.

Flodin, N.W. 1986. Atherosclerosis: an insulin-dependent disease? J. Am. Coll. Nutr. 5: 417.

Fox, I.H., John, D., DeBruyne, S., Dwosh, I. and Marliss, E.B. 1985. Hyperuricemia and hypertriglyceridemia: metabolic basis for the association. Metabolism 34: 741.

Goddard, M.S., Young, G. and Marcus, R. 1984. The effect of amylose content of insulin and glucose responses to ingested rice. Am. J. Clin. Nutr. 39: 388.

Hallfrisch, J., Reiser, S. and Prather, E.S. 1983a. Blood lipid distribution of hyperinsulinemic men consuming three levels of fructose. Am. J. Clin. Nutr. 37: 740.

Hallfrisch, J., Ellwood, K.C., Michaelis, IV, O.E., Reiser, S., O'Dorisio, T.M. and Prather, E.S. 1983b. Effects of dietary fructose on plasma glucose and hormone responses in normal and hyperinsulinemic men. J. Nutr. 113: 1819.

Hiramatsu, K. and Arimori, S. 1988. Increased superoxide production by mononuclear cells of patients with hypertriglyceridemia and diabetes. Diabetes 37: 832.

Holden, J.M., Wolf, W.R. and Mertz, W. 1979. Zinc and copper in self-selected diets. J. Am. Diet. Assoc. 75: 23.

Jackson, W.P.U., van Mieghem, W. and Keller, P. 1972. Insulin excesses as the initial lesion in diabetes. Lancet 1: 1040.

Jacobs, D. 1977. Hyperuricemia as a risk factor in coronary heart disease. Adv. Exp. Med. Biol. 76B: 231.

Jenkins, D.J.A., Wolever, T.M.S., Taylor, R.H., Barker, H., Fielden, H., Baldwin, J.M., Bowling, A.C., Newman, H.C., Jenkins, A.L. and Goff, D.V. 1981. Glycemic index of foods: a physiological basis for carbohydrate exchange. Am. J. Clin. Nutr. 34: 362.

Joannidis, M., Gstraunthaler, G. and Pfaller, W. 1990. Xanthine oxidase: evidence against a causative role in renal reperfusion injury. Am. J. Clin. Physiol. 258 (Renal Fluid Electrolyte Physiol. 27): F232.

Johnson, M.A. 1986. Interaction of dietary carbohydrate, ascorbic acid and copper with the development of copper deficiency in rats. J. Nutr. 116: 802.

Klevay, L.M. and Viestenz, K.E. 1981. Abnormal electrocardiograms in rats deficient in copper. Am. J. Physiol. 240: H185.

Klevay, L.M., Inman, L., Johnson, L.K., Lawler, M., Mahalko, J.R., Milne, D.B., Lukaski, H.C., Bolonchuk, W. and Sandstead, H.H. 1984. Increased cholesterol in plasma in a young man during experimental copper depletion. Metabolism 33: 112.

Lawrence, J. R., Gray, C.E., Grant, I.S., Ford, J.A., McIntosh, W.B. and Dunnigan, M.G. 1980. The insulin response to intravenous fructose in maturity-onset diabetes mellitus and in normal subjects. Diabetes 29: 736.

Little, J.A., Birchwood, B.L., Simmons, D.A., Antar, M.A., Kallos, A., Buckley, G.C. and Csima, A. 1970. Interrelationship between the kinds of dietary carbohydrate and fat in hyperlipoproteinemic patients. Part 1. Sucrose and starch with polyunsaturated fat. Atherosclerosis 11: 173.

Liu, G., Coulston, A., Hollenbeck, C. and Reaven, G. 1984. The effect of sucrose content in high and low carbohydrate diets in plasma glucose, insulin and lipid response in hypertriglyceridemic humans. J. Clin. Endocrinol. Metab. 59: 636.

Macdonald, I., Keyser, A. and Pacy, D. 1978. Some effects in man, of varying the load of glucose, sucrose, fructose or sorbitol on various metabolites in blood. Am. J. Clin. Nutr. 31: 1305.

Macdonald, I. 1978. The effects of dietary carbohydrates on high density lipoprotein levels in serum. Nutr. Rep. Int. 17: 663.

Mann, J.I., Watermeyer, G.S., Manning, E.B., Randles, J. and Truswell, A.S. 1973. Effects on serum lipids of different dietary fats associated with a high sucrose diet. Clin. Sci. 44: 601.

Miller, N.E., Forde, O.H., Thelle, D.S. and Mjos, O.D. 1977. The Tromso heart study. High density lipoprotein and coronary heart-disease; a prospective case-control study. Lancet 1: 968.

Moberg, B. and Wallentin, L. 1981. High density lipoprotein and other lipoproteins in normolipidemic and hypertriglyceridemic (type IV) men with coronary artery disease. Europ. J. Clin. Invest. 11: 433.

Moorjani, S., Gagne, C., Lupien, P.J. and Brun, D. 1986. Plasma triglycerides related decrease in high-density lipoprotein cholesterol and its association with myocardial infarction in heterozygous familial hypercholesterolemia. Metabolism 35: 311.

Nikkila, E.A. 1974. Influence of dietary fructose and sucrose on serum triglycerides in hypertriglyceridemia and diabetes. In "Sugars and Nutrition," p. 439. Academic Press, New York.

Olefsky, J.M., Farquhar, J.W. and Reaven, G.M. 1974. Reappraisal of the role of insulin in hypertriglyceridemia. Am. J. Med. 57: 551.

Porikos, K.P. and Van Itallie, T.B. 1983. Diet-induced changes in serum transaminase and triglyceride levels in healthy adult men. Role of sucrose and excess calories. Am. J. Med. 75: 624.

Reiser, S., Hallfrisch, J., Michaelis, IV, O.E., Lazar, F.L., Martin, R.E. and Prather, E.S. 1979. Isocaloric exhange of dietary starch and sucrose in humans. I. Effects on levels of fasting blood lipids. Am. J. Clin. Nutr. 32: 1659.

Reiser, S., Bickard, M.C., Hallfrisch, J., Michaelis, IV, O.E. and Prather, E.S. 1981. Blood lipids and their distribution in lipoproteins in hyperinsulinemic subjects fed three different levels of sucrose. J. Nutr. 111: 1045.

Reiser, S., Ferretti, R.J., Fields, M. and Smith, Jr., J.C. 1983. Role of dietary fructose in the enhancement of mortality and biochemical changes associated with copper deficiency in rats. Am. J. Clinc. Nutr. 38: 214.

Reiser, S., Scholfield, D.S., Trout, D.L., Wilson, A.S. and Aparicio, P. 1984. Effect of glucose and fructose on the absorption of leucine in humans. Nutr. Rep. Int. 30: 151.

Reiser, S., Smith, Jr., J.C., Mertz, W., Holbrook, J.T., Scholfield, D.J., Powell, A.S., Canfield, W.K. and Canary, J.J. 1985. Indices of copper status in humans consuming a typical American diet containing either fructose or starch. Am. J. Clin. Nutr. 42: 242.

Reiser, S. 1985. Effects of dietary sugars on metabolic risk factors associated with heart disease. Nutr. Health 3: 203.

Reiser, S., Powell, A.S., Yang, C.-Y. and Canary, J.J. 1987. An insulinogenic effect of oral fructose in humans during postprandial hyperglycemia. Am. J. Clin. Nutr. 45: 580.

Reiser, S. 1987. Uric acid and lactic acid. In "Metabolic Effects of Dietary Fructose," p. 113. CRC Press, Boca Raton, FL.

Reiser, S., Powell, A.S.,Scholfield, D.J., Panda, P., Ellwood, K.C. and Canary, J.J. 1989a. Blood lipids, lipoproteins, apoproteins, and uric acid in men fed diets containing fructose or high-amylose cornstarch. Am. J. Clin. Nutr. 49: 832.

Reiser, S., Powell, A.S., Scholfield, D.J., Panda, P., Fields, M. and Canary, J.J. 1989b. Day-long glucose, insulin and fructose responses of hyperinsulinemic and nonhyperinsulinemic men adapted to diets containing either fructose or high-amylose cornstarch. Am. J. Clin. Nutr. 50: 1008.

Schaffer, E.J., Anderson, D.W., Brewer, Jr., H.B., Levy, R.I., Danner, R.N. and Blackwelder, W.C. 1978. Plasma triglycerides in regulation of HDL cholesterol levels. Lancet ii: 391.

Scholfield, D.J., Reiser, S., Fields, M., Steele, N.C., Smith, Jr., J.C., Darcey, S. and Ono, K. 1990. Dietary copper, simple pigs, and metabolic changes in pigs. J. Nutr. Biochem. 1: 362.

Shekelle, R.B. and Stamler, J. 1989. Dietary cholesterol and ischemic heart disease. Lancet i: 1177.

Steiner, G. 1986. Hypertriglyceridemia and carbohydrate intolerance: interrelations and therapeutic implications. Am. J. Cardiol. 57: 27G.

Stout, R.W. 1985. Overview of the association between insulin and atherosclerosis. Metabolism 34 (Supp. 1) : 7.

Turner, J.L., Bierman, E.L., Brunzell, J.D. and Chait, A. 1979. Effect of dietary fructose on triglyceride transport and glucoregulatory hormones in hypertriglyceridemic men. Am. J. Clin. Nutr. 32: 1043.

US Department of Agriculture. 1985a. Nationwide food consumption survey. Continuing survey of food intakes by individuals. Women 19–50 years and their children 1–5 years, 1 day. Human Nutrition Information Service. (Report 85–1).

US Department of Agriculture. 1985b. Nationwide food consumption survey. Continuing survey of food intakes by individuals. Men 19–50 years, 1 day. Human Nutrition Information Service. (Report 85–3).

Van den Berghe, G. 1986. Fructose: metabolism and short-term effects on carbohydrate and purine metabolic pathways. In "Metabolic Effects of Dietary Carbohydrates," Progress in Biochemical Pharmacology 21: 1. Karger, Basel, Switzerland.

Witzum, J. and Schonfeld, G. 1979. High density lipoproteins. Diabetes 28: 326.

Wood, P.D.S., Stern, M.P., Silvers, A., Reaven, G.M. and Von Der Groeben, J. 1972. Prevalence of plasma lipoprotein abnormalities in a free-living population of the Central Valley, California. Circulation 45: 114.

Woods, H.F., Eggleston, L.V. and Krebs, H.A. 1970. The cause of hepatic accumulation of fructose-1-P on fructose loading. Biochem. J. 119: 501.

Woteki, C.E., Welsh, S.O., Raper, N. and Marston, R.M. 1982. Recent trends and levels of dietary sugars and other caloric sweeteners. In "Metabolic Effects of Utilizable Dietary Carbohydrate," p. 1. Marcel Dekker, Inc., New York.

Wursch, P., Del Vedovo, S. and Koellreutter, B. 1986. Cell structure and starch nature as key determinants of the digestion rate of starch in legumes. Am. J. Clin. Nutr. 43: 25.

Young, A.H. 1984. Fractionation of starch. In "Starch: Chemistry and Technology," p. 249. Academic Press, New York.

Zavaroni, I., Bonora, E., Pagliara, M., Dall'Aglio, E., Luchetti, L., Buonanno, G., Bonati, P.A., Bergonzani, M., Gnudi, L., Passeri, M. and Reaven, G. 1989. Risk factors for coronary artery disease in healthy persons with hyperinsulinemia and normal glucose tolerance. N. Eng. J. Med. 320: 702.

AMELIORATED PROTEINS

Improvement of the Nutritional Quality of Legume Proteins with Special Emphasis on Soybean Protein

John F. Thompson and James T. Madison
U.S. Department of Agriculture
Agriculture Research Service
U.S. Plant, Soil and Nutrition Laboratory
Tower Road
Ithaca, NY 14853

Legume seeds are an important source of protein for humans and domestic animals, but the protein is not efficiently utilized because of a deficiency of methionine and cyst(e)ine. Efforts to improve the nutritional quality of legume seed protein has focused on the storage proteins and protease inhibitors. Possible ways of improving legume seed quality include: 1) finding high-methionine cultivars; 2) increasing the level of the larger storage protein and decreasing the amount of the smaller storage protein; 3) inserting storage proteins genes modified to code for more methionine into the genome; 4) inserting a high-methionine gene into the genome. A search for high-methionine cultivars has not been successful. It has been possible to increase the content of the larger storage protein which has more methionine with a concomitant decrease in the smaller storage protein. No one has reported inserting high-methionine genes into a legume genome, though it has been accomplished with tobacco. It has also been possible to raise the content of Bowman-Birk protease inhibitor in soybean and thus increase cyst(e)ine content.

Introduction

Humans need to have food and warmth. Their food must include many specific components including vitamins, essential amino acids and fatty acids. In addition, they need a source of energy (from carbohydrates, lipids and proteins). Quantitatively, energy sources must be the major component of the diet while a source of protein to supply essential amino acids is next most important. Worldwide, grains are the major supplier of protein with legume seeds second. Meat and milk are significant suppliers of protein in a few countries. Since amino acids are not stored in the body, the essential amino acids must be present at the same time in the proper proportions in the diet to be utilized efficiently. The amino acids arising from the diet or from tissue protein degradation which are not incorporated into protein are degraded and thus serve as a source of energy. Although meat and milk are good sources of essential amino acids, they are too expensive for most people who must rely on plant sources of protein. It is generally known that grains and legume seeds are not ideal sources of protein in the diet because they do not have the proper balance of essential amino acids. The problem with legume seeds is that the sulfur amino acids, methionine and cyst(e)ine are proportionally lower than the other essential amino acids (e.g., soybean in Table 1). Although strictly speaking, cyst(e)ine is not an essential amino acid because it can be formed from methionine, methionine and cyst(e)ine are commonly considered together in evaluating nutritional value (Food and Nutrition Board, 1989) because cysteine can spare methionine (Rose and Wixom, 1955).

Over the years, there has been a great deal of interest in the total sulfur and sulfur amino acid content of legumes, both as a source of seeds and of forages (Rendig, 1986). When legumes are utilized as forages, the specific form of sulfur is not critical since they are primarily eaten by ruminants. For forage crops, sulfur supply is most important with respect to yield, not protein quality. Legume seeds are a significant source of protein, starch and lipid because they are economical sources and store well. In the United States, the soybean is by far the most important. The lipids are separated and used for food (essential fatty acids and energy supply) and for various industrial purposes. The soybean meal (after removal of lipid) is utilized primarily as a source of protein in animal feed for chickens, pigs and cattle. In the Orient, the whole bean and the extracted protein are important diet components.

A comparison of the amino acid composition of legume seed protein with amino acid requirements of humans (Food and Nutrition Board, 1989) and chicks (Subcommittee on Poultry Nutrition, 1977) makes it clear that legume seeds are deficient in the sulfur amino acids (Evans and Bandemer, 1967). Table 1 shows that more

Table 1. The amounts of soybean meal protein and egg white required to supply the daily essential amino acid requirements of humans and chicks.

Amino Acid	Humans		Chicks	
	Soybean	Egg White	Soybean	Egg White
	Grams			
Isoleucine	0.28	0.25	0.72	0.65
Leucine	0.26	0.23	1.09	0.96
Lysine	0.24	0.24	0.61	0.61
Methionine and Cyst(e)ine	0.77	0.27	1.51	0.52
Phenylalanine and Tyrosine	0.21	0.20	0.58	0.56
Threonine	0.18	0.19	0.52	0.56
Tryptophan	0.38	0.32	0.38	0.46
Valine	0.24	0.27	0.63	0.69

The underlined number in each column indicates the amount of protein required to provide that amino acid which is most deficient for that protein and species. Soybean data calculated from Yazdi-Samadi *et al.* (1977). (Reproduced with permission.)

soy meal protein is required to provide the sulfur amino acids whereas with egg white (a high quality protein) other amino acids are limiting. Table 1 also shows that less egg white than soy protein is required to provide the minimum amount of the most limiting amino acid. On the basis of feeding experiments, there is disagreement as to the quality of soy protein for humans; and this disagreement arises from the lack of an ideal method for measuring the nutritive value of a protein (McLaughlan, 1979). However, it is evident that soy protein is not as high quality as egg white protein, and this is primarily due to the low level of sulfur amino acids. The significance of this deficiency depends on several factors. These factors include: 1) the method of assay; 2) the animal species; 3) other proteins in the diet; 4) the presence of other dietary components (e.g., protease inhibitors); 5) the growth state of the animal; 6) individual variation; 7) the length of the assay; 8) digestibility of the protein; 8) level of protein supplied in a feeding assay (is the protein level near the minimum?). The best assays are based on feeding trials (Bodwell, 1979). However, the validity of these assays will depend on whether they are based on growth or nitrogen balance or plasma protein or plasma amino acid level. The length of time of the trial may be important. Feeding trials with humans tend to be of short duration, have a minimum number of subjects (which is complicated by marked individual variation) and depend on nitrogen balance rather than growth because of the expense. The problems include the unpleasantness of eating the same diet every day, the difficulty of getting satisfactory subjects, and ethical considerations (Bressani *et al.*, 1979). Animal assays usually depend on growth. They have the advantage of being able to use large numbers of genetically similar animals and the ability to use the same diet every day. The disadvantage is that the typical test animal (rat or chick) has higher sulfur amino acid requirements than humans because of hair and feathers (Steinke, 1979; Subcommittee on Poultry Nutrition, 1977) and this is particularly important when considering legume proteins. This is reflected in the low protein score of legume proteins (Evans and Bandemer, 1967).

In the United States, where most of the soy protein is used to feed animals, supplementation of the feed with methionine precursor is economically advantageous. This clearly shows the deficiency of sulfur amino acids in soy protein for this purpose. Many legume seeds have the added disadvantage of containing protease inhibitors which interfere with the utilization of protein (Kunitz and Bowman-Birk inhibitors in the soybean). The Kunitz and Bowman-Birk protease inhibitors can be inactivated by heat (DiPietro amd Liener, 1989) which is an added expense. The Bowman-Birk group of protease inhibitors have a high (20 percent) content of cyst(e)ine which could enhance the sulfur amino acid content of soybeans. The soybean has many Bowman-Birk isoinhibitors (Hwang *et al.*, 1977) that vary widely in inhibitory properties (Tan-Wilson and Wilson, 1986) leading to the suggestion that an increase in some would be nutritionally beneficial (loc. cit.).

In considering soy protein as a source of protein for humans, there are differing points of view (depending on the objective of the study). Zezulka and Calloway (1976) showed that methionine added to soy protein can increase the efficiency of utilization by humans when a portion of the amino acid requirement is supplied by nonessential amino acids (Kies and Fox, 1971). Table 2 shows that soy protein is not as good as egg white in maintaining nitrogen balance in young men, but that it is nearly as good when supplemented with methionine to raise the level to the recommended dietary allowance. This table also illustrates the importance of total amino acid supply. On the other hand, Scrimshaw and Young (1979) concluded on the basis of their experiments that "there is little nutri-

Table 2. The effect of the amount of nitrogen and added methionine on the nitrogen balance of young men fed egg white or soy protein.

Protein	Nitrogen Supplied	Supplement	Nitrogen Balance
	g/day		g/day
Egg White	3.0	"	-0.48
"	4.5	"	+0.41
"	6.0	"	+0.73
Soy Protein	3.0	"	−1.21
"	4.5	"	−0.08
"	6.0	"	+0.26
"	3.0	Methionine	+0.08
"	4.5	"	+0.64
"	6.0	"	+0.76

Methionine was added to bring its level up to the recommended daily allowance. Calculated from Zezulka and Calloway (1976).

tional or public health justification for requiring supplementation of soy protein products with methionine," even though they found that a soy protein isolate was only about 80 percent as good as egg white protein in maintaining nitrogen balance in young men and that supplemental methionine will improve the efficiency of utilization. For human use, the conclusion of Scrimshaw and Young, that enhancing the methionine content of soy protein is not justified, is probably correct because humans will have other sources of methionine. However, when soy milk protein is used for infant food, a better balance of essential amino acids would decrease the total need and avoid the problems of metabolizing extra amino acids.

The importance of the lack of sulfur amino acids in legume proteins is illustrated in Figure 1 where it is shown that the biological value of proteins (as measured by nitrogen balance with growing rats) from a number of legumes is closely related to their sulfur amino acid content. As implied from the above statements, the effect of added methionine to diets of other animals can be more dramatic. There is abundant evidence that the protein efficiency ratio (PER) (which is normally determined by rat growth) of soy protein is about 80 percent of that of casein (Patwardhan, 1962) because of the rat's higher requirement for the sulfur amino acids (Subcommittee on Laboratory Animal Nutrition, 1972). A recent report (Iowa Agriculture and Home Economics Experiment Station, 1990) stated

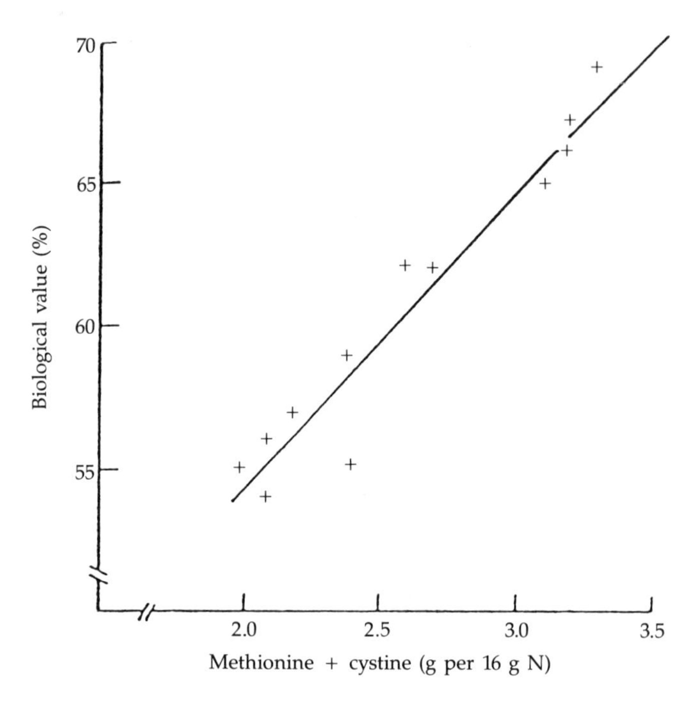

Figure 1. Relationship of sulfur amino content of various legume seed proteins and their biological value (from Khan *et al.,* 1979; Reproduced with permission.)

that it would be desirable to increase the methionine content of soybean protein because a large percentage of soybean meal in this country is utilized for feeding chicks. To improve legume proteins, the principal focus must be on the storage proteins because they constitute the bulk (>65 percent) of seed proteins and because any change in them is less likely to affect the viability of the seed than would a change in enzymes or structural proteins.

Characteristics of Legume Storage Proteins

There is justification for considering legume proteins as a group because their storage proteins are remarkably similar in size (Danielsson, 1949; Derbyshire *et al.*, 1976) and antigenicity (Kloz, 1971; Dudman and Millerd, 1976; Millerd *et al.*, 1979, Gilroy *et al.*, 1979; Doyle *et al.*, 1985; Kagawa *et al.*, 1987). In general, legumes have two major storage proteins that are stored in protein bodies (Gayler *et al.*, 1990) and are larger (100-400 kD) than most proteins. Protease inhibitors occur in protein bodies but are responsible for only a small percentage of the total protein. The protease inhibitors must be considered because of their potential adverse effect on legume seed protein digestibility.

Because the storage proteins are large, it is not surprising that they are made up of several subunits, each of which represents one polypeptide chain. The storage proteins of soybeans are typical of legume storage proteins. The smaller of the two groups has a molecular weight of 150-175 kD and is termed β-conglycinin or the 7S protein. Strictly speaking, the 7S protein is not one protein but a mixture of proteins made up primarily of three glycosylated subunits (α, α' and β) in various proportions (Figure 2) (Thanh and Shibasaki, 1976, 1978). The other major storage protein (glycinin or 11S) has a molecular weight of 350-400 kD and is composed of 12 subunits arranged in two apposed "hexagons" (Figure 3) (Badley *et al.*, 1975). Most of the common legumes such as pea, chick pea, mung bean, peanut, lima bean, soybean and pigeon pea have similar storage proteins (Danielsson, 1949). An outstanding exception is the common bean (*Phaseolus vulgaris*) which has little or none of the larger storage protein (Danielsson, 1949).

Most legume seeds have two types of protease inhibitors (2–8 percent in various legumes), the Kunitz and Bowman-Birk types (Norioka *et al.*, 1988). These inhibitors behave like storage proteins by decreasing during seed germination and by occurring in protein bodies. These protease inhibitors decrease the nutritional quality of legume seeds (Rachis and Gumbman, 1981) but their inhibitory properties can be destroyed by heating (DiPietro and Liener, 1989). The Kunitz protease inhibitor has a molecular weight of about 20,000 Daltons and an amino acid composition similar to that of the principal storage proteins (Koide and Ikenaka, 1973; Holowach *et al.*, 1984b). The Bowman-Birk protease inhibitors which have about 80 amino acids contain about 20 percent cystine (Odani and Ikenaka, 1972) so that an increase in these proteins would increase sulfur amino acid content. The sequences of these inhibitors are similar in many legumes (Ventura,1989). In the case of the Bowman-Birk type protease inhibitors, Hammond *et al.* (1984) report evidence for one or two genes for the classical Bowman-Birk type inhibitor. There are, however, about half a dozen closely related inhibitors (Hwang *et al.*, 1977; Tan-Wilson *et al.*, 1990).

Parenthetically, it might be added that the function of the protease inhibitors has been the subject of much speculation. They may serve as deterrents for insects (Steffens *et al.*, 1978; Gatehouse *et al.*,

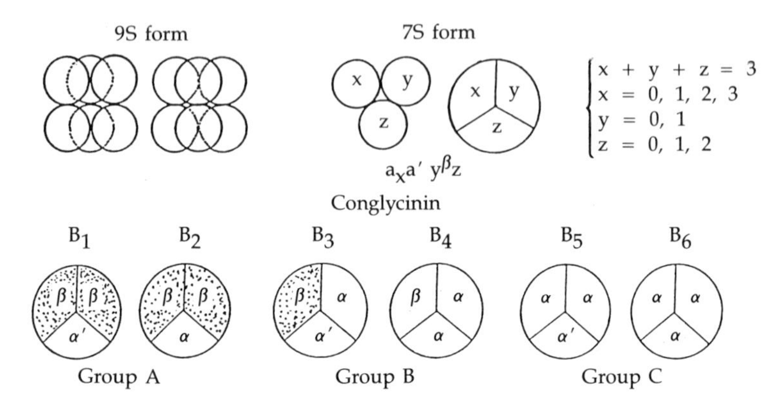

Figure 2. Models of the soybean 7S storage proteins (Thanh and Shibasaki, 1978. Reproduced with permission.)

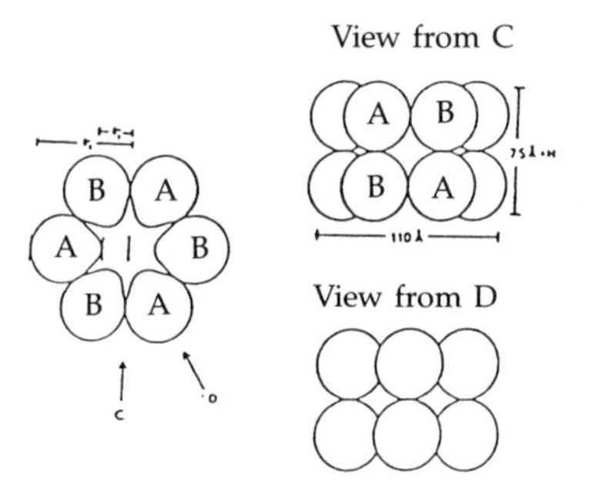

Figure 3. Model of the soybean 11S storage proteins (Bradley *et al.*, 1975. Reproduced with permission.)

1979) and may play a role in preventing various forms of cancer (Witschi and Kennedy, 1989). Nutritionally, it would be advantageous to have less Kunitz protease inhibitor, and null cultivas have been found (Orf and Hymowitz, 1979). The Bowman-Birk type protease inhibitors have a range of inhibitory activities (Tan-Wilson and Wilson, 1988) which very likely would parallel their adverse nutritional effects.

Approaches to Improving Legume Protein Quality

Several approaches have been taken to improve the efficiency of utilization of legume proteins. One approach is to add a methionine precursor and another approach is to improve the protein quality. In the United States, where a large portion of the soy protein is fed to chicks and pigs, a compound (α-hydroxy, γ-thiomethyl butyric acid) that is readily converted to methionine by the animal is often added to the feed. The problems with this approach are that the additive causes extra expense, is readily leached out, makes the diet more susceptible to spoilage, and may cause unpleasant odors and flavors. Methods of improving protein that have been considered are as follows. First, cultivar collections have been analyzed for amino acid content in an effort to discover lines that have improved nutritional quality. Analysis of a number of soybean cultivars has revealed a narrow range of sulfur amino acid content (Kuiken and Lyman, 1949; Krober and Cartter, 1966; Kakade *et al.*, 1972) that is probably related to the relatively narrow genetic base. A similar situation is manifest with chick peas, mung beans and cow peas (Khan *et al.*, 1979). In contrast, in the case of grains, opaque-2 corn has a significantly increased lysine content (Mertz, 1976). Second, a shift in the relative amounts of the two storage proteins can improve protein quality because the 11S protein has higher methionine than the 7S protein (Holowach *et al.*, 1984b). Third, the genes for the storage protein subunits can be modified and inserted into the plant genome so that they code for improved subunits. The problem with this approach is that there may be a fairly large number of genes for some proteins; e.g., the 7S protein of soybeans, (Harada *et al.*, 1989) which may make it difficult to modify a majority of the genes in that family. Fourth, genes coding for a methionine-rich storage protein from the brazil nut (Youle and Huang, 1981) have been isolated (Sun *et al.*, 1987) with the objective of insertion into legume genomes (Altenbach *et al.*, 1987). A concern in this case is that the genes must be expressed properly, at the right time, and in the correct tissue.

Factors Affecting the Sulfur Amino Acid Content of Legume Seeds

In the earlier studies, seeds from legumes subjected to various mineral deficiencies were analyzed for their amino acid content to determine nutritional quality (Eppendorfer, 1971; Arora and Luthra, 1971; Sharma and Bradford, 1973; Ham *et al.*, 1975; Radford *et al.*, 1977; Blagrove *et al.*, 1976). Improvement in yield as well as protein quality was an objective.

Over a range of sulfur nutrition, Sharma and Bradford (1973) observed that increased sulfur supply to soybeans increased the cyst(e)ine and methionine content of the seed protein but Ham *et al.*, (1975) did not confirm this observation. Arora and Luthra (1971) found that sulfur supply correlated positively with the sulfur amino acid content of leaves and seeds of mung bean where the range included conditions of sulfur insufficiency. A similar correlation was found with *Vicia faba* (Eppendorfer, 1971), and the nutritional importance of the analysis was confirmed by rat feeding trials. On the other hand, Eppendorfer and Bille (1974) found no variation of the sulfur amino acid content of protein with variation of K and P supply in pea seeds as Krober and Cartter (1966) did with soybeans. Radford *et al.*, (1977) found that methionine and cysteine content of soybeans correlated so well with N/S that the relatively easy determination of N/S was as good as the more difficult determination of the sulfur amino acids for assessing the nutritional adequacy of soybean lines and species. However, the results of Kakade *et al.*, (1972) indicate that the sulfur amino acid content of many soybean varieties does not correlate with nutritional quality as measured by protein efficiency ratio. At this time, it is not clear whether these discrepancies stem from differences in methodology or experimental plan. It does appear that there is little variation in protein quality of soybeans where the yield is not affected.

Factors Affecting Storage Protein Levels in Legume Seeds

As technologies developed, investigators began looking at individual proteins, especially the storage proteins. For example, varying the sulfur supply to *Lupinus angustifolius* plants changes the N/S of the seeds and the proportions of the major storage proteins (Blagrove *et al.*, 1976). The α-and γ-conglutins decrease in sulfur deficiency concomitant with a β-conglutin increase. Since the former contain more sulfur than the latter, the lack of sulfur amino acids apparently restricts α-and γ-conglutin synthesis (Gillespie *et al.*, 1978). In a similar fashion, a sulfur deficiency decreases the legumin to vicilin ratio in peas presumably because legumin has a higher content of methionine and cyst(e)ine (Randall *et al.*, 1979; Beach *et al.*, 1985). Here again, sulfur deficiency decreases legumin synthesis rather than accelerating legumin degradation (Chandler *et al.*, 1983) and increases vicilin synthesis (Chandler *et al.*, 1984). In an analogous fashion, sulfur deficiency decreases glycinin and increases conglycinin in soybeans (Gayler and Sykes, 1985) which correlates with the higher sulfur amino acid content of glycinin (Holowach *et al.*, 1984b).

To our knowledge, no one has successfully inserted a high-methionine protein gene into a legume. However, Altenbach *et al.* (1989) have successfully inserted the brazil nut storage protein gene (Altenbach *et al.*, 1987) into tobacco resulting in a 30 percent increase in seed protein methionine. De Lumen and Kno (1987) developed a method for identifying high-methionine proteins which should be useful for isolating these proteins and their genes (e.g., albumins from peas—Schroeder, 1984). There is clearly considerable interest in this approach, but the recent success of inserting a brazil nut gene into tobacco (Altenbach *et al.*, 1989) indicates it may be possible with legumes.

Effect of Methionine on Legume Seed Storage Proteins

In order to have a simple, reproducible system for studying the synthesis and degradation of soybean storage proteins, an immature seed culture procedure was developed (Thompson *et al.*, 1977). Utilizing this procedure, it was found that methionine increased dry weight and increased the methionine content of the total protein by over 20 percent (Thompson *et al.*, 1981). In searching for an explanation for the latter effect, it was found that added methionine decreases the 7S protein with a compensatory and equivalent increase in the 11S protein (Holowach *et al.*, 1984a) (Table 3). These results have been confirmed with intact soybean plants (Grabau *et al.*, 1986). Further investigation showed that the decrease in the 7S protein was due to the absence of its β-subunit (Holowach *et al.*, 1984a) (Table 4), since only those 7S proteins containing the α and α'-subunits can be formed (Figure 2). The replacement of the 7S protein by the 11S protein and the absence of methionine in the β-subunit (Holowach *et al.*, 1984a) help explain the increase in methionine content of the protein. It would be of considerable interest to know the mechanism whereby total storage protein level is maintained more or less constant and whereby methionine abolishes the β-subunit of the 7S protein. It could be related to the fact that the storage proteins assemble in the protein bodies. The increase in 11S protein could mean that protein bodies have a predetermined capacity for storage proteins. An interesting result of this work has been a correlation between methionine and development. Gayler and Sykes (1981) showed that during the development of the soybean seed, the α- and α'-subunits appear earlier than the β-subunit. The concentration of methionine required to suppress the

Table 3. Storage protein content of cultured cotyedons grown without and with added methionine.

Medium	7S Protein	11S Protein	7S+11S Proteins
		MG/G Wet Weight	
Basal	26	21	47
Basal Plus Methionine	18	32	50

Table 4. 7S protein subunit content and 7S protein content of cultured soybean cotyledons grown without and with methionine.

Medium	α' Subunit	α Subunit	β Subunit	7S Protein
		MG/G Wet Weight		
Basal	5.8	10.3	10.2	26.3
Basal Plus Methionine	6.6	11.4	0.0	17.0

β-subunit is in the range of 0.5 to 1 mM (Creason *et al.*, 1985). The concentration of free methionine in soybean seeds drops markedly during development. The point at which the methionine concentration drops below 1 mM is about the time that the β-subunit starts to appear (Figure 4). This result suggests that methionine may be acting as a developmental regulator. Bray and coworkers (1987) have manipulated the leader sequences of the α-, α-' and β-subunit genes but have not identified the regions that are responsible for the developmental difference in expression between the α- and β-genes.

We are interested in how methionine acts in eliminating the β-subunit of the 7S protein and thus increasing the methionine content of soy protein. The first question is whether methionine affects transcription, translation or stability of the β-subunit. Methionine does not promote the degradation of the β-subunit (Holowach *et al.*, 1986). Soybean seeds cultured in the presence of methionine contain no β-mRNA (Creason *et al.*, 1983) indicating that the methionine effect is prior to translation. Studies using run-on transcription with isolated nuclei show that methionine brings about a small (ca. 25 percent) decrease in the transcription of the β-gene (Singer *et al.*,

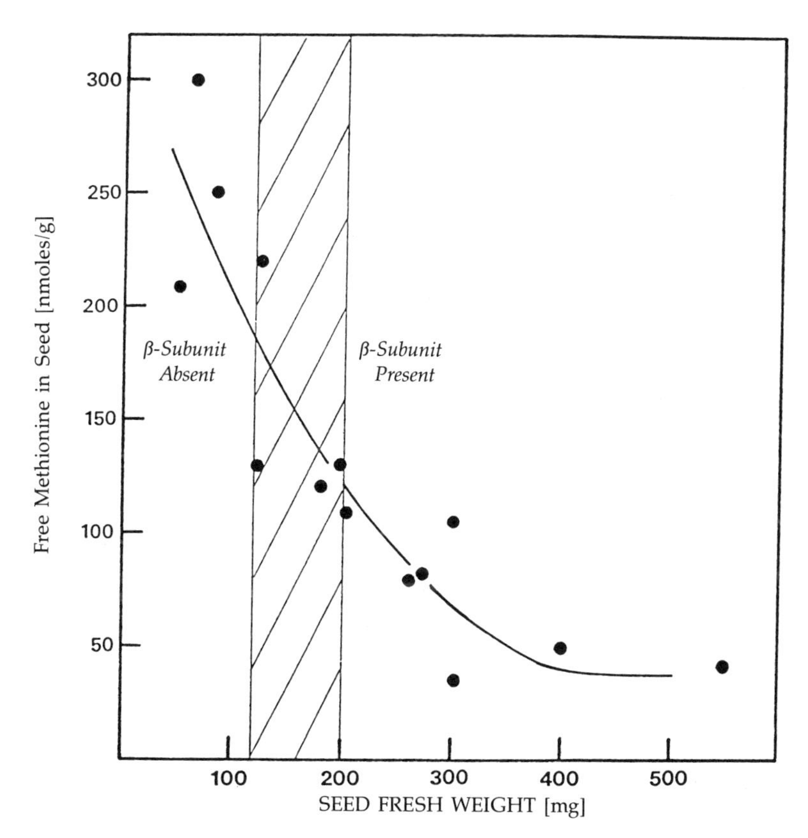

Figure 4. Free methionine content of developing soybeans compared to the concentrations of methionine required to inhibit the synthesis of the β-subunit in cultured cotyledons (Creason *et al.,* 1985; Reproduced with permission.)

1989). A test of the effect of methionine on the stability of the β-mRNA was tested and the results showed that methionine approximately doubled the degradation rate. While these effects could conceivably account for a majority of the methionine effect, calculations show that there still must be significant additional effects. Other possibilities include an effect on 5′-capping, length of the poly-A tail, splicing and transport from the nucleus to the cytoplasm. Several groups (Beach *et al.*, 1985; Higgins *et al.*, 1986; Walling *et al.*, 1986; Thompson *et al.*, 1989) found that both transcriptional and translational controls affected the rate of storage protein synthesis in soybeans and peas so that the observed effects of methionine on transcription and post-transcriptional processes is not unreasonable.

In the course of the work on the 7S subunits, we learned that methionine not only turns off the β-gene but increases the level of the Bowman-Birk protease inhibitor mRNA by several (>10) fold and the level of the protein by 4 fold (Figure 5) (Biermann *et al.*, 1990) in cultured soybean seeds. Further investigation revealed that sulfate enhanced the methionine effect on Bowman-Birk type protease inhibitors by many fold (Figure 5). As a result the cyst(e)ine content of cultured soybean seed protein was increased by about 40 percent. Since sulfate might increase the level of sulfur compounds, a number of sulfur-containing compounds have been tested as the true effector. To date, we have not been able to find another sulfur compound that has the same effect as methionine plus sulfate. This is disappointing because it is desirable to have one compound with which to test mechanisms. The Bowman-Birk type protease inhibitors contain one methionine residue per molecule and approximately 20 percent cysteine residues (as disulfides). From these facts, it is obvious that cyst(e)ine would be a logical effector. Unfortunately, cyst(e)ine is not significantly absorbed. In this case, it is clear that the methionine-plus-sulfate effect is on mRNA level. What needs to be determined is whether the effect is on transcription or on some processing step.

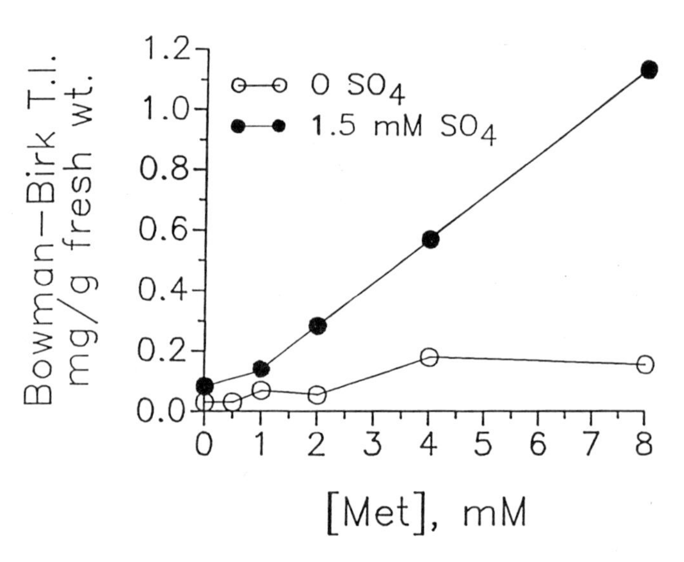

Figure 5. Bowman-Birk type protease inhibitor content of soybean cotyledons cultured with various concentrations of methionine with and without sulfate (Biermann *et al.*, 1990; Reproduced with permission.)

Our work shows that methionine suppresses the β-mRNA and the β-subunit and, as a result, the methionine content of soybean protein is increased. At present, we are producing a β-antisense gene to introduce into the soybean with the goal of suppressing the β-subunit and increasing protein methionine. A further hope is that these genes may also suppress the α- and α'-subunits and thus eliminate the 7S protein since the three genes are so homologous (Schuler *et al.*, 1982a,b; Harada *et al.*, 1989). If so, the methionine content may be still further increased, possibly to the point where another amino acid would limit nutritional quality.

Conclusions

It is known that methionine increases the methionine content of soybean protein by decreasing transcription of the β-gene and increasing the rate of degradation of the β-mRNA. The methionine content of the soybean seed is correlated with development, but whether it is a normal developmental regulator is unknown. Methionine increases the level of the Boman-Birk protease inhibitors mRNA and the level of these inhibitors. This effect is enhanced by sulfate. This should increase the cyst(e)ine content of the soybean protein because Bowman-Birk inhibitors are about 20 percent cyst(e)ine. The prospects for increasing protein methionine significantly in the soybean appear to be good. In all legumes, the prospect of raising protein methionine are enhanced by the success of inserting a high-methionine gene into tobacco.

References

Altenbach, S.B., Pearson, K.W. Leung, F.W., and Sun, S.S.M. 1987. Cloning and sequence analysis of a cDNA encoding a brazil nut protein exceptionally rich in methionine. Plant Mol. Biol. 8: 239.

Altenbach, S.B., Pearson, K.W., Meeker, G., Staraci, L.C. and Sun, S.S.M. 1989. Enhancement of the methionine content of seed proteins by the expression of a chimeric gene encoding a methionine-rich protein in transgenic plants. Plant Mol. Biol. 13: 513.

Arora, S.K., Luthra, Y.P. 1971. Relationship between sulphur content of leaf with methionine, cysteine and cysteine contents in the seeds of *Phaseolus aureus* L. as affected by S,P, and N application. Plant and Soil 34: 91.

Badley, R.A., Atkinson, D., Hauser, H., Oldani, D., Green, J.P. and Stubbs, J.M. 1975. The structure, physical and chemical properties of the soy bean protein glycinin. Biochim. Biophys. Acta 412: 214.

Beach, L.R., Spencer, D., Randall, P.J., and Higgins, T.J.V. 1985. Transcriptional and post-transcriptional regulation of storage protein gene expression in sulfur-deficient pea seeds. Nucleic Acids Res. 13: 999.

Blagrove, R.J., Gillespie, J.M. and Randall, P.J. 1976. Effect of sulphur supply on the seed globulin composition of *Lupinus angustifolius*. Aust. J. Plant Physiol. 3: 173.

Biermann, B.J., de Banzie, J., Thompson, J.F. and Madison, J.T. 1990. Regulation of Bowman-Birk proteinase inhibitor mRNA levels in cultured soybean cotyledons by exogenous methionine. Plant Physiol. In press.

Bodwell, C.E. 1979. Human versus animal assays. In "Soy Protein and Human Nutrition" p. 331. Academic Press, New York.

Bray, E.A., Naito, S., Pan, N.S., Anderson, E., Dube, P. and Beachy, R.N. 1987. Expression of the α-subunit of β-conglycinin in seeds of transgenic plants. Planta 172: 364.

Bressani, R., Navarrete, D.A., Elias, L.G. and Braham, J.E. 1979. A critical summary of a short-term nitrogen balance index to measure protein quality in adult human subjects. In "Soy Protein and Human Nutrition" p. 313. Academic Press, New York.

Chandler, P.M., Higgins, T.J.V., Randall, P.J. and Spencer, D. 1983. Regulation of legumin levels in developing pea seeds under conditions of sulfur deficiency. Plant Physiol. 71: 47.

Chandler, P.M., Spencer, D., Randall, P.J. and Higgins, T.J.V. 1984. Influence of sulfur nutrition on developmental patterns of some major pea seed proteins and their mRNAs. Plant Physiol. 75: 651.

Creason, G.L., Holowach, L.P., Thompson, J.F., Madison, J.T. 1983. Exogenous methionine depresses level of mRNA for a soybean storage protein. Biochem. Biophys. Res. Commun. 117: 658.

Creason, G.L., Thompson, J.F. and Madison, J.T. 1985. Methionine analogs inhibit production of β-subunit of soybean 7S protein. Phytochemistry 24: 1147.

Danielsson, C.E. 1949. Seed globulins of the Gramineae and Leguminosae. I. Seed globulins of the most common species of the Gramineae and their differentiation in the seed. Biochem. J. 44: 387.

de Lumen, B.O. and Kho, C-J. 1987. Identification of methionine-containing proteins and quantitation of their methionine contents. J. Agr. Food Chem. 35: 688.

Derbyshire, E., Wright, D.J., Boulter, D. 1976. Legumin and vicilin, storage proteins of legume seeds. Phytochemistry 15: 3.

DiPietro, C.M. and Liener, I.E. 1989. Heat inactivation of the Kunitz and Bowman-Birk soybean protease inhibitors. J. Agr. Food Chem. 37: 39.

Doyle, J.J., Ladin, B.F. and Beachy, R.N. 1985. Antigenic relationship of legume seed proteins to the 7S seed storage protein of soybean. Biochem. Syst. Ecol. 13: 123.

Dudman, W.F. and Millerd, A. 1975. Immunochemical behaviour of legumin and vicilin from *Vicia faba*: a survey of related proteins in the Leguminosae subfamily Faboideae. Biochem. Syst. Ecol. 3: 25.

Eppendorfer, W.H. 1971. Effects of S, N and P on amino acid composition of field beans (*Vicia fab*) and responses of the biological value of the seed protein to S-amino acid content. J. Sci. Food Agr. 22: 501.

Eppendorfer, W.H. and Bille, S.W. 1974. Amino acid composition as a function of total-N in pea seeds grown on two soils with P and K additions. Plant and Soil 41: 33.

Evans, R.J. and Bandemer, S.L. 1967. Nutritive value of legume seed proteins. J. Agr. Food Chem. 15: 439.

Food and Nutrition Board. 1989. "Recommended Dietary Allowances" 10th ed. p. 52.

Gatehouse, A.M.R., Gatehouse, J.A., Dobie, P., Kilminster, A.M. and Boulter, D. 1979. Biochemical basis of insect resistance in *Vigna unguiculata*. J. Sci. Food Agr. 30: 948.

Gayler, K.R. and Sykes, G.E. 1981. β-Conglycinins in developing soybean seeds. Plant Physiol. 67: 958.

Gayler, K.R. and Sykes, G.E. 1985. Effects of nutritional stress on the storage proteins of soybeans. Plant Physiol. 78: 582.

Gayler, K.R., Wachsmann, F., Kolivas, S., Nott, R. and Johnson, E.D. 1990. Isolation and characterization of protein bodies in *Lupinus angustifolius*. Plant Physiol. 91: 1425.

Gillespie, J.M., Blagrove, R.J. and Randall, P.J. 1978. Effect of sulfur supply on the seed globulin composition of various species of Lupin. Aust. J. Plant Physiol. 5: 641.

Gilroy, J., Wright, D.J. and Boulter, D. 1979. Homology of basic subunits of legumin from *Glycine max* and *Vicia faba*. Phytochem. 18: 315.

Grabau, L.J., Blevins, D.G. and Minor, H.C. 1986. Stem infusions enhanced methionine content of soybean storage protein. Plant Physiol. 82: 1013.

Ham, G.E., Liener, I.E., Evans, S.D., Frazier, R.D. and Nelson, W.W. 1975. Yield and composition of soybean seed as affected by N and S fertilization. Agron. J. 67: 293.

Hammond, R.W., Foard, D.E. and Larkins, B.A. 1984. Molecular cloning analysis of a gene coding for the Bowman-Birk protein. J. Biol. Chem. 259: 9883.

Harada, J.J., Barker, S.J. and Goldberg, R.B. 1989. Soybean β-conglycinin genes are clustered in several DNA regions and are regulated by transcriptional and post-transcriptional processes. Plant Cell 1: 415.

Higgins, T.J.V., Chandler, P.M., Randall, P.J., Spencer, D., Beach, L.R., Blagrove, R.J., Kortt, A.A. and Inglis, A.S. 1986. Gene structure, protein structure, and regulation of the synthesis of a sulfur-rich protein in pea seeds. J. Biol. Chem. 261: 11124.

Holowach, L.P., Madison, J.T. and Thompson, J.F. 1986. Studies on the mechanism of regulation of the mRNA level for a soybean storage protein subunit by exogenous L-methionine. Plant Physiol. 80: 561.

Holowach, L.P., Thompson, J.F. and Madison, J.T. 1984a. Effect of exogenous methionine on storage protein composition of soybean cotyledons cultured *in vitro*. Plant Physiol. 74: 576.

Holowach, L.P., Thompson, J.F. and Madison, J.T. 1984b. Storage protein composition of soybean cotyledons grown *in vitro* in media of various sulfate concentrations in the presence and absence of exogenous L-methionine. Plant Physiol. 74: 584.

Hwang, D.L.R., Lin, K.T.D., Yang, W. and Foard, D.E. 1977. Purification, partial characterization, and immunological relationships of multiple low molecular weight protease inhibitors of soybean. Biochim. Biophys. Acta. 195: 369.

Iowa Agriculture and Home Economics Experiment Station. 1990. Special report "Economic Implications of Modified Soybean Traits." 88 pp.

Kagawa, H., Yamauchi, F. and Hirano, H. 1987. Soybean basic 7S globulin represents a protein widely distributed in legume species. FEBS Lett. 226: 145.

Kakade, M.L., Simons, N.R. Liener, I.E. and Lambert, J.W. 1972. Biochemical and nutritional assessment of different varieties of soybeans. J. Agr. Food Chem. 20: 87.

Khan, M.A., Jacobsen, I. and Eggum, B.O. 1979. Nutritive value of some improved varieties of legumes. J. Sci. Food Agr. 30: 395.

Kies, C. and Fox, H.M. 1971. Comparison of the protein nutritional value of TVP, methionine enriched TVP and beef at two levels of intake for human adults. J. Food Sci. 36: 841.

Kloz, J. 1971. Serology of the leguminosae. In "Chemotaxonomy of the Leguminosae" p. 309. Academic Press, London.

Koide, T. and Ikenaka, T. 1973. Studies on soybean trypsin inhibitors 3. Amino acid sequence of the carboxyl-terminal region and the complete amino-acid sequence of soybean trypsin inhibitor (Kunitz). Eur. J. Biochem. 32: 417.

Krober, O.A. and Cartter, J.L. 1966. Relation of methionine content to protein levels in soybeans. Cereal Chem. 43: 320.

McLaughlan, J.M. 1979. Critique of methods for evaluation of protein quality. In "Soy Protein and Human Nutrition" p. 281. Academic Press, New York.

Mertz, E.T. 1976. Case histories of existing models. In "Genetic Improvement of Plants" p. 57. National Academy of Sciences. Washington, D.C.

Millerd, A., Thomson, J.A., Randall, P.J. 1979. Heterogeneity of sulphur content in the storage proteins of pea cotyledons. Planta 146: 463.

Norioka, N., Hara, S., Ikenaka, T. and Abe, J. 1988. Distribution of the Kunitz and the Bowman-Birk family proteinase inhibitors in leguminous seeds. Agr. Biol. Chem. 52: 1245.

Odani, S., and Ikenaka, T. 1972. Studies on soybean trypsin inhibitors. IV. Complete amino acid sequence and the anti-proteinase sites of Bowman-Birk soybean proteinase inhibitor. J. Biochem. (Tokyo) 71: 839.

Orf, J.H. and Hymowitz, T. 1979. Inheritance of the absence of the Kunitz trypsin inhibitor in seed protein of soybeans. Crop Sci. 19: 107.

Patwardhan, V.N. 1962. Pulses and beans in human nutrition. Am. J. Clin. Nutr. 11: 12.

Rachis, J.J. and Gumbmann, M.R. 1981. Protease inhibitors: Physiological properties and nutritional significance. In "Anti-nutrients and Natural Toxicants in Foods," p. 203. Food and Nutrition Press, Westport, CT.

Radford, R.L., Chavengsaksongkram, C. and Hymowitz, T. 1977. Utilization of nitrogen to sulfur ratio for evaluating sulfur-containing amino acid concentrations in seed of *Glycine max* and *G. soja*. Crop Sci. 17: 273.

Randall, P.J., Thomson, J.A. and Schroeder, H.E. 1979. Cotyledonary storage proteins in *Pisum sativum*. IV. Effects of sulfur, phosphorus, potassium and magnesium deficiencies. Aust. J. Plant Physiol. 6: 11.
Rendig, V.V. 1986. Sulfur and crop quality. In "Sulfur in Agriculture," p. 635. ASA,CSSA,SSSA, Madison, WI.
Rose, W.C. and Wixom, R.L. 1955. The amino acid requirements of man. XIII. The sparing effect of cystine on the methionine requirement. J. Biol. Chem. 216: 763.
Schroeder, H.E. 1984. Major albumins of *Pisum* cotyledons. J. Sci. Food Agr. 35: 191.
Schuler, M.A., Ladin, B.F., Polacco, J.C., Freyer, G. and Beachy, R.N. 1982a. Structural sequences are conserved in the genes coding for the α, α' and β-subunits of the soybean 7S seed storage protein. Nucleic Acids Res. 10: 8245.
Schuler, M.A., Schmitt, E.S. and Beachy, R.N. 1982b. Closely related families of genes code for the α and α' subunits of the soybean 7S storage protein complex. Nucleic Acids Res. 10: 8225.
Scrimshaw, N.S. and Young, V.R. 1979. Soy protein in adult human nutrion. A review with new data. In "Soy Protein and Human Nutrition," p. 121. Academic Press, New York.
Sharma, G.C. and Bradford, R.R. 1973. Effect of sulfur on yield and amino acids of soybeans. Comm. Soil Sci. Plant Anal. 4: 77.
Singer, M.E., Thompson, J.F. and Madison, J.T. 1989. Transcriptional and post-transcriptional regulation of a soybean storage protein subunit. Plant Physiol. 89: 62S.
Steffens, R., Fox, F.R. and Kassell, B. 1978. Effect of trypsin inhibitors on growth and metamorphosis of corn borer larvae *Ostrinis nubilalis* (Hubner). J. Agr. Food. Chem. 26: 170.
Steinke, F.H. 1979. Measuring protein quality of foods. In "Soy Protein and Human Nutrition," p. 307. Academic Press, New York.
Subcommittee on Laboratory Animal Nutrition. 1972. "Nutrient Requirements of Laboratory Animals." 117 pp. National Academy of Sciences, Washington D.C.
Subcommittee on Poultry Nutrition. 1977. "Nutrient Requirements of Poultry." 62 pp. National Academy of Sciences, Washington, D.C.
Sun, S.S.M., Leung, F.W. and Tomic, J.C. 1987. Brazil nut (*Bertholletia excelsa* J.B.K.) proteins: Fractionation, composition, and identification of a sulfur-rich protein. J. Agr. Food Chem. 35: 232.

Tan-Wilson, A.L., Chen, J.C., Duggan, M.C., Chapman, C., Obach, R.S. and Wilson, K.A. 1988. Soybean Bowman-Birk trypsin isoinhibitors: Classification and report of a glycine-rich trypsin inhibitor class. J. Agr. Food Chem. 35: 974.

Tan-Wilson, A.L., Chen, J.C., McGrain, A.K., Kiakagawang, D., and Wilson, K.A. 1990. Temporal pattern of trypsin inhibitor accumulation in developing soybean seeds. J. Agr. Food Chem. 37: (In press).

Tan-Wilson, A.L. and Wilson, K.A. 1986. Relevance of multiple soybean trypsin inhibitor forms to nutritional quality. In "Nutritional and Toxicological Significance of Enzyme Inhibitors in Foods," p. 391. Plenum Press, New York.

Thanh, V.H. and Shibasaki, K. 1976. Heterogeneity of β-conglycinin. Biochim. Biophys. Acta 439: 326.

Thanh, V.H. and Shibasaki, K. 1978. Major proteins of soybean seeds. Subunit structure of β-conglycinin. J. Agr. Food Chem. 26: 692.

Thompson, A.J., Evans, I.M., Boulter, D., Croy, R.R.D. and Gatehouse, J.A. 1989. Transcriptional and posttranscriptional regulation of seed storage-protein gene expression in pea (*Pisum sativum* L.). Planta 179: 279.

Thompson, J.F., Madison, J.T. and Muenster, A.E. 1977. *In vitro* culture of immature cotyledons of soya bean (*Glycine max* L. Merr.). Ann. Bot. 41: 29.

Thompson, J.F., Madison, J.T., Waterman, M.A. and Muenster, A.E. 1981. Effect of methionine on growth and protein composition of cultured soybean cotyledons. Phytochemistry 20: 941-945.

Ventura, M.M. 1989. Bowman-Birk type protease inhibitors: two-dimensional vector representation for their sequences. An. Acad. Bras. Ci. 61: 215.

Walling, L., Drews, G.N. and Goldberg, R.B. 1986. Transcriptional and post-transcriptional regulation of soybean seed protein mRNA levels. Proc. Nat. Acad. Sci. 83: 2123.

Witschi, H. and Kennedy, A.R. 1989. Modulation of lung tumor development in mice with the soybean-derived Bowman-Birk protease inhibitor. Carcinogenesis 12: 2275.

Yazdi-Samadi, B., Rinne, R.W. and Seif, R.D. 1977. Components of developing soybean seeds: Oil, protein, sugars, starch, organic acids, and amino acids. Agron. J. 69: 481.

Youle, R.J. and Huang, A.H. 1981. Occurrence of low molecular weight and high cysteine-containing albumin storage proteins in oilseeds of diverse species. Am. J. Bot. 68: 44.

Zezulka, A.Y. and Calloway, D.H. 1976. Nitrogen retention in men fed varying levels of amino acids from soy protein with or without added methionine. J. Nutr. 106: 212.

Transcriptional and Targeting Determinants Affecting Phaseolin Accumulation

M.M. Bustos
Department of Biological Sciences
University of Maryland-Baltimore County
Baltimore, MD 21228
F.A. Kalkan, D. Begum, M.J. Battraw and T.C. Hall
Department of Biology
Texas A&M University
College Station, TX 77843-3258

In higher plants, as in higher animals, fertilization marks the beginning of a new generation. A set of closely regulated developmental processes ensures the programmed differentiation of embryonic tissues that will form the new shoot and root system while vigorous accumulation of reserve materials synthesized by the parental organism provides an inheritance to be expended during the early stages of new growth. Reserve proteins must be capable of rapid assembly and accumulation in forms that avoid generating high osmotic pressures, they must be stable during overwintering or other unfavorable times and yet be readily available for controlled release of amino acids providing carbon and nitrogen sources for the fast growing seedling. These factors place similar constraints on the nature of the capital to be stored in monocotyledonous and dicotyledonous spermatophytes. In this article, these constraints will be examined in light of current molecular and cellular knowledge of phaseolin, the major storage protein of the common bean, *Phaseolus vulgaris*.

Regulation of Phaseolin Gene Transcription

The rapidity with which plants can accumulate storage protein is well exemplified by the development of legume seeds. The physiological importance of coordinating the onset of storage protein synthesis with the emergence and differentiation of storage organelles has been emphasized by the decreased stability of β-conglycinin subunits in early stages of soybean seed development

(Shattuck-Eidens and Beachy, 1985). Sun *et al.* (1978) recognized that bean (cv. Tendergreen) seeds commence accumulation of phaseolin when the cotyledons are 8 mm in length (roughly 13 days after pollination). Electrophoretic profiles and rocket immuno-electrophoresis revealed this protein to be dominant in 14 mm long cotyledons (20 DAP), where it represents over 50 percent of the total protein being synthesized. Ultrastructural studies of bean (Hall *et al.*, 1979) and pea (Craig *et al.*, 1980) cotyledons at this time show a dense rough endoplasmic reticulum with numerous Golgi stacks surrounding the cell vacuole, destined to be subdivided into the electron-dense protein bodies that are repositories of the seed storage proteins. Large deposits of phaseolin in a proteinaceous matrix can be revealed by indirect immuno-localization using anti-phaseolin antibody and colloidal gold particles (Figure 1; Greenwood and Chrispeels, 1985a).

The high proportion of phaseolin mRNA in the polyadenylated RNA of developing bean seeds facilitated its isolation, translation *in vitro* (Hall *et al.*, 1978), cloning (Sun *et al.*, 1981) and sequencing (Slightom *et al.*, 1983). Southern genomic analysis (Talbot *et al.*, 1984) of three cultivars (Sanilac, grown for its dry seeds, and green bean varieties Tendergreen and Contender) revealed the presence of a limited number (7–9) of phaseolin gene copies per haploid genome. Subsequent analyses of other crop legumes (Table 1) showed similarly low copy numbers for the major seed storage proteins. These low gene copy numbers contrast markedly with those for prolamin storage proteins in cereals, such as zein in corn (110–130 copies, Viotti *et al.*, 1979; DeRose *et al.*, 1989), wheat storage proteins (100 copies, Pernollet and Vaillant, 1984; Kreis *et al.*, 1985) and kafirin in sorghum (20 copies, DeRose *et al.*, 1989). In the absence of any evidence that certain members of the corn or sorghum storage protein genes express at higher levels than others, it appears that the strategies developed for storage protein production differ greatly between these representatives of monocot and dicot crops: whereas monocots utilize many genes (yielding a spectrum of closely-related proteins) expressing at moderate levels, storage proteins of legumes and many other dicots are vigorously expressed from a small number of genes. These contrasts impose varying constraints (e.g. dilution by the high number of endogenous genes and competition for *trans*-acting factors) on attempts to confer nutritional improvement through bioengineering.

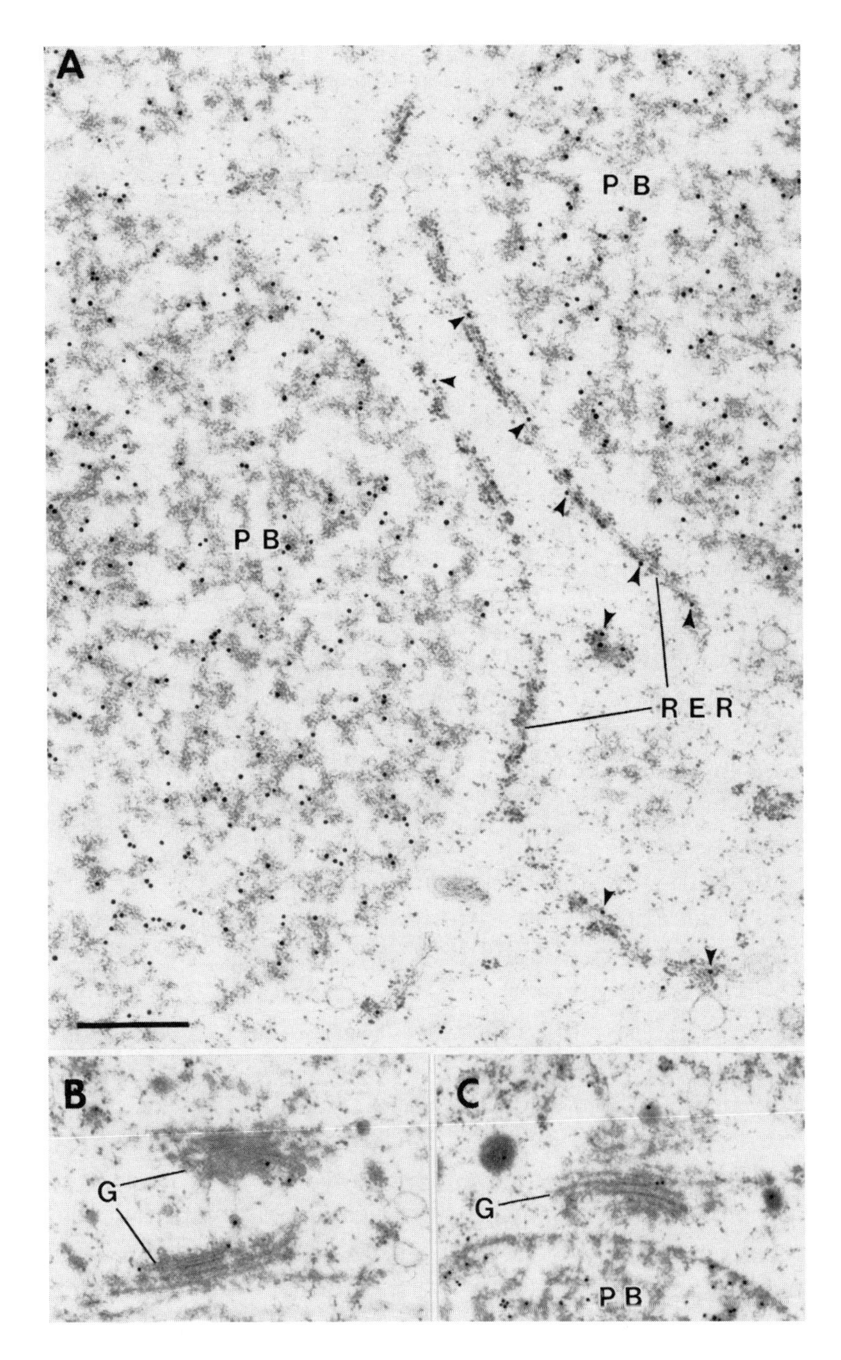

Figure 1: Immunogold labeling of phaseolin on thin sections of bean cotyledons embedded in LR White. 30,000X. Bar = 0.5 μm. A. Eighteen days after flowering, phaseolin is detectable in rough endoplasmic reticulum (RER), as indicated by the arrowheads, and in protein bodies (PB). B and C. Labeling also reveals the presence of phaseolin in Golgi bodies (G).

Table 1. Size of legume storage protein gene families

Gene Family	Gene copies per haploid genome	Reference
Phaseolin (*Phaseolus vulgaris,* 'Tendergreen')	6.9–7.5	Talbot *et al.*, 1984
Phaseolin (*Phaseolus vulgaris,* 'Sanilac')	8	"
Phaseolin (*Phaseolus vulgaris,* 'Contender')	9	"
(α')β-conglycinin (*Glycine max)*	5	Goldberg *et al.*, 1981
47 kDa Vicilin (*Pisum sativum)*	2–3	Gatehouse *et al.*, 1983
	5–7	Domoney and Casey, 1985
50 kDa Vicilin (*Pisum sativum)*	5	Gatehouse *et al.*, 1983
	4–6	Domoney and Casey, 1985
Legumin (*Pisum sativum)*	10	Casey *et al.*, 1986
Glycinin (*Glycine max)*	5	Fisher and Goldberg, 1982; Scallon *et al.*, 1985

The ability of phaseolin genes to direct high levels of transcriptional and translational expression was established definitively by the demonstration that a single copy of a β-phaseolin gene yields some 2 percent of the total seed protein in transgenic tobacco (Sengupta-Gopalan *et al.*, 1985). These early studies using transgenic hosts also revealed evolutionary conservation of many processes associated with storage protein transcription and expression. Tissue-specificity was retained, and temporal expression during tobacco seed development paralleled that of the endogenous storage proteins. These results on seed protein expression in transgenic plants have been subsequently confirmed for soybean β-conglycinin in transgenic petunia and tobacco (Beachy *et al.*, 1985; Barker *et al.*, 1988), legumin in tobacco (Ellis *et al.*, 1988) and zein in tobacco (Schernthaner *et al.*, 1988). Splicing of the β-phaseolin introns in tobacco is efficient and the resulting mRNA is translated to yield a pre-protein that undergoes correct cleavage of the transit sequence and glycosylation to yield mature glycoforms having two-dimensional electrophoretic properties indistinguishable from those of native bean phaseolin. The structure of tobacco seeds differs markedly from that of bean: mature seeds retain a high proportion of endosperm and the contents of the protein bodies is separated in two distinct phases, crystalloid and matrix. Nevertheless, phaseolin is specifically targeted to the matrix (Greenwood and Chrispeels, 1985b; Bustos *et al.*, 1990a) where it accumulates in large quantities. Hydrolysis of phaseolin in germinating tobacco seeds parallels that in bean, the full-length (48 kDa) protein being initially cleaved into peptides of 27-26 kDa and subsequently degraded, presumably to free amino acids (Sengupta-Gopalan, 1985).

The focus of gene expression studies has now moved to a determination of sequence elements controlling intensity, coordination, timing and tissue-specificity. Functional analysis of *cis* regulatory elements from a soybean gene encoding the α' subunit of β-conglycinin (Chen *et al.*, 1986) identified a region within the first upstream 257 nucleotides, essential for high-level expression in petunia seeds. Initial analysis of the first 800 bp 5′ of the phaseolin transcriptional start site by gel retardation and footprinting techniques revealed binding of bean cotyledon nuclear protein(s) to a 55 bp A/T-rich motif that was shown to enhance transcription from a heterologous CaMV-35S promoter in tobacco (Bustos *et al.*, 1989). This and similar motifs found in many other plant genes function as quantitative expression elements (Goldberg *et al.*, 1989) but apparently play no role in determining tissue-specificity. Indeed, the ability of this short sequence to promote high levels of GUS expression in roots, leaves and stems in addition to seeds, suggested that other features of the phaseolin upstream region are responsible for seed specific expression. Deletion mapping in transgenic tobacco plants and bean transient expression assays has revealed the presence of an array of positive and negative *cis*-regulatory DNA domains in the phaseolin 5′-flanking region (Bustos *et al.*, 1991b). The main seed-specific domain UAS-1, located between nucleotide positions -295 and -109, is sufficient for high-levels of expression of a bacterial reporter gene for β-glucuronidase (GUS) in the embryonic cotyledons and endosperm of tobacco. An additional domain, UAS-2, is required for expression in the embryonic axis and a pair of negative regulatory domains, NRS-1 and NRS-2, are necessary to achieve correct temporal regulation during seed development. Synergistic interactions between various upstream domains and the proximal promoter region are apparent in this system and have been reported for the viral CaMV-35S promoter complex (Benfey *et al.*, 1990b).

Analyses of the *cis*-acting DNA sequence elements responsible for regulated gene expression in legume seeds can be facilitated by transient expression of gene constructs in protoplasts from immature cotyledons (Bustos *et al.*, 1991b). As in bacteria, yeast and mammalian cells (Ptashne, 1988), the definition of *cis*-acting DNA sequences in plants has led to advances in the identification and characterization of *trans*-acting protein factors which modulate the rate of gene transcription by RNA polymerase II. Several genes have been cloned that code for DNA-binding proteins involved in the regulation of maize storage protein expression (Hartings *et al.*, 1989), flower development (Sommer *et al.*, 1990; Yanofski *et al.*, 1990), CaMV-35S transcription (Katagiri *et al.*, 1989) and histone gene expression (Tabata *et al.*, 1989). Protein:DNA binding studies indicated the presence of putative *trans*-acting factors for the seed proteins phaseolin (Bustos *et al.*, 1989), β-conglycinin (Allen *et al.*, 1989), soybean lectin (Jofuku *et al.*, 1987) and sunflower helianthinin (Jordano *et al.*, 1989) genes. Conserved upstream DNA sequence motifs, in 11S ("legumin box," Bäumlein *et al.*, 1986) and 9S ("vicilin box," Gatehouse *et al.*, 1986) storage proteins and in soybean lectin and Kunitz trypsin inhibitor genes ("AACACA" motif, Goldberg, 1986), suggest that an ensemble of proteins may be required to coordinately regulate phaseolin gene expression during seed development. Indeed, the combinatorial ordering of *cis*-actings elements in the promoter sequences of each gene may direct the conformation of proteins assembled into the transcriptional complex that results in their specific spatial and temporal specificity of expression.

Synthesis and Processing of Phaseolin

Regulated transcription of seed protein genes is but one requirement for high-level expression of the cognate protein products in specific cell types. Translational efficiency and stability of seed protein mRNA can be important; initial studies on the translation of phaseolin mRNA *in vitro* showed that each molecule yielded ten polypeptides (Hall *et al.*, 1978). An effect of various 3′-untranslated sequences of plant origin on the stability of nopaline synthase mRNA's in tobacco has been reported (Ingelbrecht *et al.*, 1989). Interestingly, improved accumulation of GUS activity during late stages of seed development has been observed upon replacement of the nopaline synthase terminator sequence (Nos-3′) with 1300 bp of phaseolin 3′-flanking DNA (Bustos *et al.*, unpublished). Experiments are underway to determine whether the effect is transcriptional or whether the 3′-untranslated region of the phaseolin transcript can extend the useful life span of the heterologous bacterial RNA sequence in tobacco seeds.

The extensive rough endoplasmic reticulum (RER) in cells of developing bean seeds (Figure 1) appears to contribute to active storage protein synthesis. Phaseolin mRNA is translated in membrane-bound polysomes (Sun *et al.*, 1975) with translocation of the resulting precursor polypeptides into the lumen of the RER (Bollini *et al.*, 1983). The long half-life of many seed proteins is important in maximizing their final levels of accumulation. It appears likely that the series of post-translational modifications to the phaseolin precursor, as it progresses through the endomembrane system, contribute to its overall stability (longevity). Cleavage of an N-terminal signal peptide takes place shortly after synthesis. Chemical sequencing of phaseolin purified from mature bean seeds has revealed heterogeneity of the NH_2-terminus in native phaseolin (Paaren *et al.*, 1987). This has been interpreted as the loss of one to three N-terminal amino acid residues during processing rather than alternative signal peptide cleavage sites. The presence of two canonical *N*-glycosylation signals, Asn_{252}-Leu_{253}-Thr_{254} and Asn_{341}-Phe_{342}-Thr_{343}, found on all phaseolin genes identified to date (Hall *et al.*, 1983), causes the alternative addition and processing of one or two oligo-mannose residues resulting in at least two different glycoforms per subunit, with *N*-glycans of the high-mannose and complex types (Sturm *et al.*, 1988). The reason for the asymmetric utilization of *N*-glycan attachment sites is not completely understood, although analysis of glycoforms expressed from a single β-phaseolin cDNA in insect cells (Bustos *et al.*, 1988) suggests that it is an intrinsic property of the phaseolin protein matrix.

Individual phaseolin subunits associate in a pH-dependent, non-covalent fashion to form oligomers of varying degrees of polymerization. Precise measurements of the sedimentation characteristics of phaseolin quaternary isomers at different pH (Sun *et al.*, 1974) demonstrated that the predominant form at neutral pH has a sedimentation coefficient of 9S, consistent with a trimeric arrangement. Lower pH values foster the reversible association of phaseolin trimers into hexameric and dodecameric forms thought to approximate the physiological state of aggregation in seeds. Elucidation of the 3-D structure of phaseolin polypeptides by crystal X-ray diffraction (Lawrence *et al.*, 1990) revealed that this dodecameric form predominates in the crystalline phase. It is not known where, in the cell, individual phaseolin subunits first associate into higher order structures.

In beans, phaseolin polypeptides undergo little or no proteolytic processing prior to seed germination. The situation in tobacco is quite different. Irrespective of the presence or absence of introns, single β-phaseolin genes direct the accumulation of a full-length, 48 kDa, polypeptide as well as smaller polypeptides of 29 and 25 kDa. Similar phaseolin fragments can be observed after limited trypsin and chymotrypsin digestion of purified bean phaseolin *in vitro* (Sun *et al.*, 1975), suggesting that processing in tobacco takes place at particularly labile sites. Indeed, engineered mutations that prevent the attachment of *N*-glycans at either or both asparagine residues of the *N*-glycosylation signals cause a marked decrease in phaseolin accumulation and modify the pattern of processed peptides (Bustos *et al.*, 1991a). These observations are consistent with increased processing of the phaseolin precursor as a consequence of the absence of oligosaccharide side-chains. Experiments are underway to determine the actual sites of proteolytic cleavage and whether the presence or absence of *N*-glycans correlates with changes in the rate of proteolysis at defined locations on the molecule. The ability to express wild-type and mutant forms in a variety of cellular environments, added to our new knowledge of the structure of phaseolin monomers, make phaseolin a model of choice for future studies of proteolysis of food proteins.

Structure and Targeting of Phaseolin

Examination of the primary amino acid structure of α and β phaseolins indicates a considerable degree of internal structural redundancy (Table 2). Alignment of the NH_2- and COOH-terminal halves of the mature protein (residues 27 to 219 and 222 to 418; the complete sequence is 421 amino acids long) based on the *local homology* algorithms (Lipman *et al.*, 1984) identifies regions of similarity that involve primarily hydrophobic residues. The structure of phaseolin at a resolution of 3 Å (Lawrence *et al.*, 1990) has confirmed the structural relatedness of the NH_2 and COOH phaseolin halves. Each half of the phaseolin monomer is composed of a "jelly-roll" β-barrel domain (strands A through J) and three α-helices connected by more or less flexible loops. There is a remarkably good correspondence between the alignment based on the primary amino acid sequence and the locations of regions of analogous secondary and tertiary structure, particularly involving β-strands B, C, D, E, G, I and J, as well as α-helices 1, 2, and 3 on

Table 2. Internal structural homology in the β-phaseolin protein.

```
                             A'      A                      B
 27   LREEEESQDNPFYFNSDNSWNTLFKNQYGHIRVLQRFDQQSKRLQNLEDYRLVEFRSKPE
      :. :.|  ::    ....|:..  |.: .   : . |:. ..| :|  :  : .:  |.:
222   VNIDSEQIEELSKHAKSSSRKSHSKQDNT...IGNEFGNLTERTDNSLNVLISSIEMKEG
                 4                 A'           A          B
          C         D         E           F            G        H
 87   TLLLPQ.QADAELLLVVRSGSAILVLVKPDDRREYFFLASDNPIFSDNQ..KIPAGTIFY
      .|::|:  ..| ::|||..|.| : || |.:.:| : :.| .: :|.::   |||:   |
278   ALFVPHYYSKAIVILVVNEGEAHVELVGPKGNKETLEFESYRAELSKDDVFVIPAA...Y
          C          D        E                 F       G
      -            I            J           1         2        3
144   LVNPDPKEDLRIIQLAMPVNNPQIHDFFLSSTEAQQSYLQEF..SKHILEASFNSKFEEI
       |. .:..::.:..::: :|| . :::: :.|:.  |  :.    :|.:|:..|.:. ||:
335   PVAIKATSNVNFTGFGINANNNN.RNLLAGKTDNVISSIGRALDGKDVLGLTFSGSGEEV
        H         I            J           1          2         3

      ----
202   NRVL.......FEEEGQQEEGQQEG   219
       :::       | :: :::::||.:
394   MKLINKQSGSYFVDGHHHQQEQQKG   418
      ------
```

The sequences of the NH_2-and COOH-terminal halves of the β-phaseolin molecule were aligned using the program Bestfit (UWGCG) with values of 3.0 and 0.1 for the gap and gap length weights. % similarity = 52%; % identity = 21%. The ']' symbol indicates identity; ':' and '.' are chemically related amino acids. Horizontal lines identify regions of common secondary structure determined by X-ray crystallographic analysis (Lawrence *et al.*, 1990): β-strands are labeled with letters and α-helices with numbers. The numbers on the left and right margins correspond to amino acid positions relative to the first methionine of the phasolin precursor deduced from the sequence of the corresponding cDNA (Slightom *et al.*, 1983).

each half of the molecule (Table 2). These observations strongly suggest that phaseolins arose by endo-duplication of an ancestral gene. Indeed, amino acid and nucleotide sequence comparisons between the vicilin-like globulins from French beans, jack beans, peas and soybeans have indicated a common origin for all the vicilin-like globulins (Argos *et al.*, 1985; Doyle *et al.*, 1986; Gibbs *et al.*, 1989) and possibly for the basic chain of 11S legumins (Argos *et al.*, 1985; Borroto and Dure, 1988). The 3-D structure has also shown that the phaseolin trimers consist of subunits (β chains) associated end-to-end, into a nearly planar, triangular configuration. The structure of legume storage proteins differs markedly from that of monocot storage prolamins (zeins from corn, hordein from barley and kafirins from sorghum) in that the latter show a considerably higher degree of internal redundancy, with a larger proportion of α-helical structure (Kreis *et al.*, 1985). These observations may indicate that multiple "solutions" to the set of constraints alluded to at the beginning of this article, have occurred during evolutionary time.

Phaseolin and phytohemagglutinin (PHA), both glycoproteins, are transported to the protein bodies of bean and tobacco cotyledon cells (Figure 1; Greenwood and Chrispeels, 1985a, 1985b; Sturm *et al.*, 1988). Several lines of evidence indicate that the presence of *N*-glycans on either protein is not absolutely required for correct intracellular targeting. Cotyledon slices treated with the antibiotic tunicamycin accumulate phaseolin and PHA polypeptides in the protein bodies, albeit less efficiently than untreated specimens (Bollini *et al.*, 1985). Additionally, mutations that prevent the attachment of *N*-glycans fail to abolish the accumulation of either protein in the protein body matrix of transgenic tobacco seeds (Voelker *et al.*, 1989; Bustos *et al.*, 1991a). Although these observations do not rule out more subtle effects of the *N*-glycans on the conformation and transport of either molecule, they suggest that potential targeting determinants reside on the polypeptide backbone itself, as has been demonstrated for targeting of mammalian nuclear and RER proteins. Interestingly, the insertion of a sequence coding for 15 additional amino acids into the β-phaseolin gene (Hoffman *et al.*, 1988) led to the expression of very low levels of protein in tobacco seeds, with no detectable accumulation in protein bodies. A more systematic and comprehensive mutagenesis program appears to be required to locate any potential amino acid domains that may be responsible for the targeting of phaseolin to protein bodies, although the complex nature of the phaseolin trimer may render this strategy very inefficient.

An alternative approach that was used to demonstrate the function of a KDEL peptide as a mammalian RER targeting signal (Munro and Pelham, 1987), makes use of in-frame fusions between the protein of interest and a passive reporter molecule. To explore the applicability of this strategy, several fusions between phaseolin NH_2-terminal fragments and the GUS gene were created (Figure 2), all under the control of the seed-specific phaseolin promoter and upstream region. High levels of GUS activity have been obtained with the control construct pβ+20/GUS in tobacco embryos and endosperm (Bustos *et al.*, 1989). The GUS protein lacks a recognizable signal peptide, and it is expected to accumulate in the cytoplasm of transgenic cells (Jefferson *et al.*, 1987). Fusion protein constructs coding for 31, 191 and 294 amino acids of phaseolin sequence fused to the NH_2-terminus of GUS via a short, heptapeptide linker ('VGVGQSL') resulted in approximately 20-times less GUS specific activity in mature tobacco seeds (4–6 independently transformed plants of each type). Although considerably lower, these values are

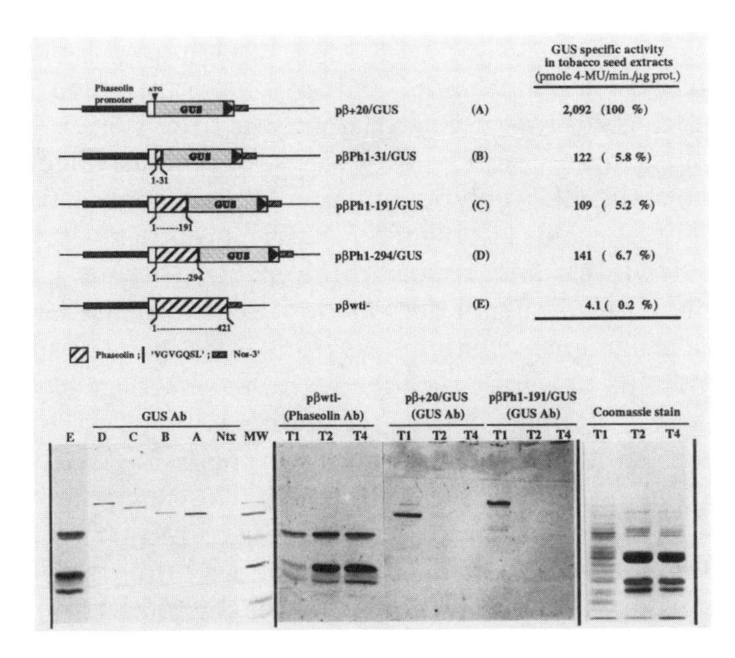

Figure 2: Expression and subcellular localization of phaseolin/GUS hybrid proteins in tobacco seeds. Top. Hybrid genes coding for 0, 31, 191 and 294 N-terminal amino acids (constructs pβ+20/GUSpβPh1-31/GUS, pβPh1–191/GUS and pβPh1–294/GUS) of phasolin fused upstream of the reporter enzyme β-glucuronidase (GUS) were constructed. The heptapeptide linker 'VGVGQSL' separates phaseolin from GUS peptide sequences. Specific GUS activities in total extracts of mature tobacco seeds are shown on the right. The presence of phaseolin peptide sequences causes an 18- to 20-fold reduction in GUS activity. Bottom. Immunodetection of GUS and phaseolin polypeptides in total seed extracts and in purified protein body fractions. Left: Total extract from seeds containing the β-phaseolin control gene pβwti-(lane E) reacted with a phaseolin specific IgG. Total extracts from seeds containing phaseolin/GUS hybrid genes (lanes A through D, corresponding to diagrams A through D on top) reacted with a GUS specific antiserum, indicate similar levels of expression and stability of all GUS proteins (both unfused and fused). Middle and right: Subcellular fractions T1, T2 and T4 from control pβwti- gene, illustrates accumulation of phaseolin in the protein body fractions T2 and T4. A similar experiment with seeds of pβ+20/GUS and pβPh1–191/GUS reacted with the GUS antiserum. Both the unfused GUS and the phaseolin/GUS fusion proteins remain in the light membrane/soluble fraction T1. A Coomassie stained gel of pβPh1–191/GUS proteins illustrates the abundance of endogenous storage proteins in the T2 and T4 fractions. Identical profiles were observed in seeds of the other gene constructs (data not shown).

still 25 to 35 times higher than those seen in non-transformed plants or in plants containing a control phaseolin gene (pβwti-). Immunodetection was used to demonstrate the presence of GUS polypeptides in total extracts of tobacco seed proteins (Figure 2). The mobilities of the GUS polypeptide products are consistent with the presence of phasolin fragments. Furthermore, they accumulate to very similar levels and appear to be stable (there is little or no indication of processing or degradation). The decreased enzymatic activity is in agreement with results from expression of patatin/GUS fusion proteins in transgenic tobacco callus (Iturriaga *et al.*, 1989). In that case, GUS transport into the endomembrane system also led to a loss of enzymatic activity, possibly due to *N*-glycosylation of the bacterial molecule at a cryptic site. This suggested that the phaseolin/GUS fusion proteins were being correctly targeted to the RER by the phaseolin signal peptide (residues 1 through 24). Subsequent transport into protein bodies, as for the phaseolin molecule, probably depends on the presence of positive information in the processed molecule. Our analysis of the subcellular distribution of phaseolin and GUS products in desiccated, mature seeds (Figure 2) indicates, so far, that the latter fail to accumulate appreciably in protein bodies. Relatively pure preparations of the storage organelles, as ascertained from the enrichment for endogenous storage proteins, can be obtained by isopicnic centrifugation in non-aqueous media (Voelker *et al.*, 1989). Immunodetection reveals that, while phaseolin polypeptides accumulate very efficiently in protein body fractions T2 and T4, the control (A) and fusion GUS proteins (C; phaseolin amino acids 1 to 191) remain in the soluble fraction T1 (analogous results have been obtained with fusion proteins B and D, data not shown). We are presently localizing the GUS fusion polypeptides by direct EM immuno-gold labeling, in the hope of establishing their final fate. These results could mean that the protein body targeting domain(s) is localized within the C-terminal 141 amino acids of phaseolin. Alternatively, it is possible that the structure of the fusion proteins is incompatible with the normal function of signal(s) present within the NH_2-terminal 294 amino acids used in these experiments. This emphasizes a major caveat inherent to mutational analysis of protein biology, namely, the need to analyze a large number of molecular variants in order to distinguish local from global structural effects.

Approaches for Nutritional Enhancement of Seed Proteins Through Protein Engineering

The relative nutritional merit for a protein is typically quantified by its protein efficiency ratio (PER), originally derived from rat feeding trials but also obtained by tests *in vitro* (Satterlee *et al.*, 1977). The presence of trypsin inhibitor in legume seed meal is nutritionally very detrimental (Liener, 1981), and excision of genes encoding such inhibitors (or inactivation of their expression) would be very helpful. Haseloff and Gerlach (1988) developed a "molecular shears" concept in which ribozyme activity is targeted to cleaving a defined mRNA sequence, but the functional application of this strategy has yet to be demonstrated. For proteins that are not readily susceptible to total digestion, sequence modification to improve digestibility may substantially improve PER values. The carbohydrates on some proteins may act as recognition markers, contributing to the protein's susceptibility to attack by intra- or extra-cellular proteases. In a study on the digestibility by trypsin, pepsin and chymotrypsin, phaseolin was found to display a marked resistance to proteolysis (Liener and Thompson, 1980). However, the glycans may have an important role in stabilizing the protein's tertiary structure, as indicated by the effect of elimination of the Asn acceptor residues on phaseolin accumulation described above.

Cereal prolamins, such as the zeins in corn and the sorghum kafirins (DeRose *et al.*, 1989), are often totally deficient in lysine. Consequently, procedures to increase the amount of these deficient amino acids would greatly improve their PER values (see Larkins *et al.*, this volume). Storage proteins of cereal seeds are frequently expressed form large gene families. This implies that their promoters are relatively weak and not well suited for expression of modified storage protein genes; whether alternative promoters can alleviate this problem awaits testing in transgenic monocots. In contrast, legume storage protein promoters are often very powerful: well over one percent of the total seed protein in the seed of transgenic tobacco containing a single phaseolin gene was phaseolin (Sengupta-Gopalan *et al.*, 1985). Consequently, modification of storage protein composition in dicots may be relatively facile in regard to transcriptional constraints.

The low methionine content of phaseolin (3–4 residues/polypeptide) reflects the deficiency of this essential amino acid in legume seed proteins. As cloning and sequencing of storage protein genes became feasible, the opportunity for increasing the proportion of methionine in legume storage proteins, such as phaseolin, through modification of the gene sequence, was apparent. However, recent results from experiments with transgenic plants containing nutritionally-modified sequences show generally very low levels of protein accumulation. Hoffman *et al.* (1988) inserted a 15 amino acid sequence containing six methionine residues into the β-phaseolin gene. In transgenic tobacco, the resulting "himet" gene yielded similar levels of mRNA to those obtained from the unmodified gene, but only 0.2 percent of the expected protein accumulation was detected. The himet phaseolin was glycosylated and assembled into trimers, but was not sequestered into the protein bodies of transgenic tobacco seed. On the basis of experiments with bean (*Phaseolus vulgaris*) phytohemagglutinin expression in yeast, Tague *et al.* (1990) postulated that a short domain (including residues 34–43 from the N-terminus) were sufficient for vacuolar targeting. The amino acid sequence resulting from the himet phaseolin insertion should not have affected such a simplistic signal, and it is possible that targeting determinants in the storage proteins are complex and related to overall structure. This conclusion is supported by the results of Saalbach *et al.*, (1990) who introduced four methionine residues into the C-terminus of a *Vicia faba* legumin B gene by a frameshift mutation. The modified legumin could not be detected in seed of transgenic tobacco and interference with intracellular transport in yeast was documented. These results support the belief that seed storage proteins, their genes and mRNA transcripts, contain complex and evolutionarily conserved features intrinsic to their temporal and spacial distribution and accumulation. Despite these reservations, interesting concepts such as the design of artificial proteins (Kim *et al.*, 1990) and expression of seed proteins in leaves and other unconventional tissues (Wandelt *et al.*, 1990) deserve experimental evaluation. Additionally, approaches based on the introduction of naturally-occurring, nutritionally-balanced storage proteins (i.e. methionine-rich), such as the Brazil nut 2S albumin (Altenbach *et al.*, 1987), are attractive and have proven effective in the tobacco model, yielding an increase of 30 percent in the methionine level in transgenic seed (Altenbach *et al.*, 1989). Protein engineering based on a combination of modifications designed with the knowledge of storage protein structure and expression of the variant proteins in transgenic plants offers exciting opportunities and challenges for the future.

Acknowledgements

We especially thank Kate VandenBosch for the electron micrograph of tobacco seed containing phaseolin. Studies on transcriptional regulation of phaseolin were supported by Grant DCB-8904886 from NSF and Texas Advanced Technology Program Grant 999902-202; phaseolin processing studies were supported by Grant DCB-8802057 from NSF.

References

Allen, R.D., Bernier, F., Lessard, P.A. and Beachy, R.N. 1989. Nuclear factors interact with a soybean Beta conglycinin enhancer. The Plant Cell. 1: 623–632.

Altenbach, S.B., Pearson, K., Leung, F.W. and Sun, S.S.M. 1987. Cloning and sequence analysis of a cDNA encoding a Brazil nut protein exceptionally rich in methionine. Plant Mol. Biol. 8: 239–250.

Altenbach, S.B., Pearson, K., Meeker, G., Staraci, L.C. and Sun, S.S.M. 1989. Enhancement of the methionine content of seed proteins by the expression of a chimeric gene encoding a methionine-rich protein in transgenic plants. Plant Mol. Biol. 13: 513–522.

Argos, P., Narayana, S.V.L. and Nielsen, N.C. 1985. Structural similarity between legumin and vicilin storage proteins from legumes. EMBO J. 4: 1111–1117.

Barker, S.J., Harada, J.J. and Goldberg, R.B. 1988. Cellular localization of soybean storage protein mRNA in transformed tobacco seeds. Proc. Natl. Acad. Sci. USA 85: 458–462.

Bäumlein, H., Wobus, U., Pustell, J. and Kafatos, F.C. 1986. The legumin gene family: structure of a B type gene of *Vicia faba* and a possible legumin gene specific regulatory element. Nucl. Acids Res. 14: 2707–2720.

Benfey, P.N., Ren, L. and Chua, N.-H. 1990a. Tissue-specific expression from CaMV 35S enhancer subdomains in early stages of plant development. EMBO J. 9: 1677-1684.

Benfey, P.N., Ren, L. and Chua, N.-H. 1990b. Combinatorial and synergistic properties of CaMV 35S enhancer subdomains. EMBO J. 9: 1685-1696.

Beachy, R.N., Chen, Z.-L., Horsch, R.B., Rogers, S.G., Hoffmann, J. and Fraley, R.T. 1985. Accumulation and assembly of soybean β-conglycinin in seeds of transformed petunia plants. EMBO J. 4: 3047-3053.

Bollini, R., Vitale, A. and Chrispeels, M. 1983. *In vivo* and *in vitro* processing of seed reserve protein in the endoplasmic reticulum: evidence for two glycosylation steps. J. Cell Biol. 96: 999-1007.

Bollini, R., Ceriotti, A., Daminati, M.G. and Vitale, A. 1985. Glycosylation is not needed for the intracellular transport of phytohemagglutinin in developing *Phaseolus vulgaris* cotyledons and for the maintenance of its biological activities. Physiol. Plant. 65: 15–22.

Borroto, K. and Dure, L. 1988. The globulin seed storage proteins of flowering plants are derived from two ancestral genes. Plant Mol. Biol. 8: 113–131.

Bustos, M.M., Luckow, V.A., Griffing, L.R., Summers, M.D. and Hall, T.C. 1988. Expression, glycosylation and secretion of phaseolin in a baculovirus system. Plant Mol. Biol. 10: 475-488.

Bustos, M.M., Guiltinan, M.J., Jordano, J., Begum, D., Kalkan, A. and Hall, T.C. 1989. Regulation of GUS expression in transgenic tobacco plants by an A/T-rich *cis*-acting sequence found upstream of a French bean β-Phaseolin gene. The Plant Cell 1: 839-853.

Bustos, M.M., Kalkan, F.A., VandenBosch, K.A. and Hall, T.C. 1991a. Differential accumulation of four phaseolin glycoforms in transgenic tobacco. Plant Mol. Biol. 16: 381-395.

Bustos, M.M., Battraw, M.J., Kalkan, F.A., Begum, D. and Hall, T.C. 1991b. Temporal and spatial regulation of gene expression by a seed storage protein promoter requires positive and negative *cis*-acting 5′-flanking DNA sequences. EMBO J. 10(6): (In press).

Casey, R., Domoney, C. and Ellis, N. 1986. Legume storage proteins and their genes. In "Oxford Surveys of Plant Molecular and Cell Biology," Miflin, B.J. (Ed.), Vol. 3, pp. 1-95, Oxford University Press, Oxford.

Chen, Z.-L., Schuler, M.A. and Beachy, R.N. 1986. Functional analysis of regulatory elements in a plant embryo-specific gene. Proc. Natl. Acad. Sci. USA 83: 8560–8564.

Craig, S., Goodchild, D.J. and Miller, C. 1980. Structural aspects of protein accumulation in developing pea cotyledons. II. Three-dimensional reconstructions of vacuoles and protein bodies from serial sections. Austr. J. Plant Physiol. 7: 329–337.

DeRose, R.T., Ma, D.-P., Kwon, I.-S., Hasnain, S.E., Klassy, R.C. and Hall, T.C. 1989. Characterization of the kafirin gene family from sorghum reveals extensive sequence homology with zein from maize. Plant Mol. Biol. 12: 245–256.

Domoney, C. and Casey, R. 1985. Measurement of gene number for seed storage proteins in *Pisum*. Nucleic Acids Res. 13: 687-699.

Doyle, J.J., Schuler, M.A., Godette, W.D., Zenger, V. and Beachy, R.N. 1986. The glycosylated seed storage proteins of *Glycine max* and *Phaseolus vulgaris*. J. Biol. Chem. 261: 9228-9238.

Ellis, J.R., Shirsat, A.H., Hepher, A., Yarwood, J.N., Gatehouse, J.A., Croy, R.R.D. and Boulter, D. 1988. Tissue-specific expression of a pea legumin gene in seeds of *Nicotiana plumbaginifolia*. Plant Mol. Biol. 10: 203–214.

Fischer, R.L. and Goldberg, R.B. 1982. Structure and flanking regions of soybean seed protein genes. Cell 29: 651–660.

Gatehouse, J.A., Lycett, G.W., Delauney, A.J., Croy, R.R.D. and Boulter, D. 1983. Sequence specificity of the post-translational proteolytic cleavage of vicilin, a seed storage protein of pea (*Pisum sativum* L.). Biochem. J. 212: 427–432.

Gatehouse, J.A., Evans, I.M., Croy, R.R.D. and Boulter, D. 1986. Differential expression of genes during legume seed development. Phil. Trans. R. Soc. Lond. B314: 367–384.

Gibbs, P.E.M., Strongin, K.B. and McPherson, A. 1989. Evolution of legume seed storage proteins - A domain common to legumins and vicilins is duplicated in vicilins. Mol. Biol. Evol. 6: 614–623.

Goldberg, R.B., Hoschek, G., Ditta, G.S. and Breidenbach, R.W. 1981. Developmental regulations of cloned superabundant embryo mRNA's in soybean. Dev. Biol. 83: 218–231.

Goldberg, R.B. 1986. Regulation of plant gene expression. Phil. Trans. R. Soc. Lond. B314: 343–353.

Goldberg, R.B., Barker, S.J. and Perez-Grau, L. 1989. Regulation of gene expression during plant embryogenesis. *Cell* 56: 149–160.

Greenwood, J.S. and Chrispeels, M.J. 1985a. Immunocytochemical localization of phaseolin and phytohemagglutinin in the endoplasmic reticulum and Golgi complex of developing bean cotyledons. Planta 164: 295–302.

Greenwood, J.S. and Chrispeels, M.J. 1985b. Correct targeting of the bean storage protein phaseolin in the seeds of transformed tobacco. Plant Physiol. 79: 65–71.

Hall, T.C., Ma, Y., Buchbinder, B.U., Pyne, J.W., Sun, S.M. and Bliss, F.A. 1978. Messenger RNA of G1 protein of French bean seeds: cell-free translation and product characterization. Proc. Natl. Acad. Sci. USA 75: 3196–3200.

Hall, T.C., Sun, S.M., Ma, Y., McLeester, R.C., Pyne, J.M., Bliss, F.A. and Buchbinder, B.U. 1979. The major storage protein of French bean seeds: characterization *in-vivo* and translation *in-vitro*. In "The Plant Seed," Rubenstein, I., Phillips, R.L., Green, C.E. and Gengerbech, B.G. Eds., pp. 3–26. Academic Press Inc., New York.

Hall, T.C., Slighton, J.L., Ersland, D.R., Murray, M.G., Hoffman, L.M., Adang, M.J., Brown, J.W.S., Ma, Y., Matthews, J.A., Cramer, J.H., Barker, R.F., Sutton, D.W. and Kemp, J.D. 1983. Phaseolin:Nucleotide sequence explains molecular weight and charge heterogeneity of a small multigene family and also assists vector construction for gene expression in alien tissue. In "Structure and Function of Plant Genomes," Ciferri, O. and Dure, L. Eds., pp. 123–142. Plenum Press.

Hartings, H., Maddaloni, M., Lazzaroni, N., Di Fonzo, N., Motto, M., Salamini, F. and Thompson, R. 1989. The O2 gene which regulates zein deposition in maize endosperm encodes a protein with structural homologies to transcriptional activators. EMBO J. 8: 2795–2802.

Haseloff J. and Gerlach, W.L. 1988. Simple RNA enzymes with new and highly specific endoribonuclease activities. Nature 334: 585–591.

Hoffman, L.M., Donaldson, D.D. and Herman, E.M. 1988. A modified storage protein is synthesized, processed, and degraded in the seeds of transgenic plants. Plant Mol. Biol. 11: 717–729.

Ingelbrecht, I.L.W., Herman, L.M.F., Dekeyser, R.A., Van Montagu, M.C. and Depicker, A.G. 1989. Different 3′ end regions strongly influence the level of gene expression in plant cells. The Plant Cell 1: 671–680.

Iturriaga, G., Jefferson, R.A. and Bevan, M.W. 1989. Endoplasmic reticulum targeting and glycosylation of hybrid proteins in transgenic tobacco. The Plant Cell 1: 381-390.

Jefferson, R.A., Kavanagh, T.A. and Bevan, M.W. 1987. GUS fusions: beta glucuronidase as a sensitive and versatile gene fusion marker in higher plants. EMBO J. 6: 3901-3908.

Jofuku, K.D., Okamuro, J.K. and Goldberg, R.B. 1987. Interaction of an embryo DNA binding protein with a soybean lectin gene upstream region. Nature. 328: 734–737.

Jordano, J., Almoguera, C. and Thomas, T.L. 1989. A sunflower helianthinin gene upstream sequence ensemble contains an enhancer and sites of nuclear protein interaction. The Plant Cell 1: 855–866.

Katagiri, F., Lam, E. and Chua, N.-H. 1989. Two tobacco DNA-binding proteins with homology to the nuclear factor CREB. Nature. 340: 727–730.

Kim, J.H., Cetiner, M.S., Blackmon, W.J. and Jaynes, J.M. 1990. The design, construction, cloning and integration of a synthetic gene encoding a novel polypeptide to enhance the protein quality of plants. J. Cellular Biochem. Supp. 14E: 349.

Kreis, M., Shewry, P.R., Forde, B.G., Forde, J. and Miflin, B.J. 1985. Structure and evolution of seed storage proteins and their genes with particular reference to those of wheat, barley and rye. In "Oxford Surveys of Plant Molecular and Cell Biology," Miflin, B.J. Ed. pp. 253–317. Oxford University Press.

Lawrence, M.C., Suzuki, E., Varghese, J.N., Davis, P.C., Van Donkelaar, A., Tulloch, P.A. and Colman, P.M. 1990. The three-dimensional structure of the seed storage protein phaseolin at 3 Å resolution. EMBO J. 9: 9-15.

Liener, I.E. and Thompson, R.M. 1980. *In vitro* and *in vivo* studies on the digestibility of the major storage protein of the navy bean *(Phaseolus vulgaris)*. Qual. Plant Foods Hum. Nutr. 30: 13–25.

Liener, I.E. 1981. Factors affecting the nutritional quality of soya products. J. Am. Oil Chem. Soc. 58: 406–415.

Lipman, D.J., Wilbur, W.J., Smith, T.F. and Waterman, M.S. 1984. On the statistical significance of nucleic acid similarities. Nucl. Acids Res. 12: 215–26.

Munro, S., and Pelham, H.R.B. 1987. A C-terminal signal prevents secretion of luminal ER proteins. Cell 48: 899–907.

Paaren, H.E., Slightom, J.L., Hall, T.C., Inglis, A.S. and Blagrove, R.J. 1987. Purification of a seed glyoprotein: N-terminal and deglycosylation analysis of phaseolin. Phytochemistry 26: 335–343.

Pernollet, J.C. and Vaillant, V. 1984. Characterization and complexity of wheat developing mRNA's. Plant Physiol. 76: 187–190.

Ptashne, M. 1988. How eukaryotic transciptional activators work. Nature 335: 683–689.

Saalbach, G., Jung, R., Kunze, G., Manteuffel, R., Saalbach, I. and Müntz, K. 1990. Expression of modified legume storage protein genes in different systems and studies on intracellular targeting of *Vicia faba* legumin in yeast. Nottingham Symposium. pp. 151–158.

Satterlee, L.D., Kendrick, J.G. and Miller, G.A. 1977. Rapid *in vitro* assays for estimating protein quality. Nutritional Reports Intl. 16: 187–199.

Scallon, B., Thanh, V.H., Floener, L.A. and Nielsen, N.C. 1985. Identification and characterization of DNA clones encoding group-II glycinin subunits. Theor. Appl. Genet. 70: 510–519.

Schernthaner, J.P., Matzke, M.A. and Matzke, A.J.M. 1988. Endosperm-specific activity of zein gene promoter activity in transgenic tobacco plants. EMBO J. 7: 1249–1255.

Sengupta-Gopalan, C., Reichert, N.A., Barker, R.F., Hall, T.C. and Kemp, J.D. 1985. Developmentally regulated expression of the bean β-phaseolin gene in tobacco seed. Proc. Natl. Acad. Sci. USA 82: 3320-3324.

Shattuck-Eidens, D.M. and Beachy, R.N. 1985. Degradation of β-conglycinin in early stages of soybean embryogenesis. Plant Physiol. 78: 895–898.

Slightom, J.L., Sun, S.M. and Hall, T.C. 1983. Complete nucleotide sequence of a French bean storage protein gene: Phaseolin. Proc. Natl. Acad. Sci. USA. 82: 3320–3324.

Smith, T.F. and Waterman, M.S. 1981. Comparison of biosequences. Advances in Applied Mathematics. 2: 482–489.

Sommer, H., Beltrán, J.-P., Huijser, P., Pape, H., Lönnig, W.-E., Saedler, H. and Schwarz-Sommer, Z. 1990. *Deficiens,* a homeotic gene involved in the control of flower morphogenesis in *Antirrhinum majus:* the protein shows homology to transcription factors. EMBO J. 9: 605–613.

Stern, D.B., Jones, H. and Gruissem, W. 1989. Function of plastid messenger RNA 3′ inverted repeats: RNA stabilization and gene-specific protein binding. J. Biol. Chem. 264: 18742–18750.

Sturm, A., Van Kuik, J.A., Vliegenthart, J.F.G. and Chrispeels, M.J. 1987. Structure, position, and biosynthesis of the high mannose and the complex oligosaccharide side chains of the bean storage protein phaseolin. J. Biol. Chem. 262: 13392–13403.

Sturm, A., Voelker, T.A., Herman, E.M. and Chrispeels, M.J. 1988. Correct glycosylation, Golgi-processing, and targeting to protein bodies of the vacuolar protein phytohemagglutinin in transgenic tobacco. Planta 175: 170–183.

Sun, S.M., McLeester, R.C., Bliss, F.A. and Hall, T.C. 1974. Reversible and irreversible dissociation of globulins from *Phaseolus vulgaris* seed. J. Biol. Chem. 249: 2118–2121.
Sun, S.M., Buchbinder, B.U. and Hall, T.C. 1975. Cell-free synthesis of the major storage protein of the bean, *Phaseolus vulgaris* L. Plant Physiol. 56: 780–785.
Sun, S.M., Mutschler, M.A., Bliss, F.A. and Hall, T.C. 1978. Protein synthesis and accumulation in bean cotyledons during growth. Plant Physiol. 68: 918–923.
Sun, S.M., Slightom, J.L. and Hall, T.C. 1981. Intervening sequences in a plant gene-comparison of the partial sequence of cDNA and genomic DNA of French bean phaseolin. Nature 289: 37–41.
Tabata, T., Takase, H., Takayama, S., Mikami, K., Nakatsuka, A., Kawata, T., Nakayama, T. and Iwabuchi, M. 1989. A protein that binds to *cis*-acting elements of wheat histone genes has a leucine zipper motif. Science. 245: 965–967.
Tague, B.W., Dickinson, C.D. and Chrispeels, M.J. 1990. A short domain of the plant vacuolar protein phytohemagglutinin targets invertase to the yeast vacuole. The Plant Cell 2: 533–546.
Talbot, D.R., Adang, M.J. Slightom, J.L. and Hall, T.C. 1984. Size and organization of a multigene family encoding phaseolin, the major seed storage protein of *Phaseolus vulgaris* L. Mol. Gen. Genet. 198: 42–49.
Viotti, A., Sala, E., Marotta, R., Alberi, P., Balducci, C. and Soave, C. 1979. Genes and mRNA's coding for zein polypeptides in *Zea mays.* Eur. J. Biochem. 102: 211–212.
Voelker, T.A., Herman, E.M. and Chrispeels, M.J. 1989. *In vitro* mutated phytohemagglutinin genes expressed in tobacco seeds: role of glycans in protein targeting and stability. The Plant Cell 1: 95–104.
Wandelt, C., Knibb, W., Schroeder, H.E., Rafiqul, M., Khan, I., Spencer, D., Craig, S. and Higgins, T.J.V. 1990. The expression of an ovalbumin and a seed protein gene in the leaves of transgenic plants. In "Plant Molecular Biology," Herrmann, R. and Larkins, B. Eds. (In press.)
Yanofski, M.F., Ma, H., Bowman, J.L., Drews, G.N., Feldmann, K.A. and Meyerowitz, E.M. 1990. The protein encoded by the *Arabidopsis* homeotic gene *agamous* resembles transcription factors. Nature. 346: 35–39.

Approaches for Enhancing the Lysine Content of Maize Seed

Mauricio A. Lopes and Brian A. Larkins
Department of Plant Sciences
University of Arizona
Tucson, AZ 85721

Maize, like other cereals, is deficient in amino acids such as lysine and tryptophan, which are required for the nutrition of monogastric animals. The low concentrations of these amino acids is a consequence of their absence in the storage proteins, which constitute more than 50 percent of the seed protein. It has been possible to improve the nutritional value of maize seed protein through the use of mutations that decrease the synthesis of zeins. However, these mutants typically have poor agronomic characteristics and are not commercially useful. As an alternative approach to increase the concentration of these essential amino acids we used site-directed mutagenesis to modify a zein coding sequence. Using a *Xenopus* oocyte protein synthesis system, we found that the addition of one or more lysine residues to an alpha-zein protein did not alter its synthesis or processing into protein bodies. We have investigated the regulation of zein gene expression to identify mechanisms that will allow high levels of synthesis of this genetically engineered protein.This has led to the identification of genotypes that enhance synthesis of a specific endosperm storage protein. With this system, as much as 35 percent of the seed storage protein is encoded by one or two genes.

Introduction

Maize is an important source of carbohydrate and protein for human and animal nutrition. However, a major limitation of maize as a food source is the very low levels of the essential amino acids lysine and tryptophan. This is due to the fact that zeins, the storage proteins of maize seed, are completely lacking in lysine and nearly so in tryptophan (Table 1). Because zeins constitute about 50 percent of the total seed protein, the result is a very poor composition of amino acids (Wallace *et al.*, 1990a).

Maize seed proteins can be sequentially fractionated into four major solubility classes (Osborne and Mendel, 1914). The water-soluble fraction is regarded as albumin, while the proteins extracted with dilute salt solutions from the remaining residue are referred to as globulins. Many globulins and albumins are probably enzymes and are synthesized early in development (Misra *et al.*, 1975; Murphy and Dalby, 1971). Zeins, the prolamine fraction, are subsequently extracted with aqueous alcohol. The remaining insoluble proteins are referred to as glutelins, and these are extracted with dilute alkali or acids. Glutelins are heterogeneous and their origin is difficult to define. Albumins, globulins and glutelins are normally referred to as non-zein proteins; these contain practically all the lysine and tryptophan of the maize kernel. Consequently, the ratio of non-zein to zein protein defines the nutritional quality of maize.

Zeins consist of four structurally distinct alcohol-soluble polypeptides that are synthesized by membrane-bound polyribosomes during endosperm development. These proteins are cotranslationally transported into the lumen of the endoplasmic reticulum, where they assemble into structures known as protein bodies (Lending *et al.*, 1988; Lending and Larkins, 1989). Molecular cloning and DNA sequencing revealed the primary amino acid sequences of the various types of zeins, which have allowed their classification based on structural relationships (Table 1). The major class, the alpha-zeins (Esen, 1986), is a complex group of closely related polypeptides of 19 and 22kD that is soluble in aqueous alcohols. The 14 kD beta-zein and 10 kD delta-zein are soluble in aqueous alcohols containing reducing agent. The 16 and 27 kD gamma-zeins are soluble in aqueous and alcoholic solvents but also require reducing conditions.

Several mutations cause dramatic decreases in zein content, which are accompanied by increases in the proportion of the other protein fractions along with a relative increase in the percentage of essential amino acids. Some of these mutations are thought to be regulatory based on their capacity to simultaneously affect the synthesis of several zein polypeptides (Soave and Salamini, 1984; Schmidt, 1987). The *opaque-2 (o2)* mutation is the most widely studied of these. Its chalky and lusterless phenotype is thought to be a consequence of the reduced amount of storage protein in the endosperm, although the basis for these phenotypic relationships is not well understood. Much research and plant breeding has been focused on the use of *opaque-2*, but the inherent phenotypic deficiencies, like soft endosperm texture, lower yield, increased seed susceptibility to pathogens and mechanical damage, have limited its use.

Table 1. Amino acid composition of maize zein proteins

	γ-Zein[1] M_r27,000	α-Zein[2] M_r22,000	β-Zein[3] M_r14,000	δ-Zein[4] M_r10,000
Leu	19	42	15	15
Gin	30	31	28	15
Ala	10	34	18	7
Pro	51	22	13	20
Ser	4	18	11	8
Phe	2	8	0	5
Asn	0	13	2	3
Ile	4	11	1	3
Tyr	4	6	16	1
Val	15	17	4	5
Gly	13	2	12	4
Thr	9	7	5	5
Arg	5	4	7	0
His	16	3	4	3
Cys	14	1	6	5
Glu	2	1	5	0
Met	2	5	11	29
Asp	0	0	4	1
Lys	0	0	0	0
Trp	0	0	1	0
TOTAL	200	225	163	129

[1]Prat *et al.*(1985) *Nucl. Acids Res.* **13,** 1493-1504.
[2]Marks and Larkins (1982) *J. Biol. Chem.* **260,** 16445-16450.
[3]Pedersen *et al.* (1986) *J. Biol. Chem.* **261,** 6279-6284.
[4]Kirihara et al. (1988) *Mol. Gen. Genet.* **211,** 477-484.

Since genes encoding prolamine proteins that contain lysine do not normally exist in the genome, it is impossible by conventional genetic means to increase the lysine content of the storage protein fraction. However, the development of methods to isolate genes and modify them by site-directed mutagenesis has provided a novel approach to introduce lysine into these proteins. Since the only known function for zeins is to serve as a source of nitrogen and sulfur for the germinating seedling, they should be ideal targets for amino acid modification. Although these experiments are technically feasible, it is unknown whether such proteins would be processed and accumulated in the seed. It is therefore important to know the structure of the storage proteins, their organization within protein bodies, and the mechanisms regulating their synthesis and deposition. We must also identify mechanisms that will allow synthesis of large amounts of the genetically engineered proteins in the endosperm.

Determining Structural Relationships Among Zein Proteins

The structural differences between classes of zein proteins are reflected in their solubility properties. Figure 1 shows a comparison of proteins extracted from mature endosperm of the inbred line W64A with water (Fig. 1, lane 1), with an aqueous solution of 20 percent 2-mercaptoethanol (Fig. 1, lane 2), or with 20 percent 2-mercapto-ethanol containing from 10 to 80 percent ethanol (Fig. 1, lanes 3 through 10). The zein fraction is commonly extracted with 70 percent ethanol in the presence or absence of reducing agent. This analysis illustrates that all the different zeins are soluble in 40 to 50 percent ethanol when a reducing agent is present. The alpha-, beta-, and delta-zeins require a solution of at least 40 percent ethanol for solubility, but the gamma-zeins (27, 16, and 12 kD) are soluble in aqueous solution if a reducing agent is present.

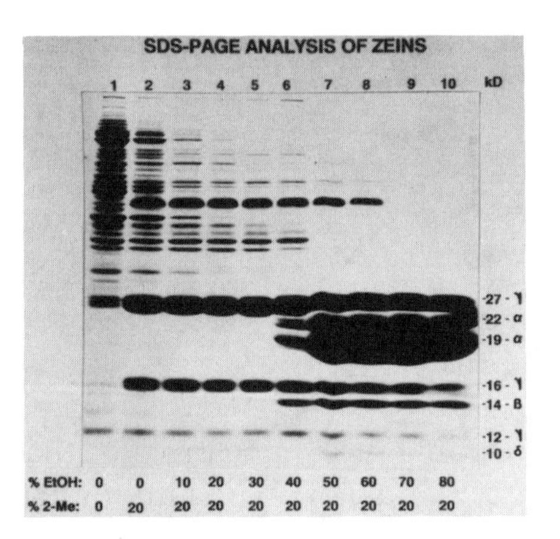

Figure 1: SDS-polyacrylamide gel electrophoresis analysis of proteins extracted from 30-day-old endosperm tissue from the maize inbred line W64A. 100-mg samples of frozen endosperm tissue were individually extracted with two volumes of the following solvents: lane 1, H_2O; lane 2, 20% 2-mercaptoethanol; lane 3, 20% 2-mercaptoethanol plus 10% ethanol, lanes 4–10, same as lane 3, but with stepwise 10% increases in ethanol concentration. Aliquots of protein samples were boiled in sample buffer (62 mM Tris-HC1, pH 6.8, 1% SDS, 5% 2-mercaptoethanol,and 10% glycerol) and analyzed by SDS-polyacrylamide gel electrophoresis on a 12.5% on a 12.5% gel (from Larkins *et al.,* 1989. Reproduced with permission.)

We purified the alpha-, beta-, and gamma-zeins and used them to generate antibodies in rabbits. Each of these zeins differs sufficiently in structure so that non-crossreacting antisera can be produced. Subsequently, the antibodies were used to localize the corresponding proteins in protein bodies as well as to study their distribution in different parts of the maize endosperm (Lending *et al.*, 1988; Lending and Larkins, 1989). These immunocytochemical studies showed that the different classes of zeins are not distributed uniformly in protein bodies. Figure 2 shows an analysis of protein bodies from endosperm cells at 18 days after pollination. For this study we compared protein bodies in the subaleurone layer (Fig. 2A-D), the first starchy endosperm layer (Fig. 2E-H), and the fifth starchy endosperm layer (Fig. 2I-L). In sections stained with uranyl acetate and lead citrate, protein bodies showed light- and dark-staining regions. There are variations in the proportion and distribution of these regions in protein bodies from different parts of the endosperm (Fig. 2A,E,I), but generally the light-staining material is in the center and the dark-staining material is at the periphery of the protein body (Lending *et al.*, 1988; Lending and Larkins, 1989). The distribution of zeins within protein bodies was revealed by immunogold staining. In sections reacted with alpha-zein antibodies, the colloidal gold label was uniformly distributed over the central light-staining region (Fig. 2B,F,J). With the beta-zein antiserum, the labelling was preferentially located over the dark-staining region, near the interface between the light- and dark-staining material (Fig. 2C,G,K). There was a small amount of labelling within the central region of the protein bodies, which was associated with patches of dark-staining material. The immunostaining with the gamma-zein antiserum was similar to that of beta-zein antiserum, although the frequency of the gold particles was reduced (Fig. 2D,H,L). We also found that the distribution of the different zein types is determined by the timing of zein deposition during endosperm development. Protein bodies in the youngest endosperm cells, those just beneath the aleurone layer, are small and contain mostly beta- and gamma-zeins (Fig. 2A-D), while those in the central, more mature endosperm cells are filled with alpha-zeins (Fig. 2I–L). Early in development, the alpha-zeins form small accretions within a matrix of beta- and gamma-zeins. As more alpha-zeins accumulate, these aggregates fuse, and the beta- and gamma-zeins become progressively displaced to the periphery of the protein body. These sulphur-rich zeins may thus form a framework upon which the alpha- and delta-zeins accumulate. As a result of this pattern of synthesis, the cells at the surface of the endosperm, which is more vitreous in the normal genotypes, have a higher concentration of the sulphur-rich beta- and gamma-zeins (Lending and Larkins, 1989).

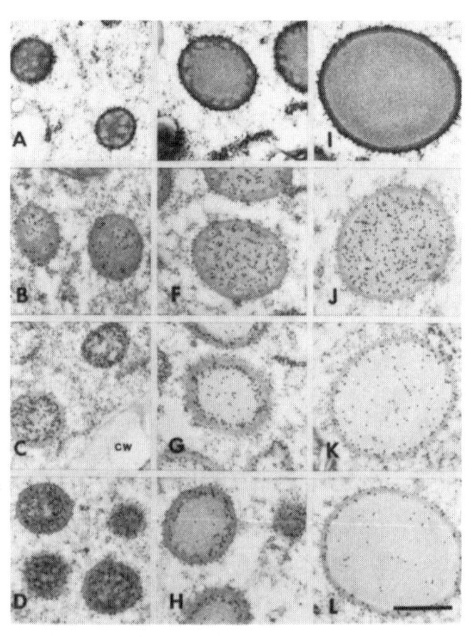

Figure 2. **Electron micrographs of protein bodies in the endosperm cells of the maize inbred line W64A at 18 days after pollination. Micrographs are representative of the morphology and size within the cell layer indicated. Maize seeds were fixed in 1% (v/v) gluteraldehyde and 4% (w/v) freshly prepared formaldehyde in phosphate buffer (50 mM KH_2PO_4/Na_2HPO_4, pH 7.0) and prepared for immunocytochemistry as described by Lending and Larkins, 1989. For immunostaining, sections were incubated in the indicated primary antibody and then in 10-nm goat anit-rabbit IgG conjugated to colloidal gold. Immunostained sections were post-stained with uranyl acetate for 5 min and with lead citrate for 2 min. Bar = 0.5 μM. Ato D are protein bodies from the subaleurone layer, E to H are protein bodies from the first starchy endosperm layer, and I to L are protein bodies from the fifth starchy endosperm layer. A, E, and I are sections post-stained with uranyl acetate and lead citrate. Note the light-staining central region surrounded by a peripheral region that stains darkly. B, F and J are sections immunostained for alpha-zein. The colloidal gold label is uniformly distributed over the central, light-staining region. C, G, and K are sections immunostained for beta-zein. The labelling is preferentially distributed over the dark-staining region, near the interface with the light-staining material. Sparse labelling is observed within the central region of the protein bodies; this is often associated with patches of dark-staining material. D, H and L are sections immunostained for gamma-zein. Immunostaining is similar to that observed for beta-zein although fewer gold particles are seen (from Lending and Larkins, 1989. Reproduced with permission.)**

Engineering a Zein Protein with High Lysine Content

Zeins perform no known enzymatic function and thus appear to be ideal candidates for amino acid modification by genetic engineering. However, as discussed above, they may have important roles in protein body formation, and ultimately in endosperm structure. The absence of lysine in these proteins may reflect the fact that charged amino acids would adversely affect protein aggregation and thus formation of normal protein bodies (Wallace *et al.*, 1988). A major concern in engineering these proteins is the possibility that insertion of charged amino acids may affect formation of protein bodies and ultimately protein stability in the endosperm of transformed plants.

Protein bodies are dense structures, 1.22 - 1.25 mg/cm3 (Hurkman *et al.*, 1981), which reflect the fact that zeins pack very tightly. A structural model for alpha-zeins which explains this was proposed by Argos *et al.* (1982). The alpha-zeins have a highly conserved primary amino acid sequence. After the first 40 amino acids, the polypeptides contain a series of repeated peptides of about 20 amino acids (Fig. 3). These repeated sequences are thought to exist as antiparallel alpha helices separated by glutamine-rich turn regions. Hydrogen bonding between glutamine residues at the ends of the repeated peptides and polar amino acids on the surface of the repeats, as well as nonpolar interactions, are proposed to be responsible for packing these proteins into protein bodies (Argos *et al.*, 1982).

Modification of DNA sequences encoding alpha-zeins. On the basis of this structural model for alpha-zeins, we used site-directed mutagenesis to modify the coding sequence of a 19 kD alpha-zein and substituted lysine residues at a number of positions. Single lysine replacements were made in the amino-terminal sequence preceding the repeated peptides, at the ends of the repeated peptides, and within the repeated peptides (Fig. 3, constructs 1 to 5). Following these changes, we recombined sequences so that double lysine substitutions were created (Fig. 3, combination constructs 1 to 5). Using newly created restriction enzyme sites, we inserted short oligonucleotides coding for several lysine and tryptophan residues (Fig. 3, constructs a, b, and c). Lastly, to significantly perturb the structure of the molecule, we inserted a sequence encoding a 17 kD fragment of the SV40 VP2 protein (Fig. 3, construct*).

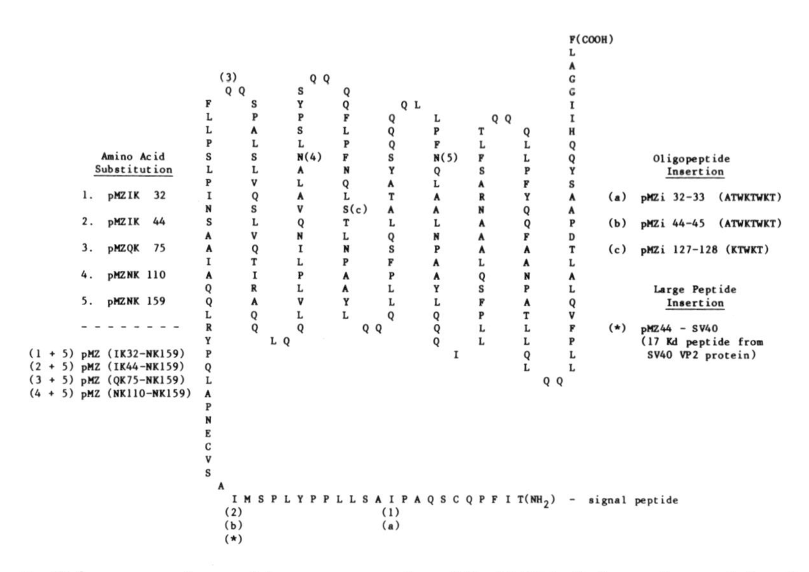

Figure 3: Primary amino acid sequence of an M_r 19,000 alpha-zein protein showing position of lysine substitutions, oligopeptide insertions, and a large peptide insertion. The plasmid constructs listed at the sides of the figure indicate the site of the substitution or insertion. Synthesis of the mutagenized sequence and *in vitro* transcription of the mRNA were previously described (Wallace *et al.*, 1988; Reproduced with permission).

Since a method to transform maize cells with engineered genes and obtain regenerated plants was not available, we used a heterologous system to analyze the synthesis and processing of the modified zein proteins. Injection of zein messenger RNA into *Xenopus* oocytes resulted in the synthesis and processing of zein proteins into membrane-enclosed structures with the physical characteristics of maize protein bodies (Larkins *et al.*, 1989; Hurkman *et al.*, 1981; Wallace *et al.*, 1988). With this system it was possible to study the consequences of lysine substitutions and other modifications to zein proteins on protein body formation (Wallace *et al.*, 1988). Synthetic mRNAs for the modified sequences were transcribed *in vitro* and injected into oocytes. To detect the zein proteins synthesized in oocytes, ^{3}H-leucine was injected into the oocytes several hours after mRNA injection. Figure 4A shows an analysis of zein proteins synthesized in oocytes injected with total zein mRNA. A mixture of radioactive oocyte proteins remained near the top of the gradient, but the majority of the zein proteins sedimented in a broad zone beneath the band of yolk platelets. Protein bodies from maize endosperm were also found to sediment in this region. When oocytes were injected with the unmodified alpha-zein mRNA, the distribution of radioactive proteins was similar to that with total zein mRNA, although the radioactive protein sedimenting deep in the gradient was in less dense vesicles than when total native mRNA was injected (Fig. 4B). However, the density of these vesicles was significantly greater than when the protein was added exogenously to the oocyte homogenate (Fig. 4C).

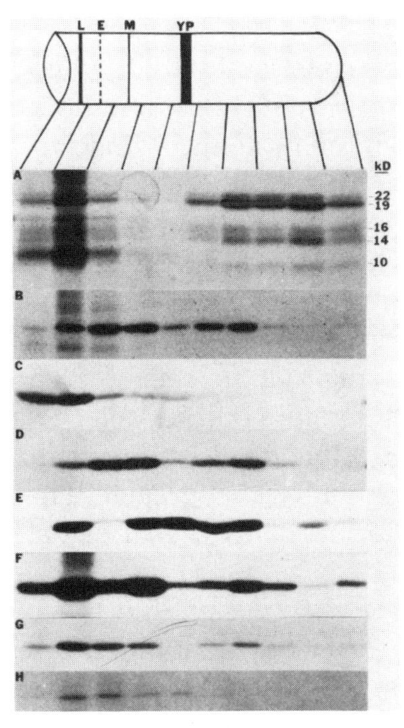

Figure 4. Density gradient separation of protein bodies from *Xenopus* oocytes injected with zein mRNAs. Oocytes were injected with synthetic mRNA transcripts and ^{3}H-leucine as previously described (Wallace *et al.,* 1988). Following homogenization, extracts were separated by centrifugation in 4-ml gradients of 10–50% metrizamide. Gradients were fractionated manually, and the distribution of radioactive alcohol-soluble proteins determined by SDS-polyacryamide gel electrophoresis and fluorography. Panels correspond to the following mRNAs (see Fig. 3): A) native total zein mRNA; B) wild-type M_r 19,000 alpha-zein; C) exogenous ^{3}H-zein added to oocyte homogenate; D) MZIK 32; E) MZ (NK110-NK159); F) MZi 32–33; G) MZi 127–128; H) MZ44–SV40. The M_r of the zeins is shown in panel A; other zeins migrated at M_r 19,000 except the MZ44-SV40 protein (H), which had an M_r of 35,000 (Wallace *et al.,* 1988; Reproduced with permission).

The synthesis of alpha-zein protein was not affected by the addition of single lysine residues (Fig. 4D), two lysine residues (Fig. 4E), or short oligopeptides containing several lysine and tryptophan residues (Fig. 4F, G). Furthermore, the density of vesicles containing the modified zeins was similar to the unmodified alpha-zein (Fig. 4B). However, in all cases in which the alpha-zein mRNA was injected alone, the vesicles were slightly less dense than when the total zein mRNAs were injected (cf. Fig. 4A and Fig. 4B to G). The addition of the 17 kD fragment of the SV40 VP2 protein interfered with the aggregation of the alpha zein-protein. Most of the labeled protein remained at the top of the gradient, and little or no aggregation occurred (Fig. 4H).

These results are consistent with the hypothesis that the association of zein polypeptides into a protein body results primarily from interactions between the proteins themselves. Aggregation of alpha-zeins can occur in the absence of the beta- and gamma-zeins, although the protein bodies that form are somewhat less dense than those formed with a mixture of zein proteins. In subsequent experiments we found that the protein bodies formed following injection of a mixture of synthetic mRNAs sedimented deeper in the gradient, suggesting that the beta- and gamma-zeins are required to form a fully dense protein body (Wallace *et al.*, 1990b). Our results indicate that the introduction of lysine residues into the alpha-zeins does not affect their ability to aggregate. However, more significant alterations may interfere with their normal interactions.

Synthesis of modified zeins in transgenic plants. We have also analyzed the synthesis of these modified zeins in transgenic tobacco and petunia plants. We found that the promoter sequences flanking these genes did not function efficiently in the transgenic *Solanaceous* species (Ueng *et al.*, 1988). Therefore, we used a promoter from a dicot storage protein to direct expression of the zein coding sequence. High levels of mRNA transcripts were produced from these gene constructs, but the proteins were barely detectable (Williamson *et al.*, 1988; Wallace *et al.*, 1990b; Ohtani *et al.*, 1990). The small amount of alpha-zeins accumulated in dicot seeds appears to result from instability of the proteins, since both the normal and modified zein polypeptides were degraded with a half-life of about 4 hours. In mature seeds, the proteins were observed as aggregates appressed against the cell wall (Wallace *et al.*, 1990b).

Searching for Mechanisms to Overexpress Modified Zein Proteins

An analysis of alpha-zein proteins by SDS-polyacrylamide gel electrophoresis reveals two major components of 19 and 22 kD.

However, isoelectric focusing resolves these into a large number of polypeptides, indicating substantial charge heterogeneity for this protein class (Fig. 5). The 19 and 22 kD alpha-zeins appear to be coded by a large multigene family (Rubenstein and Geraghty, 1985). Therefore, the transformation of maize plants with a modified gene containing an alpha-zein promoter would not be expected to impact on the nutritional quality of the grain protein. Before the modified genes can be genetically engineered in plants, mechanisms leading to high levels of expression must be identified. We found a potential solution to this when we characterized a group of endosperm modifier genes that appear to enhance transcription of certain zein genes.

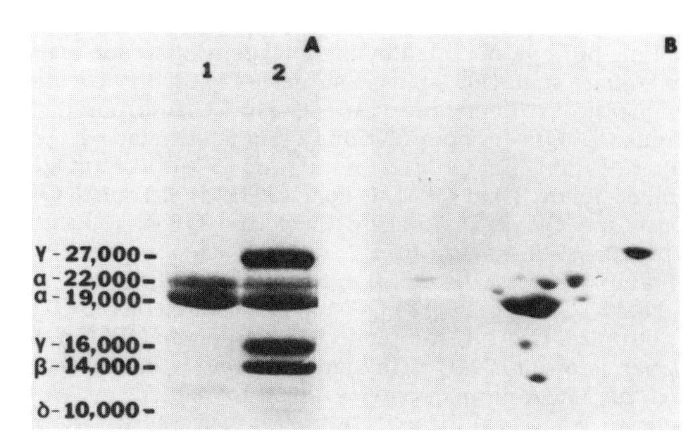

Figure 5. Analysis of maize proteins by one and two-dimensional gel electrophoresis. Panel A: one-dimensional separation by SDS-PAGE of zeins extracted from endosperm with 70% ethanol (lane 1) and with 70% ethanol plus 1% beta mercaptoethanol (lane 2). Panel B: two-dimensional separation of total zein extracted with 70% ethanol plus 1% beta mercaptoethanol. SDS-PAGE from top to bottom and isoelectric focusing from left to right. Apparent molecular weights are indicated on the left.

Quality protein maize (QPM) synthesizes large amounts of gamma-zein. Recently we characterized a group of "modified" hard endosperm *opaque-2* maize genotypes called quality protein maize (QPM). Seeds of these "modified" opaque genotypes contain two to four times more of the 27 kD gamma-zein as standard *opaque-2* mutants. This quantity of this protein is second only to the alpha-zein fraction in the endosperm of normal maize; it is the largest storage protein fraction in the endosperm of normal maize; it is the largest storage protein fraction in the endosperm of the "modified" *opaque-2* mutants (Fig. 6).

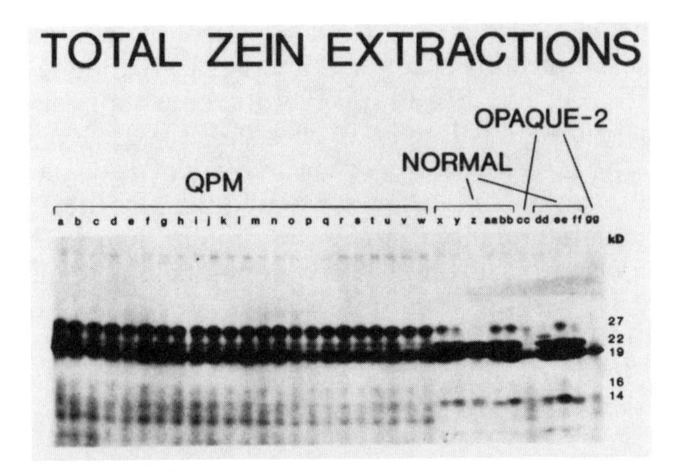

Figure 6. Total zein protein from several genotypes of maize. Equivalent extracts of 1 mg of flour, obtained by the total zein extraction method described by Wallace *et al.* (1990a), were subjected to SDS-PAGE and stained with coomassie brilliant blue. Lane a, extract from population 63-Blanco Dentado-1 QPM; b, population 64-Blanco Dentado-2 QPM; c, population 62-White Flint QPM; d, populaiton 65-Yellow Flint QPM; e, population 66-Yellow Dent QPM; f, Pool 25 QPM; g, Pool 26 QPM; h, Blanco Cristalino QPM; i, Amarillo Cristalino QPM; j, La Posta QPM; k, Obregon 7940; l, Poza Rica 7940; m, Guanacaste 7940; n, population 69-Templado Amarillo QPM; o, population 70-Templado Amarillo QPM; p, Pool 33 QPM; q, Pool 34 QPM; r, Amarillo del Bajio QPM; s, Amarillo Subtropical QPM; t, Templado Blaco Dentado QPM; u, Obregon 7941; w, San Jeronimo 7941; x, Tuxpeno nomral;; y, B73 normal; aa, W22 normal; bb, W64A normal; cc, W64A*o2*; dd, W64A*fl2;* ee, W64A*xx*; ff, Oh43 normal; gg, Poza Rica*o2* (Wallace *et al.*, 1990a; Reproduced with permission).

The "modified" mutants were developed at the International Maize and Wheat Improvement Center, CIMMIT/Mexico, by selecting for hard, vitreous kernels with a high protein content, while maintaining the recessive *opaque-2* mutation (Vasal *et al.*, 1980). Since gamma-zein gene transcription is not significantly affected by *opaque-2* (Kodrzycky *et al.*, 1989), this selection procedure appears to have brought together genes that increase gamma-zein expression. Although homozygous for the *opaque-2* gene, the QPM varieties are easily distinguished from the unmodified *opaque-2* materials by the phenotype of the kernels: kernels from QPM genotypes have translucent, vitreous endosperm, like normal maize kernels (Wallace *et al.*, 1990a; Lopes and Larkins, 1990). Like typical *opaque-2* genotypes, the QPM varieties have small amounts of alpha- and beta-zeins in comparison to normal maize (Fig. 6 and 7) (Wallace *et al.*, 1990a).

It remains to be established whether there is a correlation between kernel modification (vitreousness) and high gamma-zein content. A recent analysis of reciprocal crosses of QPM genotypes by unmodified *opaque-2* inbred lines demonstrated a relationship between vitreousness of the endosperm and content of gamma-zein. Similar correlations were observed when QPM genotypes were crossed into normal, *floury-2,* and *floury-2 opaque-2* double mutants, as well as in F2 progenies segregating for the modified genotype (Lopes and Larkins, 1990).

There appears to be a relationship between the increased synthesis of the gamma-zein protein and the activity of "modifier genes" accumulated by QPM selection. Therefore, the use of promoter sequences of the highly expressed gamma-zeins to regulate expression of zein genes encoding essential aminio acids may provide a method for generating large amounts of genetically engineered proteins in the future.

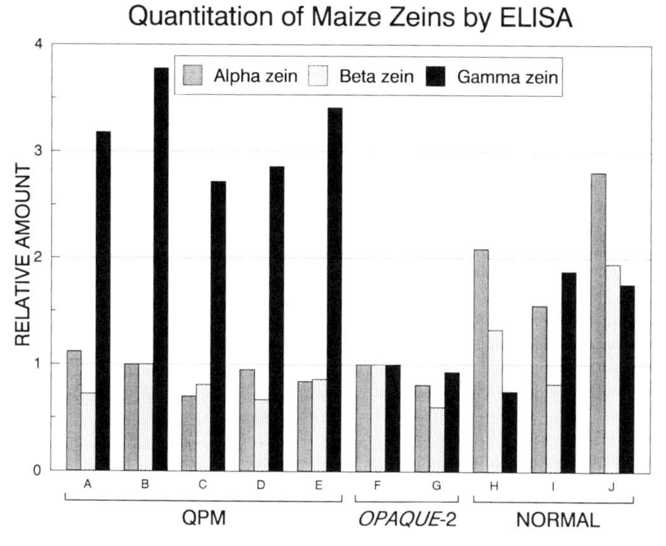

Figure 7. Quantitation of alpha-, beta-, and gamma-zeins. ELISA assays were performed as described by Wallace *et al.* (1990). For all assays, absorbance at 410 nm was graphed vs relative antigen concentration. Regression analysis was performed on the region which formed a straight line; the slope of that line is proportional to relative antigen concentration. Results were normalized to those from extract from meal of the *opaque-2* version of the inbred W64A (F), which was included in all assays. All results reported are averages of duplicate assays and did not differ from the mean by more than 10%. A, population 64-Blanco Dentado-2 QPM; B, population 65-Yellow Flint QPM; C, population 66-Yellow Dent QPM; D, Pool 25 QPM; E, Templado Blanco Dentado QPM; F, W64A*o2*; G, Poza Rica *o2*; H, B73 normal; I, W22 normal; J; W64A normal (Wallace *et al.,* 1990a; Reproduced with permission).

Mechanisms affecting gamma-zein gene expression in QPM. It is unclear whether the increased synthesis of the gamma-zein in QPM endosperms is due to gene amplification, cis- or trans-acting factors that increase gene transcription, or a combination of these. Except for few differences, the restriction maps we have determined for the gamma-zein genes in the "modified" *opaque-2* mutants coincide with those described for "unmodified" genotypes (Prat *et al.*, 1985; Das and Messing, 1987). Based on Southern blots and polymerase chain reaction (PCR) analysis, there are two similar genes (called A and B) in all highly modified genotypes we examined. Many of the normal inbred lines we have examined, including W64A, Oh43, Pa91, Mo17 and B73, contain only the A gene. We have cloned and sequenced the A gene from genomic libraries of a normal inbred, W64A, and a modified *opaque-2* mutant, Blanco Dentado. The coding and flaking sequences of these genes are highly homologous. However, one interesting difference between the two A genes is a very AT-rich sequence between -800 and -700, which is longer in Blanco Dentado QPM than W64A (Larkins *et al.*, in preparation). It is not clear if the duplication of the gene and the increase in size of this AT-rich sequence in the A gene plays a role in increasing the amount of protein synthesized in the endosperm.

Summary and Conclusions

Zeins, the storage proteins of maize, play a major role in determining both the nutritional quality and physical properties of the grain. Since zeins contain no lysine and very little tryptophan, their large amounts in the endosperm lower the concentration of these essential amino acids in the seed. We found that zeins are deposited in an organized way in the protein bodies. Analysis of several maize mutants with defective storage protein synthesis suggest zeins play an important role in protein body assembly and that they affect the physical properties of the grain. We found that it is possible to modify an alpha-zein gene, so that the protein it encodes contains lysine. These lysine-containing proteins assemble into protein bodies in a manner similar to native zeins. However, the expression of one modified gene may be sufficient for improving seed protein quality, since it will be diluted by the expression of the other members of the alpha-zein multigene family. Before the modified genes can be effectively genetically engineered into plants, mechanisms must be identified that will lead to high levels of gene expression and protein synthesis. A potential solution may be through the use of modified *opaque-2* mutants called quality protein maize (QPM). As is typical of *opaque-2* mutants, QPM seeds contain lower amounts of alpha-zeins. But unlike typical *opaque-2* mutants, QPM seeds contain very high concentrations of the 27 kD gamma-zein. The action of the "modifier" genes causes the gamma-zein to account for as much as 35 percent of the total storage protein of the endosperm of the modified *opaque-2* mutants. This protein is encoded by only one or two genes; the feasibility of using their regulatory sequences to increase the expression of modified zein genes encoding essential amino acids is a potential mechanism for generating large amounts of genetically engineered proteins in the future.

References

Argos, P., Pedersen, K., Marks, M.D. and Larkins, B.A. 1982. A structural model for maize zein proteins. J. Biol. Chem. 257: 9984-9990.

Das, O.P., and Messing, J. 1987. Allelic variation and differential expression at the 27 - kilodalton zein locus in maize. Mol. Cell. Biol. 7: 4490-4489.

Esen, A. 1986. Separation of alcohol-soluble proteins (zeins) from maize into three fractions by differential solubility. Plant Physiol. 80: 623-627.

Hurkman, W.J., Smith, L.D., Richter, J. and Larkins, B.A. 1981. Subcellular compartmentalization of maize storage proteins in *Xenopus* oocytes. J. Cell Biol. 89: 292-299.

Kodrzycky, R., Boston, R. and Larkins, B.A. 1989. The *opaque-2* mutation of maize differentially reduces zein gene transcription. The Plant Cell 1: 105-114.

Larkins, B.A., Lending, C.R., Wallace, J.C., Galili, G., Kawata, E.E., Geetha, K.B., Kriz, A.L., Martin, D.N. and Bracker, C.E. 1989. Zein gene expression during maize endosperm development. In "The Molecular Basis of Plant Development," Goldbert R.B. (Ed), Alan R. Liss, Inc., pp. 109–120.

Lending, C.R. and Larkins, B.A. 1989. Changes in the zein composition of protein bodies during maize endosperm development. The Plant Cell 1: 1011–1023.

Lending, C.R., Kriz, A.L., Bracker, C.E. and Larkins, B.A. 1988. Structure of maize protein bodies and immunocytochemical localization of zeins. Protoplasma 143: 51–61.

Lopes, M.A., and Larkins, B.A. 1990. Relationships between gamma-zein content and endosperm modification in quality protein maize (QPM). (Submitted for publication.)

Misra, P.S., Mertz, R.T. and Glover, D.V. 1975. Studies on corn proteins. VII. Developmental changes in endosperm proteins of high-lysine mutants. Cereal Chem. 52: 734–739.

Murphy, J.J. and Dalby, A. 1971. Changes in the protein fractions of developing normal and *opaque-2* maize endosperm. Cereal Chem. 48: 336–349.

Ohtani, T., Galili, G., Wallace, J.C. and Larkins, B.A. 1990. Normal and lysine-containing zeins are unstable in transgenic tobacco seeds. (Submitted for publication.)

Osborne, T.B. and Mendel, L.B. 1914. Nutritive properties of proteins of the maize kernel. J. Biol. Chem. 18: 1–6.

Prat, S., Cortadas, J., Puigdomenech, P. and Palau, J. 1985. Nucleic acid (cDNA) and amino acid sequences of the maize endosperm protein glutelin-2. Nucl. Acids Res. 13: 1493–1504.

Rubenstein, I. and Geraghty, D.E. 1985. The genetic organization of zeins. In, "Advances in Cereal Science and Technology," Pomeranz, Y. (Ed.), Vol. VIII, pp. 298–315.

Wallace, J.C. Gallili, G. Kawata, E.E., Cueller, R.E., Shotwell, M.A. and Larkins, B.A. 1988. Aggregation of lysine-containing zeins in protein bodies in *Xenopus* oocytes. Science 240: 662-664.

Wallace, J.C., Lopes, M.A., Paive, E. and Larkins, B.A. 1990a. New methods for extraction and quantitation of zeins reveal a high content of gamma-zein in modified *opaque-2* maize. Plant Physiol. 92: 191-196.

Wallace, J.D., Ohtani, T., Lending, C.R., Lopes, M., Williamson, J.D., Shaw, K.L., Gelvin, S.B. and Larkins, B.A. 1990b. Factors affecting physical and structural properties of maize protein bodies. In, "Plant Gene Transfer, UCLA Symposium on Molecular and Cellular Biology, New Series," Lamb, C. and Beachy, R.N. (Eds.), Alan R. Liss, Inc., Vol. 181(4): 467-474.

Williamson, J., Galili, G. Larkins, B.A. and Gelvin, S.B. 1988. The synthesis of a 19 kilodalton zein protein in transgenic *Petunia* plants. Plant Physiol. 88: 1002-1007.

Manipulation of Potato Protein: Biotechnological Approaches and Lessons from Evolution

William D. Park
Department of Biochemistry and Biophysics
Texas A&M University
College Station, TX 77843

Patatin accounts for 30–40 percent of the protein in potato tubers. Unlike most plant storage proteins, patatin is extremely well balanced nutritionally with an amino acid composition similar to casein. By using the promoter from the patatin gene and a high efficiency transformation system that we have developed, we can introduce any foreign gene into potato plants and have it specifically expressed in tubers. To determine how to best increase the amount of protein in tubers, we are investigating the regulation evolution of major "tuber-specific" genes such as patatin. We have found that all of the trans-acting regulatory factors necessary for high levels of patatin gene expression are present in both tuberizing and nontuberizing plants. While both the patatin structural gene and the cis-acting elements necessary for tuber-specifc expression are also present in nontuberizing species, they are arranged differently and in some cases are even on different chromosomes.

Introduction

Most of the work on manipulation of the nutritional quality of plants has focused on cereal and legume seed proteins. However, much of the world's food comes directly from root and tuber crops such as potato, sweet potato and casava. Not only are root and tuber crops important dietary sources of carbohydrate, they can also be significant dietary sources of nutritionally well balanced protein. In the case of potato, for example, protein accounts for only 1–2 percent of the fresh weight, but the amount of protein produced per acre is second only to soybean (Johnson and Lay, 1974). Also, unlike most of the major cereals and legumes, the protein of potato is of excellent nutritional quality. In fact, patatin, which accounts for 30–40 percent of the total tuber protein in most potato varieties, has almost exactly the same amino acid composition as the milk protein, casein (Liedl *et al.*, 1987).

Potatoes are ideally suited for the use of biotechnology since they can be easily manipulated by tissue culture and are amenable to genetic transformation using Ti plasmid vectors. Not only do potatoes provide a very favorable system to investigate how to best use genetic engineering to enhance the nutritional quality of plant products, they also provide an opportunity to examine how nature has gone about "genetic engineering" in the evolution of tubers from components found in nontuberizing species.

Somatic Storage Tissue Differentiation in Potato

A potato tuber is simply an underground stem which has differentiated into a storage organ by expanding radially and accumulating large amounts of starch and a characteristic set of tuber proteins (Artschager, 1924). This differentiation is influenced by a variety of environmental and hormonal factors (reviewed in Cutter, 1978; Ewing, 1990; Krauss, 1985; and Vreugdenhil and Struik, 1989). High levels of photosynthate and a high carbon-to-nitrogen ratio favor tuberization. High levels of available nitrogen can completely block tuberization in hydroponically grown potato plants under conditions that were otherwise favorable for tuberization. When nitrogen is withdrawn, tubers form immediately, but frequently revert to stolon-like growth if the plants are again exposed to high levels of nitrogen.

Photoperiod is another factor important in controlling the process of tuberization. Whereas exposure to a long photoperiod favors vegetative growth, exposure to a short photoperiod favors the development of tubers. In a classic set of experiments (Kumar and Waring, 1973), it was shown that grafting leaves or stems from a long photoperiod induced tuberization. A great deal of work has been done to identify the nature of this graft-transmissible "tuberization stimulus." In the late 1960's it was demonstrated that stolon tips could be induced to form tubers *in vitro* by exposure to cytokinin and high concentrations of sucrose (Palmer and Smith, 1970). Subsequently, a number of compounds, including ABA, coumarin, auxins, and ethylene and jasmonic acid derivatives, have been shown to stimulate tuberization under various conditions. While it has not been proven that any of these compounds are the actual "tuberization stimulus," it has become clear that the control of tuberization includes both positive and negative factors and that gibberellic acid appears to play an important role as an inhibitor of tuberization.

Use of Patatin as a Biochemical Marker for Tuberization

To study tuberization at the molecular level, one must be able to recognize a tuber biochemically. The marker we have used in most of our work is patatin, a family of glycoproteins that accounts for approximately 40 percent of the soluble protein in potato tubers (Racusen and Foote, 1980; Racusen, 1983; Paiva *et al.*, 1983). Typically, commerical potato cultivars contain 12–15 patatin species (Racusen and Foote, 1980; Park *et al.*, 1983). Although they show extensive charge heterogeneity and differ between cultivars, the apparent molecular weights of all of the forms of patatin are approximately 40,000. All of the patatin proteins in tubers appeared to be immunologically identical by crossed immunoelectrophoresis and have homologous amino-terminal sequences (Park *et al.*, 1983). This conclusion is also supported by work at the nucleic acid level since the coding regions of all of the patatin cDNA and genomic clones that have been isolated are extremely homologous (Mignery *et al.*, 1984; Bevan *et al.*, 1986; Nakamura *et al.*, 1986; Pikaard *et al.*, 1986; Rosahl *et al.*, 1986a and 1986b; Twell and Ooms, 1988; Mignery *et al.*, 1988).

Unlike most other storage proteins, patatin has enzymatic activity (Racusen, 1984; Rosahl *et al.*, 1987; Andrews *et al.*, 1988). Using a baculovirus sytem to express the patatin cDNA, pGM01, Andrews *et al.* showed that patatin has lipid acyl hydrolase and transferase activity. It is active with phospholipids, monoacylglycerols, and *p*-nitrophenol fatty acid esters, moderately active with galactoplipids, but apparently inactive with di- and tri-acyl glycerols. Patatin's lipid acyl activity may be involved in the tuber's response to wounding. Patatin is normally sequestered in the vacuole (Sommewald *et al.*, 1989) and thus would be inactive. However, upon mechanical or pathogen damage, patatin would be released and could remove fatty acids from damaged membranes for use in production of suberin and cytotoxic waxes. It may also be involved in phytoalexin production since the twenty-carbon fatty acid, arachadonic acid, is a potent elicitor of phytoalexin production in potato tubers (Bostock *et al.*, 1981).

Patatin normally is not detectable (at least 100-fold less abundant) in leaves or stems from either tuberizing or nontuberizing plants (Paiva *et al.*, 1983) and yet it accounts for approximately 40 percent of the protein in mature tubers. Patatin is not detectable in stolon tips from non-induced plants, but increases vary rapidly as tubers develop, accounting for 5–7 percent of the soluble protein in a one gram tuber. As would be expected, patatin mRNA is very abundant in developing tubers, but is not normally detected in either stems or leaves in either tuberizing or nontuberizing plants (Lee *et al.*, 1983; Rosahl *et al.*, 1986a).

While tubers normally form on underground stolons, tubers can also be formed from axillary buds of stem cuttings exposed to the proper photoperiod or from stolon tips or axillary buds cultured *in vtiro* on medium containing cytokinin and 8–10 percent sucrose. As expected, these tubers contain large amounts of patatin and the other major tuber proteins (Paiva *et al.*, 1982; Bourque *et al.*, 1987). When axillary buds are cultured in the light in basal medium containing 1–2 percent sucrose, they develop as shoots. The leaves and stems of these shoots do not normally contain significant amounts of patatin (Bourque *et al.*,1987).

Molecular Characterization of the Patatin Multigene Family in *S. Tuberosum*

Patatin is encoded by 10–15 genes per monohaploid genome, giving 40–60 copies in a typical tetraploid potato cultivar. All of the patatin genes that have been isolated (Bevan *et al.*, 1986; Pikaard *et al.*,1986; Rosahl *et al.*, 1986b; Twell and Ooms, 1988; Mignery *et al.*, 1988) have highly homologous coding regions that contain six introns. These genes can be divided into two classes by the presence (Class-II) or absence (Class-I) of a 22 bp insertion in their 5′ flanking regions (Pikaard *et al.*, 1987; Mignery *et al.*, 1988) and by differences in their 5′ flanking regions. The flanking regions of the two classes are 90 percent homologous to position -87, relative to the transcription start site. This conserved region contains CAAT and TATA homologies as well as a region homologous to a core enhancer sequence. Beyond -87 the homology between the two classes essentially ends. There are additional large regions of homology within a class, but differences exist which allow the patatin genes to be further divided into a number of discrete subclasses (Mignery *et al.*, 1988).

The Class-I patatin genes encode 98–99 percent of the patatin mRNA in tubers as shown by S1 nuclease protection and primer extension experiments (Mignery *et al.*, 1988). The Class-II patatin genes encode only 1–2 percent of the patatin mRNA in tubers and, unlike the Class-1 genes, are also normally expressed at low levels in roots (Pikaard *et al.*, 1987; Mignery *et al.*, 1988). Perhaps due to difference in post translational processing, the patatin protein in roots is immunologically and electrophoretically distinct from that in tubers. Although they are equal to or greater in number than Class-1, many Class-II genes appear to be pseudogenes (Pikaard *et al.*, 1986; Twell and Ooms, 1988).

All copies of the patatin multigene family in potato appear to be located at the end of chromosome eight on a single 1.4 million base pair DNA fragment based on RFLP analysis and pulsed field gel electrophoresis (Ganal *et al.*, 1991). While the detailed structure of this locus is not known, pulsed field gel electrophoresis indicates physical separation of at least a considerable number of the Class-I and Class-II genes.

Expresssion of Patatin Promoters in Transgenic Potato Plants

To determine which parts of the patatin genes are responsible for their tissue specific pattern of regulation, we attached the 5′ flanking sequences from both Class-I and Class-II patatin genes to the GUS reporter gene in the binary Ti-plasmid vector pBl101.1 using the Dra-1 site at position +10 of both classes (Wenzler *et al.*, 1989a). These constructs were transferred into *Agrobacterium* strain LBA4404 and were used to produce transgenic plants using the leaf disc transformation method and the potato cultivar FL1607.

Under normal conditions, the GUS reporter gene was not expressed at significant levels in roots, stems or leaves of either tuberizing or nontuberizing plants containing Class-I patatin/GUs fusions. In contrast, extracts of tubers from these plants had high leves of GUS activity (approximately 3000-fold higher than those seen in stolon tips before tuberization.)

Tuber slices stained with the histochemical substrate 5-bromo-4-chloro-3-indolyl-b-D-glucuronic acid (X-gluc) showed a relatively uniform pattern of GUS expression except in the outermost 5–6 cell layers which did not have detectable activity. Similar results have been reported for other Class-I patatin/GUS constructs in transgenic potato plants (Rosha-Sosa *et al.*, 1989; Jefferson *et al.*, 1990).

Constructs containing the 5′ flanking region of the Class-II patatin clone PS22 were expressed at much lower levels in tubers (1–2 percent of that seen for Class-I). Unlike the Class-I constructs, they also were normally expressed in the rhizodermis of roots (Koster-Topfer, 1989).

Although the regions of the Class-I and Class-II patatin promoters that are required for activity have not yet been completely defined, recent data have shown that sequences between -40 and -400 relative to the translation start site are sufficient to direct "tuber-specific" expression of a truncated 35S promoter in an orientation independent fashion (Jefferson *et al.*, 1990). The sequences required have been further delimited by the work of Liu *et al.*, (1990), who showed that sequences downstream of -228 were sufficient to direct "tuber-specific" expression. This region contains two notable repeats of 33 and 37 nucleotides which are also present in the region between -400 and -600. At least two pieces of evidence suggest that these repeats may be involved in directing "tuber-specific" expression. Deletion of the sequences from -228 to -195, which removes most of the 33 bp repeat, greatly reduces expression in tubers (Liu *et al.*, 1990). In contrast, addition of the sequences between -400 and -600, which contain the second copy of both repeats, markedly increases expression (Jefferson *et al.*, 1990). A conserved AT-rich region contained within the repeat binds nuclear proteins present in leaf and tuber extracts (Liu *et al.*, 1990). The DNA fragment between -183 and -143 that contains the binding site is not, however, able to enhance the expression of a truncated 35S CaMV promoter.

While the repeats may be necessary for "tuber-specific" expression, several additional elements also appear to be necessary to get high levels of expression. The trans-acting factors which interact with some of these elements may be cultivar specific since the expression of Class-I patatin gene fusions in transgenic potato plants varies with both the gene and the cultivar (Blundy *et al.*, 1991).

Introduction of Class-I Patatin Expression without the Morphology of Tuberization

Although Class-I patatin is normally expressed at significant levels only in tubers and in stolons attached to developing tubers, under certain conditions it can be induced to express at high levels in stems and leaves. When potato plants were placed under conditions of sink limitation, significant amounts of patatin where shown to accumulate in the stem and petioles (Paiva *et al.*, 1983). Using single node stem cuttings from which the axillary buds had been removed, the levels of patatin, starch and the other tuber proteins in the petiole and lower section of the rachis could reach those normally seen in mature underground tubers.

To determine what factors were responsible for the induction of patatin accumulation, internodal stem segments and leaf explants from transgenic plantlets containing the Class-I patatin/GUS chimeric gene were placed on tissue culture media supplemented with various sugars (Wenzler *et al.*, 1989a). Class-I patatin/GUS is normally not expressed in significant levels in leaves or stems of micropropagated plantlets, but could be induced to express at levels comparable to those seen in tubers by placing the stem or leaf explants on media containing 300-400 mM sucrose. The induction of patatin expression is not an osmotic effect since mannitol, or mannitol in the presence of low levels of sucrose, did not induce expression. However, induction is not completely sucrose-specific since glucose, fructose and maltose also gave lower levels of induction.

It is not yet known if the sequences responsible for the sucrose induction of Class-I patatin are exactly the same as those responsible for its normally "tuber-specific" pattern of expression. However, the level of sucrose in developing tubers appears to be sufficient to account for the expression of Class-I patatin genes (Jefferson *et al.*, 1990) and it has been shown that 228 bp of flanking sequence is sufficient for either "tuber-specific" expression or sucrose-inducible expression in leaves and stems (Liu *et al.*, 1990). Even though the Class-I and Class-II patatin genes are highly conserved to position –87, constructs containing 5′ flanking sequences from Class-II patatin genes are not sucrose inducible in either stem or leaf explants.

Leaf and stem explants placed on high concentrations of sucrose do not show noticeable tuber-like swelling or cell proliferation, but they do accumulate large amounts of starch as well as patatin and other tuber proteins. It has recently been shown that a number of key enzymes involved in carbohydrate metabolism, including sucrose synthase (Salanoubat and Belliard, 1989) and ADP-glucose pyrophosphorylase (Muller-Rober *et al.*, 1990), are also sucrose inducible in potato.

Modulation of the Sucrose Induction of Patatin Gene Expression by Factors Known to Influence Tuberization

The accumulation of starch in somatic tissues is consistent with the classical observation that high levels of photosynthate favor tuberization. To determine whether other factors known to influence tuberization also affect the sucrose induction of Class-I patatin/GUS constructs, we examined the effects of GA and available nitrogen (Park, 1990). While the maximum level of induction was not significantly altered, the amount of sucrose needed for induction of patatin constructs in leaf explants was altered by these treatments in a fashion consistent with their effect on the overall process of tuberization. While 80mM sucrose gave only a slight induction of Class-I patatin gene expression in media with normal Murashige and Skoog salts, 80mM sucrose gave almost maximal expression in explants cultured on media without nitrogen. In contrast, gibberellic acid increased the amount of sucrose required to give maximal levels of induction. None of the treatments had a significant effect on expression of a CaMV 35S/GUS construct in leaf or stem explants cultured on either high or low sucrose.

While some of the factors which regulate tuberization also modulate Class-I patatin expression in the expected fashion in the absence of morphogenesis, the regulation of starch and tuber protein accumulation appears to differ somewhat from that of the overall process of tuberization. Tuberization is strongly influenced by photoperiod. However, large amounts of patatin and starch can accumulate in the petioles of cuttings from which the axillary buds have been removed under either a long or a short photoperiod (Paiva *et al.*, 1983). Also, while cytokinins promote tuberization *in vitro* and increase when plants are placed under conditions favoring tuberization, cytokinin is not required for the sucrose induction of patatin in leaf and stem explants and does not reduce the amount of the sucrose required. Rather than being one part of the tuberization stimulus *per se*, cytokinins may be required for the cell division that normally occurs during tuberization. As Ewing (1990) has pointed out, the kinetics of cytokinin accumulation are not consistent with it playing a major role in the induction of tuberization. Also, application of cytokinin to stolon tips of intact plants causes them to develop as negatively geotrophic shoots rather than as tubers (Kumar and Waring, 1972).

Evolution of Tuberization

In contrast to the clear evoluntionary relationships which exist between the seed storage proteins of different species (Kreis *et al.*, 1985), no clear evolutionary relationship exists in tuber storage proteins or between plants which have the ability to form tubers. Dahlia (*Dahlia pinnata*) and Jerusalem artichoke (*Helianthus tuberosum*) form tubers, but are in different families than potato. While family Solanaceae contains over 3500 species, including a wide range of famililar plants such as tobacco, petunia,and tomato, only approximately 160 wild and 7 cultivated species within the section petota of the genus Solanum have the ability to form tubers (D'arcy, 1979; Hawkes, 1979; 1990). Most tuber-bearing potato species can be crossed with comparative ease and form highly fertile offspring. This has been taken to suggest that the tuber-bearing species are a relatively young group that has evolved recently (Hawkes, 1979; 1990).

Patatin Trans-acting Factors in Nontuberizing Species

If tuberization evolved only recently, one would expect that all of the cis- and trans-acting factors involved in tuberization also would be present in nontuberizing species. By understanding what roles these factors play in nontuberizing species and how they have been altered during the evolution of tuberization, we may be able to gain insight into how tuberization is regulated and also determine how to best go about modifying it for our own purposes.

To determine if the trans-acting factors necessary for the sucrose induction of patatin gene expression are present in nontuberizing species, we first looked at tobacco. While tobacco contains patatin genes, they are sufficiently divergent to not cross-hybridize to a patatin cDNA clone at the usual stringencies of hybridization (Rosahl *et al.*, 1987). However, tobacco still contains all of the trans-acting factors necessary for sucrose induction of Class-I patatin gene expression (Wenzler *et al.*, 1989b). Not only could a Class-I patatin/GUS chimeric gene from potato be induced to express efficiently by sucrose after being inserted into tobacco plants, the sucrose dose response curve was virtually identical to that seen in potato. The induction of the Class-I patatin construct occurred without morphogenesis or cell proliferation and was accompanied by the accumulation of large amounts of starch, just as in potato. Since similar results can also be obtained with transgenic tomato plants containing a potato Class-I/GUS construct, it appears that all of the trans-acting factors necessary for the sucrose induction of patatin are present in a wide variety of nontuberizing species. This suggests that the mechanism that controls patatin gene expression in potato tubers evolved from a widely distributed mechanism in which gene expression is regulated by the level of available photosynthate.

Patatin Genes in Tomato

Tomato plants contain endogenous Class-II patatin genes which are very similar to those in potato. As expected, these are expressed in roots. The situation with the cis-acting factors that comprise the Class-I patatin promoter, however, is more subtle. Tomato contains sequences homologous with the functionally important part of the Class-I patatin promoter and, as was discussed above, also contains all of the trans-acting factors necessary to express Class-I patatin promoter from potato. However, the endogenous patatin genes in tomato are not sucrose inducible. The reason for this has become apparent in recent experiments done in collaboration with Steve Tanksley of Cornell Univeristy. In tomato, all of the structural genes for patatin are on the end of chromosome eight, just as in potato (Ganal *et al.*, 1991). The 5′ flanking sequences of the Class-II patatin genes also map to this same locus. In tomato, however, we detected only one region homologous to Class-I specific patatin promoter sequences. This is located near the middle of chromosome three and is not linked to a patatin structural gene.

We have cloned the Class-I promoter homologous region from chromosome 3 of VFN-8 tomato. Other than three short insertions, it is greater than 90 percent homologous to the 33 and 37 bp repeats in the potato Class-I patatin promoter that appear to be involved in sucrose induction. It is also greater than 90 percent homologous to sequences adjacent to both the proximal and distal copies of the 33 and 37 bp direct repeats (Figure 1).

The Class-I promoter homology in tomato is adjacent to a series of five open reading frames which are separated by putative introns which follow the GT/AG rule. Transcripts which contain these open reading frames are detectable, but they are expressed in a wide range of tissues and do not appear to be sucrose inducible.

It is not yet known if the sequences homologous to Class-I patatin promoters are actually functionally important in expression of the

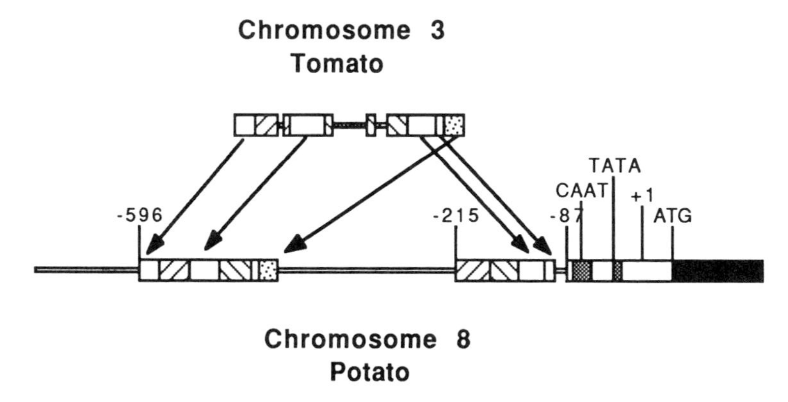

Figure 1: Sequences on chromosome 3 of VFN-8 tomato which are homologous to the Class-I patatin promoter from *S. tuberosum*. The 33 bp and 37 direct repeats are indicated by upward and downward diagonal striping, respectively. The TATA and CAAT homologies are shown by dark stippling and the coding region of exon 1 is solid black. As indicated by the arrows, the same sequences are found flanking the two repeats in both tomato and potato, but they are arranged slightly differently.

presumptive open reading frame on chromosome 3 of tomato. However, if they are, they may be used very differently. For example, the 33 nucleotide repeat, which is present twice in the 5′ flanking region of Class-I patatin at -170 and -570, contains the sequence TATATAATA. This sequence is also present on chromosome 3 of tomato, but is surrounded by two short insertions and is in approximatley the correct location to serve as the TATA box for the open reading frame. Work is currently in progress to directly test this region for enhancer activity, both with a truncated 35S promoter and a patatin promoter truncated at -87.

Patatin Genes in Tobacco

The chromosome 3 fragment from tomato contains sequences with striking homology to parts of the Class-I patatin promoter that are thought to be essential for "tuber-specific" and sucrose-inducible expression. It does not, however, contain regions homologous to the sequences between +1 and -87 which are common to both Class-I and Class-II or the Class-I specific sequences between the two sets of 33 and 37 bp repeats. Interestingly, sequences with strong homologies to both of these are found in tobacco genomic DNA.

Using PCR, we isolated a sequence from *Nicotiana tabacum* cv xanthi which is very similar to the proximal part of the Class-I patatin promoter. It contains both the 33 and 37 bp repeats and also contains their flanking sequences. However, these are arranged differently than in the Class-I promoter from *S. tuberosum.* Interestingly, the order of the "Class-I specific" proximal sequences in tobacco is the same as on chromosome 3 of tomato, but in tobacco there are no insertions and the sequences are adjacent to a patatin structural gene. The tobacco clone contains the sequences between +1 and -87, but unlike the Class-I genes in *S. tuberosum,* it also contains the 22 bp insertion at +14 and additional sequences at -87 that are normally found only in Class-II genes.

As shown in Figure 2, tobacco also contains sequences more than 95 percent homologous to the region between -227 and -596 of the potato Class-I patatin clone PS20. This contains the distal set of 33 and 37 bp repeats. Unlike potato, in tobacco the two sequences homologous to the Class-I promoter are not attached to each other. Either region can easily be isolated by PCR, but we have not been able to isolate any PCR products that contain both regions. Which chromosome(s) these homologous regions are located on in tobacco is not yet known.

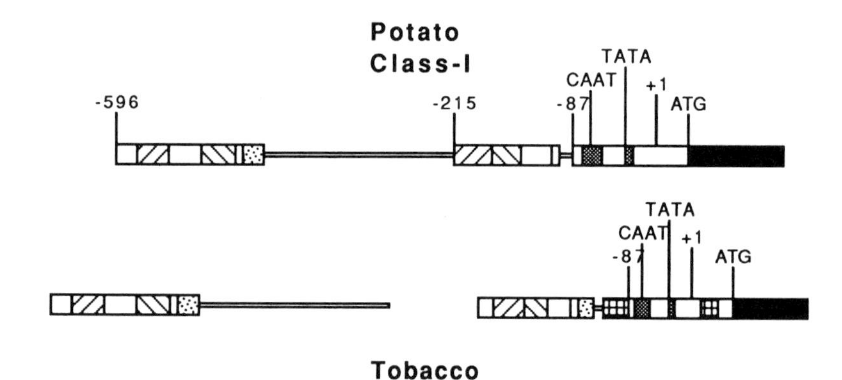

Figure 2. Sequences from *Nicotiana tabacum* cv xanthi which are homologous to the Class-I patatin promoter from *S. tuberosum*. Cross-hatched regions indicate the 22 bp insertion at +14 and additional sequences inserted at -87 which are normally found in Class-II patatin genes, but not in Class-I.

Even though tobacco contains sequences which are very homologous to the regions that are thought to be required for sucrose induction and these are attached to a patatin structural gene, the endogenous patatin genes are not sucrose inducible. The reason for this is not known, but it can not be due to a lack of the proper trans-acting factors since, as discussed above, Class-I promoters from potato show the expected pattern of sucrose induction when introduced into tobacco.

In addition to the patatin genes discussed thus far, there is also an "anther specific" form of patatin that has been described in pepper and potato (Vancanneyt *et al.*, 1989). The structure of the promoter responsible for expression of patatin in anthers has not been determined, but none of the Class-I or Class-II/GUS constructs tested thus far are expressed in anthers. We have found that sequences homologous to patatin mRNA are also present in tobacco flowers, but interestingly they do not appear to be present in flowers from VFN-8 tomato.

In addition to cis- and trans-acting elements which regulate Class-I patatin gene expression, other components of the systems that regulate tuberization also are present in nontuberizing species, but appear to be playing different roles. Even the graft transmissable "tuberization stimulus" may be present in species such as tobacco. It has been shown that scions of tobacco will induce tuberization when grafted onto potato rootstocks (Chailakyan *et al.*, 1981; Martin *et al.*, 1982). When scions of tobacco requiring short days for flowering were used, tuberization occurred only if the tobacco leaves received short days. When tobacco species requiring long days for flowering were used, however, tuberization occurred only under long days.

Patatin Genes in Wild Potato Species

To determine how the components present in nontuberizing species have been altered and adapted during the evolution of tuberization, we are examining patatin gene structure and function in wild potato species. We began by looking at the Mexican and Central American diploid species since these are the most evolutionarily divergent potato species and they are thought to have given rise to the more highly conserved Sourth American species (Hawkes, 1990).

The endogenous patatin genes of most of the Mexican diploid tuberizing species are sucrose inducible, but we have identified two species, *S. lesteri* and *S. polyadenium*, in which the endogenous patatin genes are not sucrose inducible. While we have not directly demonstrated that these species contain all of the necessary trans-acting factors for sucrose induction of patatin, it should be noted that both of these species form tubers and that their endogenous sucrose synthase genes are sucrose inducible, just as they are in *S. tuberosum*.

To examine patatin gene structure in the Mexican diploid species, genomic DNA was used as a template in PCR reactions with primers at -596 (just upstream of the distal repeat) and at +140 in the first exon of the patatin coding region. With *S. tuberosum*, this set of primers gives two products which correspond to PS20 and to PS3/PS27 (the variants of Class-I that contain a 478 bp insertion and 26 bp deletion), as well as a couple of other products which have not yet been characterized. With all of the Mexican species, including *S. lesteri* and *S. polyadenium*, we obtained only one product. This product contained only one copy of the 33 and 37 bp repeats and was strikingly similar to that seen in tobacco. All of the PCR products contained the 22 bp insert characteristic of Class-II as well as 30 to 100 bp of additional Class-II sequence inserted at -87 (the point at which the Class-I and Class-II regions normally diverge). There were slight differences between species in the arrangment of the sequences flanking the 33 and 37 bp direct repeats, but none of these appears to be related to sucrose-inducibility.

Using primers at -596 and -227, we were also able to detect the distal part of the patatin promoter in all of the Mexican diploid species. However, as in tobacco, we were not able to detect any PCR products in which the two regions homologous to the Class-I promoter were connected in these species.

Using the PCR primers at -596 and -227 and +140, we did not detect any difference in the gross arrangement of "Class-I specific" patatin promoter elements between those species which have sucrose inducible patatin genes and those which do not. However, if we moved the 5′ primer only a few bases further upstream, a very different story emerged.

With primers at -618 and +140, we got slightly longer versions of the same products from all of the species which were sucrose inducible. However the non-sucrose inducible species *S. lesteri* and *S. polyadenium* did not give any products with these primers nor did tobacco or any of the nontuberizing species that we tested. This suggests that the sequences 5′ of -596 are different for sucrose inducible and non-sucrose inducible patatin genes and further suggests that a critical step in patatin gene evolution may have been the removal of sequences further upstream which prevents sucrose inducibility.

Summary

The Class-I patatin genes appear to have evolved from components with different patterns of expression which are present in nontuberizing species and which are presumably serving different functions. This evolution appears to have occurred in at least two, and probably several, steps which involved rearrangement of existing cis-elements to create a new pattern of expression and further changes to direct the accumulation of the patatin protein to very high levels. The idea of a new pattern of gene expression having occurred by "shuffling" existing promoter elements is in agreement with recent findings suggesting that there is a combinatorial code underlying development in which various combinations of promoter elements give different tissue and developmental stage-specific patterns of expression (Benfey *et al.*, 1990). While this is still highly conjectural, it can be directly tested by using gene transfer technology to determine what pattern of gene expression the Class-I promoter components from nontuberizing species support, both in their current context and when rearranged as they are in different tuberizing species.

While there is still a great deal to be learned, patatin provides a powerful system to examine the stepwise process by which storage protein genes are recruited and their regulation is modified so that they are expressed in large amounts within the appropriate physiological context. Not only is this of theoretical interest, it may help us determine how to best use the tools of both traditional plant breeding and biotechnology to further improve the nutritional quality of potato tubers. In particular, it will be very interesting to look at some of the very high protein varieties, such as *S. phureja*, to determine the molecular basis for their high protein content and to determine how this trait can best be transferred into commerical cultivars.

References

Andrews, D.L., Beames, B., Summers, M.D. and Park, W.D. 1988. Characterization of the lipid acyl hydrolase activity of the major potato (*Solanum tuberosum*) tuber protein, patatin, by cloning and abundant expression in a baculovirus vector. Biochem. J. 252: 199–206.

Artschwager, E. 1924. Studies on the potato tuber. J. Agric. Res. 27: 809–835.

Bevan, M., Barker, R., Goldsbrough, A., Jarvis, M., Kavanagh, T. and Iturriaga, G. 1986. The structure and transcription start site of a major tuber protein patatin. Nucleic Acid Res. 14: 5564–5566.

Benfry, P.M., Ren, L. and Chua, N.-H. 1990. Combinatorial and synergistic properties of CaMV 35S enhancer subdomains. EMBO J. 9: 1685–1696.

Bostock, R.M., Kluc, J.A. and Laine, R.A. 1981. Eicospentaenoic acid and arachidonic acids from Phytophthora infestans elicit fungitoxic sesquiterpenes in the potato. Science 212: 67–69.

Bourque, J.E., Miller, J.C. and Park, W.D. 1987. Use of an *in vitro* tuberization system to study tuber protein gene expression. *In vitro* Cellular & Develop. Biol. 23: 381-386.

Blundy, K., Blundy, M., Carter, D., Wilson, F., Park, W. and Burrell, M. 1991. The expression of Class-I patatin gene fusions in transgenic potato varies with both gene and cultivar. Plant Molec. Biol. (In press).

Chailakyan, M.K., Yanina L.I., Devedzhyan, A.G. and Lotova, G.N. 1981. Photoperiodism and tuber formation in grafting of tobacco on to potato. Dokl. Akad. Nauk. SSSR 257: 1276–1279.

Cutter, E.G. 1978. Structure and development of the potato plant. In: Harris, P.M. (ed.), "The Potato Crop: The Scientific Basis for Improvement." pp. 70–152. Chapman and Hall, London.

D'Arcy, W.G. 1979. The classification of the Solanaceae. In: Hawkes, J.G., Lester, R.N., Skelding, A.D., (eds.), "The Biology and Taxonomy of the Solanaceae." pp. 3–47. Academic Press, London.

Ewing, E. 1990. Induction of tuberization in potato. In: Vayda, M. and Park, W. (eds.), "Cellular and Molecular Biology of the Potato." pp. 25–42. CAB International, Wallingford, Oxon, UK.

Ganal, M.W., Bonierbale, M.W., Roeder, M.S., Park, W.D. and Tanksley, S.D. 1991. Genetic and physical mapping of the patatin genes in potato and tomato. Molec. Gen. Genet. (In press).

Hawkes, J.G. 1979. Evolution and polyploidy in potato species. In: Hawkes, J.G. Lester, R.N., Skelding, A.D. (eds.), "The Biology and Taxonomy of the Solanaceae." pp. 637–645. Academic Press, London.

Hawkes, J.G. 1990. Potato evolution. In: "The Potato: Evolution, Biodiversity, and Genetic Resources." pp. 48-61. Smithsonian Insititution Press, Washington.

Jefferson, R., Goldsbrough, A. and Bevan, M. 1990. Transcriptional regulation of a patatin-I gene in potato. Plant Molec. Biol. 14: 995–1006.

Johnson, V.A. and Lay C.L. 1974. Genetic Improvement of Plant Protein. Agric. and Food Chem. 22: 558–566.

Koster-Topfer, M., Frommer, W., Rosha-Sosa, M., Rosahl, S., Schell, J. and Wilmitzer, L. 1989. A Class-II patatin promoter is under developmental control in both transgenic potato and tobacco plants. Molec. Gen. Genet. 218: 390–396.

Krauss, A. 1985. Interaction of nitrogen nutrition, phytohormones, and tuberization. In: Li, P. (ed.), "Potato Physiology." pp 209–230. Academic Press, Orlando.

Kreis, M., Forde, B.G., Rahman, S., Miflin, B.J. and Shewry, P.R. 1985. Molecular evolution of the seed storage proteins of barley, rye, and wheat. J. Mol. Biol. 183: 499–502.

Kumar, D. and Waring, P.F. 1972. Factors controlling stolon development in the potato plant. New Phytol. 71 639–648.

Kumar, D. and Waring, P.F. 1973. Studies on tuberization in *Solanum andigena*: Evidence for the existence and movement of a specific tuberization stimulus. New Phytol. 72: 283–287.

Lee, L., Hannapel, D., Mignery, G., Shumway, J. and Park, W. 1983. Control of tuber protein synthesis in potato. UCLA Symposium on Molecular and Cellular Biology, New Series 12: 355-365.

Liedl, B.E., Kosier, T. and Desborough, S.L. 1987. HPLC isolation and nutritional value of a major tuber protein. Am. Potato J. 64: 545-558.

Liu, X.J., Prat, S., Wilmitzer, L. and Frommer, W.B. 1990. Cis regulatory elements directing tuber-specific and sucrose-inducible expression of a chimeric Class-I patatin promoter/GUS-gene fusion. Molec. Gen. Genet. 223: 401–406.

Martin, C., Vernay, R. and Paynot, N. 1982. Physiologie v g tale. Photop riodisme, tub risation, floraison et ph nolamides. C R Hebd Seances Acad Sci 295: 565–568.

Mignery, G.A., Pikaard, C.S., Hannapel, D.J. and Park, W.D. 1984. Isolation and sequence analysis of cDNAs for the major tuber protein, patatin. Nucleic Acids Res. 12: 7987–8000.

Mignery, G.A., Pikaard, C.S. and Park, W.D. 1988. Molecular characterization of the patatin multigene family of potato. Gene 62: 27–44.

Muller-Rober, B.T., Kossman, J. Hannah, L.C., Wilmitzer, L. and Sommewald, U. 1990. One of the two different ADP-glucose pyrophosphorylase genes from potato responds strongly to elevated levels of sucrose. Mol. Gen. Genet. 224: 136–146.

Nakamura, K., Hattori, T., and Asahi, T. 1986. Direct immunological identification of full-length cDNA clones for plant protein without gene fusion to *E. coli* protein. FEBS 198: 16–20.

Palmer, C.E. and Smith, D.E. 1970. Effect of kinetin on tuber formation on induced stolons of *Solanum tuberosum* cultured *in vitro.* Plant Cell Physiol. 11: 303-314.

Paiva, E., Lister, R.M. and Park, W.D. 1982. Comparison of the protein in axillary bud tubers and underground stolon tubers in potato. Am. Potato J. 59: 425–433.

Paiva, E.P., Lister, R.M., and Park, W.D. 1983. Induction and accumulation of the major potato tuber protein, patatin. Plant Physiol. 71: 161–168.

Park, W.D., Blackwood, C., Mignery, G.A., Hermondson, M.A. and Lister, R.M. 1983. Analysis of the heterogeniety of the 40,000 molecular weight tuber glycoprotein of potato by immunological methods and by NH_2-terminal analysis. Plant Physiol. 71: 156–160.

Park, W. 1990. Induction of Tuberization in Potato. In: Vayda, M. and Park, W. (eds.), "Cellular and Molecular Biology of the Potato." pp. 43–56. CAB International, Wallingford, Oxon, UK.

Pikaard, C.S., Mignery, G.A., Ma, D.P., Stark, V.J. and Park, W.D. 1986. Sequence of two apparent pseudogenes of the major tuber protein patatin. Nucleic Acid Res. 14: 5564–5566.

Pikaard, C.S., Brusca, J.S., Hannapel, D.J. and Park, W.D. 1987. The two classes of genes for the major tuber protein, patatin, are differentially expressed in tubers and roots. Nucleic Acids Res. 15: 1979–1994.

Racusen, D. and Foote, M. 1980. A major soluble glycoprotein of potato. J. Food Biochem 4: 43–52.

Racusen, D. 1983. Occurrence of patatin during growth and storage of potato tubers. Can. J. Bot. 61: 370–373.

Racusen, D. 1984. Lipid acyl hydrolase activity of patatin. Can. J. Bot. 62: 2156–2165.

Rocha-Sosa, M., Sonnewald, U., Frommer, W., Stratman, M., Schell, J. and Willmitzer. L. 1989. Both developmental and metabolic signals activate the promoter of a Class-I patatin gene. EMBO J. 8: 23–29.

Rosahl, S., Eckes, P., Schell, J. and Willmitzer, L. 1986a. Organ-specific gene expression in potato: isolation and characterization of tuber-specifc cDNA sequences. Mol. Gen. Genet. 202: 368–373.

Rosahl, S., Schmidt, R., Schell, J. and Willmitzer L. 1986b. Isolation and characterization of a gene from *Solanum tuberosum* encoding patatin, the major storage protein of potato tubers. Mol. Gen. Genet. 203: 214–220.

Rosahl, S., Schell, J. and Willmitzer, L. 1987. Expression of a tuberspecific storage protein in transgenic tobacco plants: demonstration of an esterase activity. EMBO J. 6: 1155–1159.

Salanoubat, M. and Belliard, G. 1989. The steady-state level of potato sucrose synthase is dependent on wounding, anaerobiosis and sucrose concentration. Gene 84: 181–185.

Sommewald, U., Sturm, A., Chrispeels, M. and Wilmitzer, L. 1989. Targeting and glycosylation of patatin, the major potato tuber protein in leaves of transgenic tobacco. Planta 179: 171–180.

Twell, D. and Ooms. G. 1988. Structural diversity of the patatin gene family in potato cv. Desiree Mol. Gen. Genet. 212: 325–336.

Vancanneyt, G., Sonnewald, U., Hofgen, R. and Wilmitzer, L. 1989. Expression of a patatin-like protein in the anthers of potato and sweet pepper flowers. The Plant Cell 1: 533–540.

Vreugdenhil, D. and Struik, P.C. 1989. An integrated view of the hormonal regulation of tuber formation in potato (*Solanum tuberosum*). Physiol Plant. 75: 525–531.

Wenzler, H.C., Mignery, G.A., Fisher, L.M. and Park W.D. 1989a. Analysis of a chimeric Class-I patatin-GUS gene in transgenic potato plants: high level expression in tubers and sucrose-inducible expression in cultured leaf and stem explants. Plant Molec. Biol. 12: 41–50.

Wenzler, H., Mignery, G., Fisher, L. and Park, W. 1989b. Sucrose-regulated expression of a chimeric potato tuber gene in leaves of transgenic tobacco plants. Plant Molec. Biol. 13: 347–354.

Towards an Understanding of Starch Biosynthesis and Its Relationship to Protein Synthesis in Plant Storage Organs

---

Cathie Martin, Madan Bhattacharyya, Ian Dry, Cliff Hedley, Noel Ellis, Trevor Wang and Alison Smith
John Innes Institute
John Innes Centre for Plant Science Research
Colney Lane
Norwich NR4 7UH

Starch biosynthesis is one of the most important metabolic processes of developing storage organs. The importance of this process in determining seed composition is illustrated by a number of mutations affecting genes in the biosynthetic pathway such as *r* and *rb* of pea and some of the high lysine lines of maize and barley. Lesions that limit starch quantity and quality also have effects on the accumulation of lipid protein. To understand the relationship between starch biosynthesis and the production of other storage products we have begun to isolate a number of the genes involved in starch biosynthesis. These genes allow us to define particular mutants biochemically, and by genetic engineering allow us to affect particular steps in starch production in transgenic plants. By analyzing these transgenic plants we hope to develop our understanding of the regulation of starch biosynthesis and its relationship to the determination of seed composition as a whole. We anticipate that these experiments will also produce a number of transgenic plants with modifications in starch quantity and quality.

Introduction

Starch is the major digestible complex carbohydrate of most crop plants and is of considerable importance to the paper, textile and printing industries as well as for human and animal consumption. Different crop plants produce different types of starch which vary in their physical and chemical properties and have somewhat different uses within the food industry. However, although an enormous variety of starch types are produced by different plants, over 50 percent of starch isolated for industrial and food processing use comes from maize, about 15 percent from potato and the rest from wheat, rice, Manihot, sorghum and sago. This is subsequently processed chemically and enzymatically, producing starch with the desired qualities.

The biological basis for the qualitative differences between starches is relatively poorly understood. Starch is synthesized in the chloroplasts in leaves or in amyloplasts in developing storage organs and is laid down in starch grains, which have a complex physical and chemical form. Starch, biochemically, is made up of two components, the almost exclusively linear 1-4 α linked glucan chains called amylose, and branched chains consisting of 1-4 α linked glucan chains joined by 1-6 α linkages called amylopectin. Amylose and amylopectin themselves are very different physically and chemically, and within each chemical subgroup there is variation. For example, differences in the extent of branching within amylopectin molecules and the average chain length may affect physical properties such as solubility, viscosity and crystalin structure. As the starch grain in storage organs such as cotyledons, endosperm and tubers is formed over an extended period of development, the relative time of synthesis of amylose and amylopectin may also have important effects on the final properties of the starch grain. In addition, starch is phosphorylated in some plants such as potato and the degree of phosphorylation has a major influence on the properties of starch produced. Much of the qualitative variation in starch available prior to processing has arisen from the successful introduction of maize and potato varieties with altered starch metabolism that changes the proportion of amylose and amylopectin produced.

While the factors determining starch quality are poorly understood, even less is known of those factors that limit starch production in developing storage organs. Breeding has selected for highly productive lines, but there has been little accompanying biochemistry to determine the biological basis for increased starch production.

Considerable academic research is required into the biochemistry and molecular biology of starch biosynthesis to construct a picture of how the quantity and quality of starch is determined. With an understanding of the controls regulating starch production, it may be possible to modulate starch quality and quantity in crop plants to develop required charactertistics prior to processing.

Starch Biosynthesis

Within the non-photosynthetic plastids, starch is synthesized from hexose or hexose phosphate precursors, imported from the cytosol. Recent evidence suggests that, at least in some species such as pea, this transported precursor is glucose-6-phosphate (Hill and Smith, unpublished results). The actual precursor of the committed pathway of starch biosynthesis is glucose-1-phosphate and, therefore, if glucose-6-phosphate is imported it has to be converted to glucose-1-phosphate prior to use for starch biosynthesis. Beyond these steps there are three committed steps involved in starch biosynthesis (Figure 1). It is these steps that we are examining in detail in order to understand how to manipulate starch production genetically.

ADP glucose-pyrophosphorylase: The first committed step in starch biosynthesis involves the formation of precursor for synthesis of glucose chains. The precursor in plants is ADP-glucose, and it is synthesized from ATP and glucose-1-phosphate by ADP glucose pyrophosphorylase. There is evidence for metabolic regulation of this enzyme, whose activity *in vitro* is stimulated by 3-phosphoglycerate and inhibited by phosphate (reviewed by Preiss and Levi, 1980; Preiss, 1982). The activity may also be affected by fructose-6-phosphate, fructose-1, 6-bisphosphate and phosphoenolpyruvate (reviewed by Preiss, 1988). The importance of this enzyme is also highlighted by the phenotypes of mutants that appear to be lesions in ADP glucose pyrophosphorylase such as *sh2* and *bt2* in maize (Tsai and Nelson, 1965; Bhave *et al.*, 1990; Preiss *et al.*, 1990), the starchless mutant of *Arabidopsis* (Lin *et al.*, 1988) and *rb* in pea (Smith *et al.*, 1989; Smith, unpublished results). The enzyme appears to be a heterotetramer with two types of subunits, in maize encoded by *sh2* and *bt2* genes. Both subunit types are believed to contribute to the regulatory mechanism and to the catalytic function. It is not clear whether these are the only polypeptides giving rise to ADP glucose pyrophosphorylase activity. In maize, neither the *sh2*, the *bt2*, nor the double *sh2 bt2* mutations result in a complete abolition of ADP glucose pyrophosphorylase activity (Dickenson and Preiss, 1969; Preiss *et al.*, 1990; Bhave *et al.*, 1990). Similarly, the *adg2* mutation in *Arabidopsis* (Lin *et al.*, 1988) and the *rb* gene in pea (Smith *et al.*, 1989) do not completely abolish activity. It is possible that complete nulls are lethal or that there are other isoforms of ADP glucose pyrophosphorylase

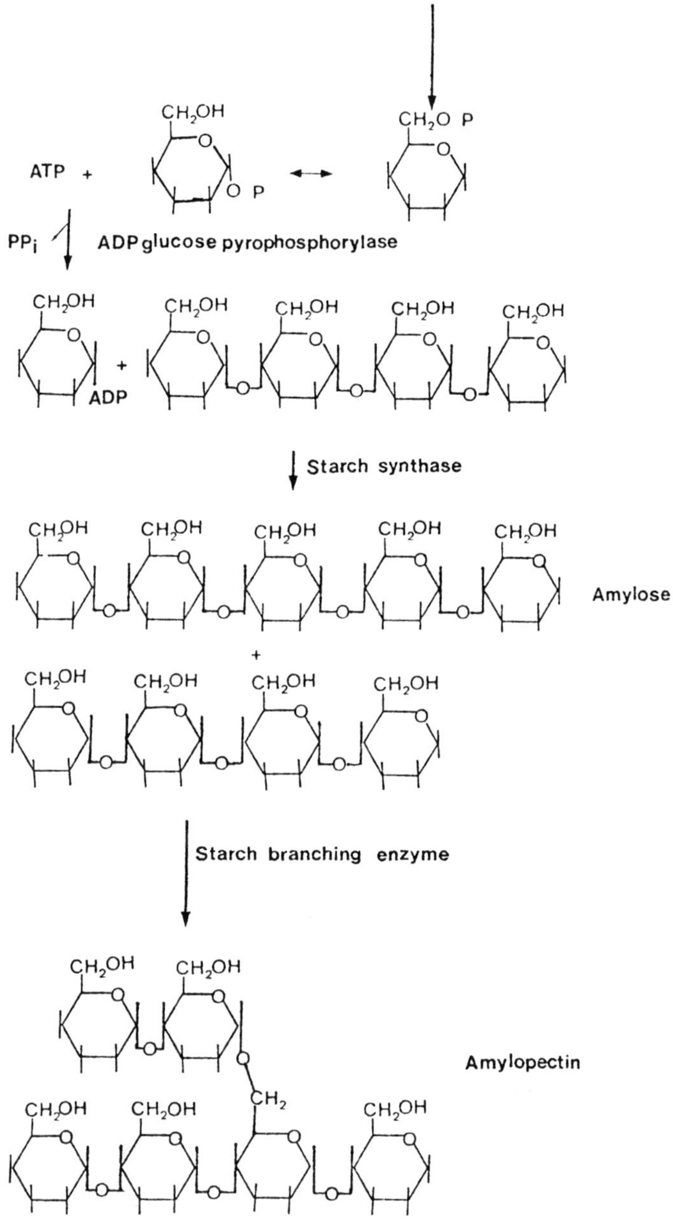

Figure 1. Diagramatic scheme of starch biosynthesis in storage organs. Glucose-6-phosphate is transported into the amyloplast where it is converted to glucose-1-phosphate. Glucose-1-phosphate is converted to ADP glucose using ATP and producing pyrophosphate by ADP glucose pyrophosphorylase. ADP glucose is added to the non-reducing end of 1-4 α-linked glucan chains by starch synthase, to produce amylose. 1-4 α-linked glucan chains are linked by starch branching enzyme by 1-6 α linkages to form amylopectin.

with low activity. In pea, the *rb* mutation completely abolishes one polypeptide of ADP glucose pyrophosphorylase causing a dramatic reduction in, although not complete loss of, activity. The phenotypic effects of mutations in these genes are primarily seen in the developing seed where starch production is reduced. It is interesting to note that in the *bt2* and *sh2* mutations enzyme activity in the endosperm is reduced to 3 percent and 6 percent respectively, while starch production is only reduced to 29 percent. In *Arabidopsis* leaves the *adg*2 mutation reduces ADP glucose pyrophosphorylase activity to 5 percent, but starch production only drops to 40 percent of wild-type (Lin *et al.*, 1988). These mutations are completely recessive, the heterozygotes being phenotypically indistinguishable from the wild-type. This evidence suggests that ADP glucose pyrophosphorylase may not always be the major rate-limiting step in starch production in developing storage organs.

The *rb* mutation of pea also alters the ratio of amylose to amylopectin produced in pea (Wang *et al.*, 1990), suggesting that this enzyme may play a role in the quality of starch made as well as the quantity, although the basis for this influence is not understood.

Starch Synthase: Glucose from ADP glucose is added via a 1-4 α linkage to the non-reducing end of the growing amylose chain by starch synthase. Activity has been found associated with the starch grain and in the stroma of the plastid. The *waxy* mutation of maize dramatically reduces the activity of granule-bound starch synthase in the endosperm (Nelson and Rhines, 1962), although activity in the embryos and seed coats is unaffected (Nelson and Tsai, 1964). This is also the effect of the *amf* mutant in potato (Hovenkamp-Hemelink *et al.*, 1987; Visser *et al.*, 1989) and the *waxy* mutants of rice (Sano, 1984) and sorghum (Hseih, 1988). Biochemical and genetic data suggest that there are at least two isoforms of granule-bound starch synthase (Nelson *et al.*, 1978). The effect of the waxy gene involves a complete loss of amylose in the endospern suggesting that its product may have a greater activity in extending linear glucan chains than the other enzymes. In the other parts of the waxy maize plant, 25 percent of starch is amylose, suggesting that the activity of the other granule-bound synthase (or synthases) is more important outside the kernel. There is also thought to be more than one form of soluble starch synthase (Ozbun *et al.*, 1971). No mutation has been found that is associated with the loss of activity of soluble starch synthase alone, although the *null* mutant of maize affects soluble starch synthase activity along with the activity of starch branching enzyme (Preiss and Boyer, 1980; Boyer and Preiss, 1981). The absence of a null mutant in soluble starch synthase could indicate the loss of this enzyme is lethal.

There is also some evidence that starch synthase may represent a major rate-limiting step in starch production. This point is considered by Lin *et al.* (1988) for starch production in *Arabidopsis* leaves where soluble starch synthase activity was 70-fold lower than ADP glucose pyrophosphorylase activity and exactly matched the rate of starch biosynthesis. Starch biosynthesis in several species is known to be temperature-sensitive and in wheat may be considerably limited at temperatures of 35°C and above, which is within the growing temperatures encountered in many countries. There is physiological evidence to suggest that it is the soluble starch synthase that is reduced in activity under such conditions and, therefore, limiting starch production (Jenner, C., personal communication).

Starch Branching Enzyme: The third committed step in starch biosynthesis forms amylopectin from the linear 1-4 α linked amylose chains via the creation of 1-6 α linkages. This is catalyzed by starch branching enzyme. There is biochemical evidence to suggest that there is more than one isoform of SBE in plants which may have somewhat different kinetic properties (Boyer and Preiss, 1978a,b, 1981; Baba *et al.*, 1982; Matters and Boyer, 1981, 1982; Boyer, 1985; Boyer and Fisher, 1984; Smith 1988). The resultant amylopectin products may also be somewhat different. In maize, one isoform of SBE is thought to be encoded by the amylose extender gene (*ae*) (Boyer and Preiss, 1978b, 1981; Preiss and Boyer, 1980) and in pea the *rugosus* (*r*) (Matters and Boyer, 1982; Edwards *et al.*, 1988; Smith, 1988) mutation dramatically reduces SBE activity. In these mutants starch biosynthesis is reduced, and the amount of amylopectin is considerably lower relative to amylose. Therefore, the effect of SBE is primarily on starch quality although in mutants where activity is greatly reduced, starch production is also affected. In heterozygotes, no effect is seen on starch production suggesting that this is not a step normally limiting starch biosynthesis.

The formation of starch in plants, therefore, involves three basic steps. However, there is good evidence that starch synthase and starch branching enzyme exist in multiple forms which may be performing different biosynthetic roles. When one also considers that in developing storage organs the relative importance of each isoform may vary with time, unravelling the individual contribution of each component and the regulation of each participating gene becomes extremely complex. In addition, from the study of mutants, it is clear that modifications of starch biosynthesis have pleiotropic effects on other elements of seed composition. In order to achieve the successful genetic manipulation of starch production, an understanding of how each component contributes to starch biosynthesis and storage organ development is essential.

Experimental

Starch Branching Enzyme: We have started our analysis of starch biosynthesis by using pea because it is very suitable for biochemical study and has a large number of genetic variants to facilitate analysis. We have extended our investigations to potato as this is a major starch-forming crop with established protocols for reproducible transformation, making it the most suitable target for the modification of starch production by genetic engineering.

The most direct route to analyze the role of individual components in starch biosynthesis is to study the effects of mutations. In pea, the *rugosus* or *r* locus originally studied by Mendel (1865) offered a suitable starting point. Near-isogenic lines had been developed (Hedley *et al.*, 1986) for the exact definition of the phenotypic effects of *r.* These are listed in Table 1.

The primary effects in mutant, wrinkled peas include reduced starch production and decreased amylopectin to amylose ratios, although *r* also has a selective effect on storage protein and lipid production and increases water uptake. The effect of *r* on starch suggested that it might affect starch branching enzyme (SBEI) activity and Smith (1988) was able to show that wrinkled seeds lack one isoform of starch branching enzyme that is active early in pea seed development. A second isoform, active later in development, is not affected. The two isoforms have distinguishable activities, SBEI giving a less highly branched, less soluble form of starch than SBEII. We decided to find out whether *r* encoded SBEI to determine whether all the phenotypic effects of *r* could be attributed to loss of SBEI.

Using an antibody to SBEI we isolated a number of cDNA clones from an expression library in λgt11. When these clones were used to probe RNA from developing round and wrinkled peas, a transcript of 3.2 kb was observed in round seeds. In wrinkled seeds an aberrant transcript of 4.2 kb was observed of lower abundance than the transcript in round seed. This suggested that wrinkled seed has a

Table 1. Compositional differences between round (*RR*) and wrinkled (*rr*) pea embryos

	Round	Wrinkled	Data derived from
Starch content % dry weight	51	29	Wang *et al.* (1990)
Amylose % dry weight	15	21	Wang *et al.* (1990)
Amylopectin % dry weight	35	8	
Storage protein components (μg per 100μg protein)			
legumin	33	23	Domoney and Casey (1985)
vicilin	54	61	
Lipid % dry weight	1.7	3.4	Wang *et al.* (1990)
Sucrose % dry weight	4.1	7.4	Wang and Hedley unpublished results
All data for mature seed from near isogenic lines (Hedley *et al.*, 1986)			
% water content at 30 days post-anthesis	60	74	Wang *et al.*, (1987)

For further details see Bhattacharyya *et al.* (1990).

mutated form of the SBEI gene. To confirm this we studied the segregation of a restriction fragment length polymorphism found in the SBEI gene between *RR* and *rr* isolines and compared it to the phenotypic segregation. The SBEI gene showed 100 percent co-segregation with the round and wrinkled phenotypes establishing that, by definition, it is at the *r* locus. We also showed that the SBEI gene in wrinkled peas carries an 800bp transposable element-like insertion within the coding region, explaining the larger transcript observed and why wrinkled seeds have no detectable SBEI activity (Bhattacharyya *et al.*,1990).

It is clear from the phenotype of wrinkled seeds that loss of SBEI activity has a profound effect on starch biosynthesis and seed composition. It would appear that reduced starch biosynthesis results from the loss of SBEI because branching normally makes additional non-reducing ends of 1-4 α linked glucan chains. As the non-reducing ends provide one substrate for the further extension of chains by starch synthase, decreased branching enzyme activity may substrate-limit starch synthase. This limitation placed on starch biosynthesis will then result in the accumulation of sucrose in wrinkled seeds as it is the ultimate precursor of starch. The increase in free sucrose in cells of wrinkled seed will cause an increase in osmotic pressure and this is probably the cause of the pleiotropic effects of *r* (Kappert, 1915; Wang *et al.* 1987). Increased cellular osmotic pressure will cause greater water uptake during seed development, and the testa of the developing pea will expand to accommodate this increase in cellular size. On maturation, as the pea desiccates, the extra water will be lost, and the expanded testa will wrinkle to give the characteristic phenotype. Increased osmotic pressure or higher sucrose levels may also affect legumin biosynthesis as pea embryos cultured on high sucrose show reduced steady-state levels of legumin mRNA. This may be an effect on message stability (Turner, 1988). This model for the role of SBEI in pea development highlights the role that osmotic pressure or levels of free sucrose may have on the development of storage organs and stresses the biological co-ordination between starch and storage protein production. In fact, a large proportion of the effects of *r* results from its limitation of starch biosynthesis. Confirmation that this model is correct may then come from comparison to other mutations that limit starch biosynthesis to see if they have similar pleiotropic effects. This option has been somewhat limited in pea due to the lack, until recently, of many other mutations that affect seed composition available as isogenic material (Wang *et al.*, 1990). The *rb* mutation does give rise to wrinkled seed; and it causes increased lipid accumulation, although its effects on storage proteins have not been well characterized. In barley and maize, a number of high-lysine lines, selected originally for their reduced hordein and zein contents, have also been shown to have major effects on starch biosynthesis (Creech, 1969; DiFonzo *et al.*, 1978; Shrewry *et al.*, 1987). Some of these mutants in maize have been biochemically characterized and shown to be lesions in starch biosynthetic enzymes such as *bt2, sh2,* (ADP glucose pyrophosphorylase), *wx* (granule-bound starch synthase) and *ae* (starch branching enzyme). Thus the influence of limiting starch production on storage protein synthesis is reflected by mutations in other steps of starch biosynthesis and in other species, supporting the model for the role of SBEI.

The *r* mutation also has a major effect on the form of the starch grains (Gregory, 1903). They are compound and highly fractured in wrinkled seed compared with the simple grains of round seed. This difference is probably the result of loss of amylopectin. However, some amylopectin is synthesized in wrinkled peas presumably through the activity of SBEII. This amylopectin is synthesized later in pea development (Smith, 1988). The effect on grain structure and physical properties might be the result of the later production of this amylopectin, or an alteration in its form, or both.

Having identified the SBEI gene from pea we have isolated a full length cDNA clone encoding this protein. We have constructed vectors using a double CaMV 35S enhancer and CaMV terminator for over expression in transgenic potato. We have also used the SBEI cDNA as a probe for the equivalent cDNA from developing potato tubers. Using this clone we are building antisense constructs to inhibit the activity of SBE in developing tubers.

Starch Synthase

In maize, one protein associated with granule-bound starch synthase activity is the product of the *waxy* gene. This is a major polypeptide of 65kD in size adhering to the starch grain. Nelson *et al.* (1978) have reported a second granule-bound starch synthase activity in waxy mutants with a lower Km for ADP glucose than the *waxy* gene product. MacDonald and Preiss (1985) reported that there are at least four proteins with granule-bound starch synthase activity. The activities of the two classes of granule-bound starch synthases did not vary synchronously during development (Nelson *et al.*, 1978). The *waxy* gene product increased in activity between 12 and 16 days after pollination. The enzyme(s) with low Km did not increase in activity over this period. An equivalent protein to the maize *waxy* protein has been observed in pea which is also antigenically related to the equivalent *waxy* protein of potato. In pea, two antigenically related polypeptides of 59 kD and 60 kD are observed on the starch grain. Purification of these polypeptides has failed to reveal any starch synthase activity associated with them. In addition, immunoprecipitation of these proteins does not result in any loss of activity from solubilized granule-bound proteins (Smith, 1990). Activity has been found associated with a less abundant polypeptide of 77 kD which is immunologically quite distinct. Using antisera, cDNAs to both the 77 kD and the 60 kD polypeptides have been isolated from pea, and sequenced. The cDNA encoding the 60 kD polypeptide is highly homologous to the *waxy* sequence from maize and homologous to glycogen synthase from *E. coli.* The derived protein sequence contains the amino acids lysinetheronine-glycine-glycine positioned some 15 amino acids from the N-terminus. These residences are thought to represent the active site of the enzyme. These data imply that this gene could encode an active granule-bound starch synthase although the biochemical data have shown no activity associated with this protein.

The derived protein sequence from the cDNA encoding the 77 kD polypeptide is similar to the waxy gene product and the *E. coli* glycogen synthase, and contains the active site motif, although it also carries an extra 15 kD of protein sequence at its amino-terminus. Both biochemical and molecular data support the role of this product as starch synthase.

Clearly the activities of these two enzyme types are not the same. Analysis of gene expression in developing pea seeds has indicated that the 77 kD starch synthase gene is expressed mostly in young embryos up to 300 mg fresh weight (fwt) while the 60 kD gene is expressed throughout seed development, expression increasing as the seed develops up to >500 mg fwt. Thus, these genes may play different roles in the timing of starch biosynthesis. While the embryo of pea and the endosperm of maize are not directly equivalent organs, it is possible that these two gene products are equivalent to the two granule-based starch synthases identified by Nelson *et al.* (1978) in maize. The lack of mutants in these genes has hampered the evaluation of their contribution in pea, but we have been able to isolate a cDNA homologous to the pea 77 kD protein cDNA from potato. The equivalent cDNA to the 60 kD protein from potato has also been isolated (Visser *et al.*, 1989; Dry and Martin, unpublished results). Using these cDNAs we are making antisense constructs for transformation into potato. By inhibiting the expression of these genes we hope to characterize their role in starch biosynthesis and the effect of modifying their activity on starch quantity and quality in developing potato tubers. The 77 kD protein may also represent one component of the soluble starch synthase activity as an enzyme of 77 kD is also found in the soluble fraction in peas. However, biochemical evidence suggests that more than one enzyme is involved (Denyer, unpublished results). We are taking similar approaches to characterize these components.

ADP glucose pyrophosphorylase: Antisera to the major polypeptides purified with ADP glucose pyrophosphorylase activity from developing peas has been produced and used to isolate cDNA clones from pea from expression libraries. Homologous cDNAs have also been isolated from potato and are at present being characterized before constructs for transformation are built.

Discussion

The analysis of mutants with reduced activity in individual components of starch biosynthesis allows for an evaluation of their qualitative contribution to starch biosynthesis. Their quantitative contribution depends on the extent to which they limit flux into starch under any particular environmental conditions or at any particular development time. This relative importance of each contribution to the flux into starch can be calculated from the control coefficient for each enzyme, provided mutants with reduced activities of the enzyme are available. The antisense transformants of potato may provide mutants for each component from which control coefficients can be calculated. Once the major rate-limiting step has been identified for any particular set of conditions, it may be possible to increase starch production in transgenic plants by over expressing the gene, encoding the enzyme, or modifying the protein structure of the enzyme.

Clearly, increased starch production by genetic engineering is something that requires considerable research if it is to be effectively achieved. At present the best candidate genes to engineer for this purpose would appear to be those encoding soluble starch synthase and possibly ADP glucose pyrophosphorylase. The gene(s) encoding soluble starch synthase need to be isolated before this approach can begin in earnest.

Qualitative changes may arise through modification of ADP glucose pyrophosphorylase activity (although this may also affect starch quantity), the granule-bound starch synthases and starch branching enzyme. In addition to over expression and inhibition of branching enzyme, changing the relative balance between different isoform types (resulting in the production of different types of amylopectin) and their time of expression during development may give added variety to the types of starch that can be produced in transgenic potato. However, one sobering thought amidst the images of variation that could be produced is the reminder of the effect of *r* and other lesions in starch biosynthesis in barley on storage protein production. One reason why the high lysine lines of barley were never commercially useful was that their decreased hordein production was always accompanied by a loss of productivity (due in part to reduced starch production) (Shrewry *et al.*, 1987). The biological link between starch and protein is important to understand not only to define the controls operational in seed development but also to manipulate starch production successfully by genetic engineering.

References

Baba, T., Arai, Y., Yamamoto, T. and Hori, T. 1982. Some structural features of amylomaize starch. Phytochemistry 21: 2291-2296.

Bhattacharyya, M.K., Smith, A.M., Ellis, T.H.N., Hedley, C. and Martin, C. 1990. The wrinkled-seed character of pea described by Mendel is caused by a transposon-like insertion in a gene encoding starch branching enzyme. Cell 60: 115-122.

Bhave, M., Lawrence, S., Barton, C. and Hannah, L.C. 1990. Identification and molecular characterization of *shrunken-2* cDNA clones of maize. The Plant Cell 2: 581-588.

Boyer, C.D. 1985. Soluble starch synthases and starch branching enzymes from developing seeds of sorghum. Phytochemistry 24: 15-18.

Boyer, C.D. and Preiss, J. 1978a. Multiple forms of starch branching enzyme in maize: evidence for independent genetic control. Biochem. Biophys. Res. Commun. 80: 169-179.

Boyer, C.D. and Preiss, J. 1978b. Multiple forms of (1-4)α D glucan (1-4)α D glucan-6-glucosyl, transferase from developing *Zea mays* L. kernals. Carbohydrate Res. 61: 321-334.

Boyer, C.D. and Preiss, J. 1981. Evidence for independent genetic control of the multiple forms of maize endosperm branching enzymes and starch synthases. Plant Physiol. 67: 1141-1145.

Boyer, C.D. and Fisher, M.B. 1984. Comparison of soluble starch synthases and branching enzymes from developing maize and *Teosinte* seeds. Phytochemistry 23: 733-737.

Creech, G.R. 1969. Carbohydrate synthesis in maize. Adv. Agron. 21: 275-322.

Dickenson, D.B. and Preiss, J. 1969. Presence of ADP glucose pyrophosphorylase in *shrunken-2* and *brittle-2* mutants of maize endosperm. Plant Physiol. 44: 1058-1062.

DiFonzo, N., Fornasari, E., Gentinetta, E., Salamini, F. and Soave, C. 1978. Proteins and carbohydrate accumulation in normal opaque-2 and floury maizes. In "Carbohydrate and Protein Synthesis", Miflin, B. and Zoschte, M. (Eds.), pp. 199-212. Commission of the European Communities, Brussels.

Domoney, C. and Casey, R. 1985. Measurement of gene number for seed storage proteins in *Pisum*. Nucl. Acids Res. 13: 687-699.

Edwards, J., Green, J.H. and Rees, T. 1988. Activity of branching enzyme as a cardinal feature of the *Ra* locus of *Pisum sativum*. Phytochemistry 27: 1615-1620.

Gregory, R.P. 1903. The seed characters of *Pisum sativum*. The New Phytologist 2: 226-228.

Hedley, C.L., Smith, C.M., Ambrose, M.J., Cook, S. and Wang, T.L. 1986. An analysis of seed development in *Pisum sativum*. II. The effect of the *r* locus on growth and development of the seed. Ann. Bot. 58: 371-379.

Hovenkamp-Hermelink, J.H.M., Jacobsen, E., Ponstein, A.S., Visser, R.G.F., Vos-Scheperkeuter, G.H., Bijmolt, E.W., de Vries, J.N., Witholt, B. and Feenstra, W.J. 1987. Isolation of an amylose-free mutant of the potato, (*Solanum tuberosum* L.). Theor. Appl. Genet. 75: 217-221.

Hseih, J-S. 1988. Genetic studies on the *Wx* gene of sorghum (*Sorghum bicolor* (L) Monench.). 1. Examination of the protein product of the *waxy* locus. Bot. Bull. Academia Sinica. 29: 293-299.

Kappert, H. 1915. Untersuchtungen on Mark-,Kneifel-and Zuckererbsen und ihren Bastarden. Z. Ind. Abst, Vererbungslehre B: 1-57.

Lin, T-P., Caspar, T., Somerville, C.R. and Preiss, J. 1988. A starch deficient mutant of *Arabidopsis thaliana* with low ADP glucose pyrophosphorylase activity lacks one of the two subunits of the enzyme. Plant Physiol. 88: 1175-1181.

MacDonald, F.D. and Preiss, J. 1985. Partial purification and characterization of granule-bound starch synthases from normal and *waxy* maize. Plant Physiol. 78: 849-852.

Matters, G.L. and Boyer, C.D. 1981. Starch synthases and starch branching enzymes from *Pisum sativum*. Phytochemistry 20: 1805-1809.

Matters, G.L. and Boyer, C.D. 1982. Soluble starch synthases and starch branching enzymes from cotyledons of smooth and wrinkled seeded lines of *Pisum sativum*. L. Biochem. Genet. 20: 833-848.

Mendel, G. 1865. Versuche über Pflanzen-Hybriden. Verh. Naturforsch. Ver. Brunn 4: 3-47.

Nelson, O.E. and Rhines, H.W. 1962. The enzymatic deficiency in the *waxy* mutant of maize. Biochem. Biophys. Commun. 9: 297-300.

Nelson, O.E. and Tsai, C.Y. 1964. Glucose transfer from adenosine diphosphate glucose to starch in preparations of *waxy* seeds. Science 145: 1194-1195.

Nelson, O.E., Chourey, P.S. and Chang, M.T. 1978. Nucleoside diphosphate sugar-starch glucosyl transferase activity of *wx* starch granules. Plant Physiol. 62: 383-386.

Ozbun, J.L., Hawker, J.S. and Preiss, J. 1971. Adenosine diphospho glucose-starch glucosyl-transferases from developing kernals of *waxy* maize. Plant Physiol. 48: 765-769.

Preiss, J. 1982. Biosynthesis of starch and its regulation. Encyclopedia of Plant Physiology, New Series 13A: 397-417.

Preiss, J. 1988. Biosynthesis of starch and its regulation. In "The Biochemistry of Plants.", Stumpf, P.K., and Conn, E.E. (Eds.) Vol. 14. pp. 181-254. Academic Press, New York.

Preiss, J. and Boyer, C.D. 1980. Evidence for the independent genetic control of the multiple forms of maize endosperm branching enzymes and starch synthases. In "Mechanisms of Saccharide Polymerization and Depolymerization." Marshall, J.J. (Ed.) Academic Press, New York.

Preiss, J. and Levi, C. 1980. Starch biosynthesis and degradation. In "The Biochemistry of Plants." Preiss, J. (Ed.) Vol. 3, pp. 371-423. Academic Press, New York.

Preiss, J., Danner, S., Morell, M., Barton, C.R., Yang, L. and Nieder, M. 1990. Molecular characterization of maize endosperm ADP glucose pyrophosphorylase subunits and implication of *Brittle-2* as a structural gene. Plant Physiol. In press 1991.

Sano, Y. 1984. Differential regulation of *waxy* gene expression in rice endosperm. Theor. Appl. Genet. 68: 467-473.

Shrewry, P.R., Williamson, M.S. and Kreis, M. 1987. Effects of mutant genes on the synthesis of storage components in developing barley endosperms. In "Developmental Mutants in Plants." Thomas, H., and Grierson, D. (Ed.) pp. 95-118. Cambridge University Press.

Smith, A.M. 1988. Major differences in isoforms of starch branching enzyme between developing embroys of round- and wrinkled-seeded peas (*Pisum sativum* L.). Planta 175: 270-279.

Smith, A.M., Bettey, M. and Bedford, I.D. 1989. Evidence that the *rb* locus alters the starch content of developing pea embryos through an effect on ADP glucose pyrophosphorylase. Plant Physiol. 89: 1279-1284.

Smith, A.M. 1990. Evidence that the *waxy* protein of pea (*Pisum sativum* L.) starch granules is not the major granule-bound starch synthase. Planta (In press.)

Tsai, C.Y. and Nelson, O.E. 1966. Starch deficient maize mutants lacking adenosine diphosphate glucose pyrophosphorylase activity. Science 151: 341-343.

Turner, S.R. 1988. The effect of the *r*-locus on the synthesis of storage proteins in *Pisum sativum*. Ph.D. Thesis, University of East Anglia.

Visser, R.G.F., Hergersberg, M., Van der Leij, F.R., Jacobsen, E., Witholt, B. and Feenstra, W.J. 1989. Molecular cloning and partial analysis of the gene for granule-bound starch synthase from wild type and amylose-free potato (*Solanum tuberosum* L.). Plant Science 64: 185-192.

Wang, T.L., Smith, C.M., Cook, S.K., Ambrose, M.J. and Hedley, C.L. 1987. An analysis of seed development in *Prisum sativum*. III. The relationship between the *r*-locus, the water content and the osmotic potential of seed tissues *in vivo* and *in vitro*. Ann. Bot. 59: 73-80.

Wang, T.L., Hadavizideh, A., Harwood, A. Welham, T.J., Harwood, W.A., Faulks, R. and Hedley, C.L. 1990. An analysis of seed development in *Pisum sativum* L. XIII. The chemical induction of storage product mutants. Plant Breeding (In press.)

ENHANCEMENT OF VITAMINS AND MINERALS

Genetic Improvement of Vegetable Carotene Content

P. W. Simon
Vegetable Crops Research
USDA, Agricultural Research Service
Department of Horticulture
University of Wisconsin
Madison, Wisconsin 53706

Much of the world dietary vitamin A is derived from vegetable carotenes. Vegetable carotene content can be improved by genetically increasing total carotene content, by genetically increasing the content of provitamin A carotenoids, or both. These approaches have been successful for tomatoes, sweet potatoes, maize, and carrots. Major genes affecting carotenoid content have also been identified in peppers and cucurbits. Genetic selection has also yielded germplasm which contains carotenes for such typically carotene-free vegetables as cauliflower, yams, cucumber, and potatoes. The genetic basis for improving vegetable carotene content is poorly understood. Even so, gain from selection can be quite rapid since total carotene content can be visually estimated quite accurately. Genetic improvement of vegetable carotene content has depended completely upon classical breeding and genetic methods. As the biochemistry and molecular biology of carotene metabolism is better understood, new methods may be useful in genetically enhancing vegetable carotene content.

Introduction

Vitamin A deficiency is a major health problem in parts of Africa, Asia, Latin America, and the Near East (USAID, 1989). In developed areas, vitamin A deficiency is rare although it is estimated that 30 percent of the U.S. population consumes less than 70 percent of their recommended vitamin A daily allowance (Briggs, 1981). Much of the world's vitamin A is derived from provitamin A carotenoids in vegetables. Carotenoids not only provide vitamin A but also have been implicated as cancer protectants and enhancers of immune function (Krinsky, 1988; Bendich and Olson, 1989). It has been estimated that Americans consume only one-fourth of the 6 mg of the beta-carotene equivalent recommended to meet dietary goals (Lachance, 1988). Thus, increased carotene consumption could provide several health benefits.

Carotene consumption can be increased by increasing consumption of the available carotene-containing foods or by dietary supplementation (e.g., carotene capsules, carotene fortification of foods). Carotene consumption can also be increased without altering or supplementing the diet directly since it is possible to genetically increase vegetable carotene content.

Vegetables as Carotene Sources

The available sources of dietary carotene provide a perspective for the potential prospects of genetic improvement of vegetable carotene content. Table 1 lists all U.S. sources of dietary carotene which contain at least 2 ppm provitamin A carotenoids. Although 21 commodities are listed, only carrot, sweet potato, and tomato provide at least 5 percent of the available U.S. vitamin A since many of the high-carotene commodities are not consumed in large quantity. Conversely, several of the vegetables and fruits consumed in high volume, such as potato, lettuce, onion, and apple, contain little carotene. The relative contributions of carrot, sweet potato, red pepper, and broccoli have increased since 1975 (Senti and Rizek, 1975) due to the availability of new higher carotene cultivars and higher per capita consumption.

Genetic Improvement of Carotenoid Content

The genetics of carotene biosynthesis have been characterized only in tomato fruit, maize fruit, and carrot root. Genetic variation for total carotenoid content and relative content of specific carotenes has been evaluated. Some genetic selection for higher carotene content in sweet potato root has occurred. Major genes controlling the carotene content of pepper, cantaloupe, squash, and pumpkin fruit have also been identified. In addition to the genes which affect carotene accumulation in fruit and roots, mutants have been identified in many plant species which limit leaf carotene content resulting in chlorotic or albino plants. Since these plants are very weak and contribute no dietary carotene, they are not discussed here.

Although preliminary genetic analysis of tomato color dates back to the 1920's, Lincoln and Porter in 1950 published the first biochemical pathway of carotene biosynthesis in tomato based upon genetic stocks available. With his co-workers, Porter continued to elaborate the biochemical basis of tomato fruit color mutants throughout his career (Porter *et al.*, 1984). As in other crops, many of the tomato fruit carotene mutants analyzed are pale orange, yellow, or colorless with low carotene levels.

Tomes and coworkers also examined tomato fruit color biochemical genetics but focused on genetic manipulations which yielded germplasm with greater nutritional value. The result was the high beta-carotene cultivar, Caro-Red (Tomes and Quackenbush, 1958). Tigchelaar has carried this work further in the development of Caro-Rich (Tigchelaar and Tomes, 1974).

More than 20 genes have been characterized in tomato which affect the type, amount, or distribution of fruit carotenoids (Darby, 1978; Kirk and Tilney-Basset, 1978). A simplified scheme of beta-carotene synthesis is: Carotene precursors ($\xrightarrow{1}$) Phytoene ($\xrightarrow{2}$) Phytofluene ($\xrightarrow{3}$) Zeta-carotene ($\xrightarrow{4}$) Neurosporene ($\xrightarrow{5}$) Lycopene ($\xrightarrow{6}$) Delta-carotene ($\xrightarrow{7}$) Beta- or Alpha-carotene. The genes *ap,* apricot, and *r,* yellow flesh, reduce pigment levels at step (1); *gh,* ghost, at step (2); *vo,* virescent orange, at step (4); *t,* tangerine, at step (5); *Del,* high delta, at step (7). Genes *y,* colorless epidermis, and *u,* uniform fruit color, affect pigment distribution.

Lycopene, which has no provitamin A activity, is the primary carotene in typical red-fruited tomatoes accounting for 50-70 percent of the total carotenes. Incorporation of the dominant allele at the *B* locus (step 7) results in a change from red fruit to deep orange fruit with 40-80 percent beta-carotene. Incorporation of the recessive *hp,* high pigment allele (step 1), increases the total carotene content 30-50 percent without markedly altering the relative percentage of beta-carotene. Unfortunately, the red color is highly preferred to orange by tomato consumers, and the *B* allele is associated with unrelated but undesirable characteristics. Consequently, production of high beta-carotene tomatoes is very limited. Another high pigment gene, og^c, crimson (step 6), also increases color but reduces provitamin A activity. Thus, some genetic changes to improve tomato nutritional value have effects that are contrary to consumer expectations of tomato appearance (Tigchelaar, 1988). Genetic manipulation to increase tomato total carotene content further would increase provitamin A carotene availability and probably be acceptable to the consumer. An increase in beta-carotene content without a concommitant reduction of lycopene may also be acceptable if red color could be maintained.

In addition to the well-characterized single genes which affect tomato carotene biosynthesis, early research indicated that certain accessions of tomato, *Lycopersicon esculentum,* and of interspecific crosses of tomato with wild species, *L. pimpinellifolium* and *L. hirsutum,* have 150-400 ppm lycopene and up to 120 ppm beta-carotene (Kohler *et al.* 1947; Porter and Lincoln, 1950). The prospects for incorporating high-carotene characteristics into commercially acceptable tomato germplasm should be examined.

Maize has not yielded many major genes affecting carotene biosynthesis. Hauge and Trost (1928) discovered the *Y,* yellow, locus, which is incompletely dominant for yellow (versus white) endosperm pigmentation. Brunson and Quackenbush (1962) found that yellow maize ranges from 20 to 40 ppm total carotenoids (dry weight basis). From 10 percent to 20 percent of these carotenoids have provitamin A activity. Variation in carotenoid type and amount is presumably controlled by many genes. Differences in relative amounts of individual carotenes and xanthophylls were noted between maize lines but inheritance was not evaluated (Grogan and Blessin, 1968). The development of high-carotene maize for human consumption could have a major positive impact. Either higher total carotenoids or a higher relative percentage of provitamin A carotenoids would benefit consumers.

Carrots were white-rooted in the Middle Ages but orange-colored carrot roots have been known since the 1700's. Since the 1960's, Gabelman and coworkers (Gabelman and Peters, 1979) gathered evidence to indicate three genes controlling major color categories: *y,* colored; *b,* light orange, and *c,* orange with *b* hypostatic to *Y* and *c* hypostatic to *Y* and *B.* Furthermore, genes Y_1 amd Y_2 affect color distribution in the xylem and phloem, whereas genes *a* and *l* condition the inhibition of alpha-carotene and lycopene biosynthesis, respectively. It is interesting to note that low carotene content is dominant in carrot roots but recessive in tomato and maize fruits. This may reflect the fact that fruits at some point have photosynthetic function and thereby require carotenoids, whereas root carotenoids serve no function to the plant.

More recently, Simon *et al.* (1989) developed very dark orange carrot strains with 300-600 ppm total carotenes using phenotypic selection. From 40 percent to 60 percent of the carotene content is comprised of beta-carotene with the balance largely alpha-carotene (Simon and Wolff, 1987). The ability to select for dark orange color in carrot roots has depended upon using genetically diverse parents of European and Asian origin. Selection for higher carotene content in carrots should be possible (Simon, 1988). A shift to higher relative beta-carotene content would increase the provitamin A activity somewhat.

In contrast to typical sweet potatoes which contain 120 ppm carotenes, genetic variation in carotene content up to 170 ppm has been noted (Collins, 1988). No genetic analysis of this variation has been reported. The large consumption of sweet potatoes in developing countries suggests a ready application for high-carotene sweet potatoes. Beta-carotene accounts for most of the sweet potato carotene makeup, so provitamin A content cannot be improved without an increased total carotene content.

Several genes control the flesh and epidermis color of typically orange-fleshed cucurbit fruits (Robinson *et al.*, 1976). Orange flesh color is dominant to both white flesh (*wf*) and green flesh (*gf*) in muskmelon. In watermelon, red flesh is dominant to one gene for yellow flesh (*y*) but recessive to another yellow flesh gene and to white flesh (*Wf*). Although some varieties of squash are quite high in carotene content, little is known about fruit flesh genetics. In cucumbers, white flesh color is dominant to yellow (*yf*) and to the double recessive orange flesh (*yf, wf*). Cucurbit carotenes are primarily beta-carotene in orange fruit and lycopene in red fruit.

High carotenoid content in *Capsicum* is conditioned by four dominant genes, B, C_1, C_2, and Y, and one recessive gene, t (Kirk and Tilney-Bassett, 1978). Alleles B_1, c_1 and t^+ especially reduce beta-carotene accumulation but all other carotenoids are also present in reduced quantities. Allele c_2 blocks the synthesis of all carotenoids whereas y reduces the level of capsanthin, the major carotenoid of red pepper fruits.

Based upon the genetic studies summarized above, there is great potential for increasing carotene intake of tomato, corn, carrot, sweet potato, peppers and cucurbits without increasing consumption of these commodities. High values for carotene content are listed in Table 2. Although genetic studies have not been performed, high carotene cultivars or breeding stocks have also been identified for nearly all of the fruits and vegetables from Table 1. These values are also included in Table 2 as are high carotene values for cabbage, orange, and banana.

Table 1. Provitamin A carotenoid content of selected vegetables and fruits and their contribution to U.S. vitamin A availability.

Commodity	Carotene Content[a]	Contribution to U.S. Vitamin A Availability[b]
Carrot	169	30 %
Sweet Potato	120	11 %
Spinach	40	2 %
Parsley	31	
Mango	23	
Cantaloupe	19	2 %
Apricot	16	
Papaya	12	
Squash	1–47[c]	1 %[d]
Pumpkin	10	
Broccoli	9	1 %
Red Pepper[e]	8	2 %
Tomato[e]	7	5 %
Tangerine	5	
Green Bean	4	
Green Pea	3	
Green Pepper	3	
Peach	3	
Lettuce	2	
Plum	2	
Sweet Corn	2	

[a]ppm; provitamin A carotenes, fresh weight basis; from Gebhardt *et al.*, 1982; Haytowitz & Matthews, 1984.
[b]From Putnam, 1988; Raper 1990, based on carotene content and per capita availability, values<1 percent not listed.
[c]Squash cultivars vary markedly in carotene content.
[d]Pumpkin and squash availability data pooled.
[e]Substantial quantities of carotenoids with no provitamin A activity occur.

Table 2. High carotenoid reported for carotene-containing vegetables and fruits and their potential contribution to U.S. diet.

Commodity	High Reported Carotene Content[a]	Reference[b]	Potential Annual U.S. per Capita Carotene Consumption[c]
Carrot	600	A	2942
Sweet Potato	170	B	332
Spinach	70	C	25
Parsley	60	C	
Mango	60	C	
Cantaloupe	120	C	926
Apricot	19	C	4
Papaya	12	D	
Squash	47	E	
Pumpkin	60	C	30
Broccoli	25	C	53
Red Pepper	248	C	
Tomato	95[d]	F	2113
Tangerine	5	D	3
Green Bean	4	E	11
Green Pea	4	E	7
Green Pepper	11	C	15
Peach	15	C	63
Lettuce	13	C	121
Plum	8	G	
Sweet Corn	5	C	50
Cucumber	6	H	22
Orange	6	G	76
Cabbage	3	C	9
Banana	2	C	3
Yams	14	H	
Cauliflower	3	H	4
Potatoes	3	C	102

[a] ppm total carotene, fresh weight basis
[b] A-Simon *et al.*, 1989
B-Collins, 1988
C-Klaui and Bauernfeind, 1981
D-Gebhardt *et al.*, 1982
E-Haytowitz and Matthews, 1984
F-Porter and Lincoln, 1950
G Polacchi *et al.*, 1982
H-Simon, 1989
[c] In mg per capita per year; based on 1985-87 per capita retail weight consumption (Putnam, 1988). Commodities consumed less than 0.5 lb. per capita are not included.
[d] Large fruited types.

New Sources of Vegetable Carotenes

Carotene sources discussed above are vegetables and fruits which typically contain some carotene. In contrast, potatoes, yams, and cauliflower typically contain no carotene, yet experimental varieties which do contain carotenes have been developed (Howard, 1970; Simon, 1990). These commodities are also included in Table 2. With their high volume of consumption in many areas, the utilization of high-carotene yams and potatoes could have a highly significant positive effect in vitamin A-poor regions of the world.

Conclusion

Genetic investigations indicate the potential for significantly increasing carotene availability in the food supply by genetic manipulations of vegetables. These genetic changes would provide several health benefits without otherwise modifying the diet. The yellow or orange pigmentation resulting from genetic improvement of carotene content may require consumer education to ensure consumption of high-carotene varieties. Molecular biological techniques may provide methods to supplement classical plant breeding to enhance vegetable carotene content.

References

Bendich, A. and Olson, J.A. 1989. Biological actions of carotenoids. FASEB J. 3: 1927-1932.

Briggs, M.H. 1981. "Vitamins in Human Biology and Medicine. " CRC Press, Boca Raton, FL.

Brunson, A.M. and Quackenbush, F.W. 1962. Breeding corn with high provitamin A in the grain. Crop Sci. 2: 344.

Collins, W.W. 1988. Genetic improvement of sweet potatoes for meeting human nutritional needs. In "Proceedings of the 1st International Symposium on Horticulture and Human Health," p. 191. Prentice Hall, Englewood Cliffs, NJ.

Darby, L.A. 1978. Isogenic lines of tomato fruit colour mutants. Hort. Res. 18: 73.

Gabelman, W.H. and Peters, S. 1979. Genetical and plant breeding possibilities for improving quality of vegetables. Acta Hort. 93: 243.

Gebhardt, S.E., Cutrufelli, R. and Matthews, R.H. 1982. Composition of foods: fruits and fruit juices. Handbook 8–9, U.S. Department of Agriculture, Washington, DC.
Grogan, C.O. and Blessin, C.W. 1968. Characterization of major carotenoids in yellow maize lines of differing pigment concentration. Crop Science 8: 730.
Hauge, S.M. and Trost, J.F. 1928. An inheritance study of the distribution of vitamin A in maize. J. Biol. Chem. 80: 107.
Haytowitz, D.B. and Matthews, R.H. 1984. Composition of foods: vegetables and vegetable products. Handbook 8–11, U.S. Department of Agriculture, Washington, DC.
Howard, H.W. 1970. "Genetics of the Potato." Springer-Verlag, New York, New York.
Kirk, J.T.O. and Tilney-Bassett, R.A.E. 1978. "The Plastids." Elsevier Press, New York, New York.
Klaui, H. and Bauernfeind, J.C. 1981. Carotenoids as food colors. In "Carotenoids as colorants and vitamin A precursors," p. 47. Academic Press, New York, New York.
Kohler, G.W., Lincoln, R.E., Porter, J.W., Zscheile, F.P., Caldwell, R.M., Harper, R.H. and Silver, W. 1947. Selection and breeding for high beta-carotene content (provitamin A) in tomato. Bot. Gaz. 109: 219-225.
Krinsky, N.I. 1988. The evidence for the role of carotenes in preventative health. Clin. Nutr. 7: 107.
Lachance, P. 1988. Dietary intake of carotenes and the carotene gap. Clin. Nutr. 7: 118.
Lincoln, R.E. and Porter, J.W. 1950. Inheritance of beta-carotene in tomatoes. Genetics 35: 206.
Polacchi, W., McHargue, J.S. and Perloff, B.P. 1982. Food composition tables for the Near East. FAO Food and Nutrition Paper 26, United Nations, FAO, Rome.
Porter, J.W., Spurgeon, S.L. and Sathyamoorthy, N. 1984. Biosysthesis in carotenoids. In "Isopentenoids in plants: biochemistry and function", p.161-183. Marcel Dekker, Inc., New York.
Porter, J.W. and Lincoln, R.E. 1950. I. Lycopersicon selections containing a high content of carotenes and colorless polyenes. II. The mechanism of carotene biosynthesis. Arch. Biochem. Biophys. 27: 390.
Putnam, J.J. 1988. Food consumption, prices, and expenditures, 1966-87. USDA Statistical Bulletin 773, Wash., DC.
Raper, N. 1990. Personal communication. USDA Health and Nutrition Info. Service, Wash., DC.

Robinson, R.W., Munger, H.M., Whitaker, T.W. and Bohn, G.W. 1976. Genes of the cucurbitaceae. HortScience 11: 554.
Senti, F.R. and Rizek, R.L. 1975. Nutrient levels in horticultural crops. HortScience 10: 243.
Simon, P.W. 1988. Genetic improvement of carrots for meeting human nutritional needs. HortScience 25: 1495-1499.
Simon, P.W. 1990. Carrots and other horticultural crops as a source of provitamin A carotenes. Hort Science 25: 1495-1499.
Simon, P.W. and Wolff, X.Y. 1987. Carotenes in typical and dark orange carrots. J. Agric. Food Chem. 35: 1017-1022.
Simon, P.W., Wolff, X.Y., Peterson, C.E., Kammerlohr, D.S., Rubatzky, V.E., Strandberg, J.O., Bassett, M.J. and White, J.M. 1989. High carotene mass carrot population. HortScience 24: 174.
Tigchelaar, E.C. and Tomes, M.L. 1974. Caro-Rich tomato. HortScience 9: 82.
Tigchelaar, E.C. 1988. Genetic improvement of tomato nutritional quality. In "Proceedings of the 1st International Symposium on Horticulture and Human Health." p. 185. Prentice Hall, Englewood Cliffs, NJ.
Tomes, M.L. and Quackenbush, F.W. 1958. Caro-Red, a new provitamin A rich tomato. Econ. Bot. 12: 256.
U.S. Agency for International Development. 1989. Vitamin A field support component of the vitamin A for health project 936-5116.

Enzymology and Genetic Regulation of Carotenoid Biosynthesis in Plants

Bilal Camara
Laboratoire de Biochimie et de Régulations Cellulaires UA, CNRS 568 33400 Talence France
Rhodolphe Schantz
Institut de Biologie Moléculaire des plantes du CNRS, 12 rue du Général Zimmer F-67000 Strasbourg France
René Monéger
Laboratoire de Biochimie et Phathologie Végétales Université Paris VI, 4 Place Jussieu 75230 Paris Cedex 05 France

Carotenoids are lipophilic pigments present in all photosynthetic and some non-photosynthetic organisms. Several aspects of carotenoid biosynthesis have been delineated through the use of chromoplasts isolated from higher plants. The sequence of reactions involves the synthesis of geranylgeranyl pyrophosphate and its dimerization into phytoene, the first C_{40} carotenoid. All these steps are catalyzed by operationally soluble enzymes localized in plastid stroma. During subsequent reactions the desaturation isomerization cyclization, and hydroxylation of phytoene, afford different xanthophylls.

The three basic enzymes of phytoene systhesis (isopentenyl pyrophosphate isomerase, geranylgeranyl pyrophosphate synthase and phytoene synthase) have been purified to homogeneity. Available evidence indicates that these enzymes are encoded by the nuclear genome and that their synthesis is specifically triggered during the massive accumulation of carotenoids which occurs in chromoplasts. Therefore, a potential mean by which the plant cell regulates its carotenoid content is displayed and open for genetic manipulations.

Introduction

Great impetus to the investigation of carotenoid biosynthesis was given by the introduction of Porter and Lincoln (1950) of the hypothesis that colored carotenoids were formed by the sequential dehydrogenation of the unsaturated polyenes detected in several tomato fruit lines. Since then, the implications of these studies have led to a massive exploration which now allows us to describe the different steps according to the basic scheme shown in Figure 1. Several reviews (Porter and Spurgeon, 1979; Camara and Monéger, 1982; Porter and Spurgeon, 1983; Jones and Porter, 1986; Bramley and Mackenzie, 1988; Britton, 1988; and Kleinig, 1989) present different pictures of our advances in the field, but they also inevitably point to persistent gaps in our understanding. A search of the literature leaves the impression that as far as the enzymology and the underlying basis of its control are concerned, little progress has been made there. The lag can be accounted for by the exceptionally great difficulties that such studies must face. In this chapter, we consider some of the recent developments in this area, focusing mainly on higher plants.

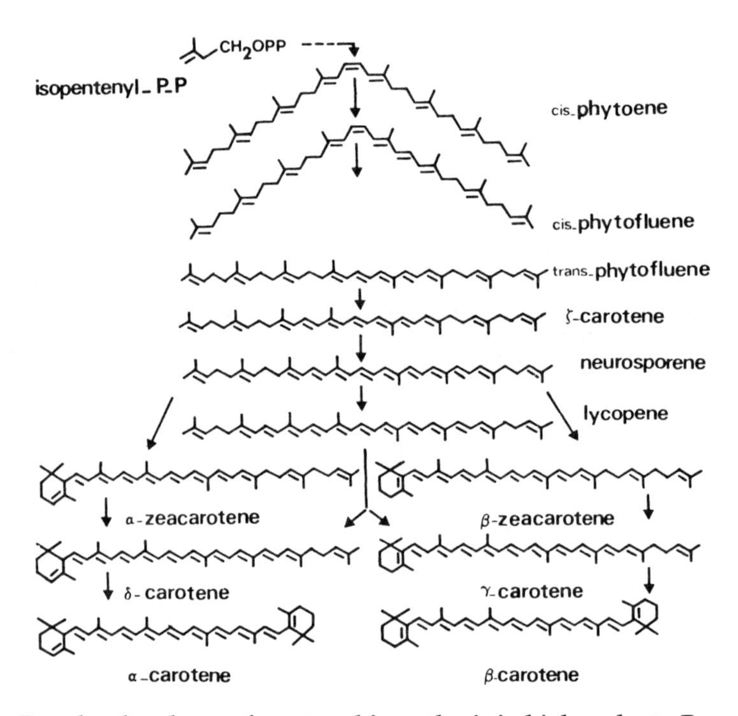

Figure 1. Postulated pathway of carotene biosynthesis in higher plants. Reproduced with permission from Camara and Monéger (1982).

Biosynthesis of the Isopentenyl Pyrophosphate Unit

In comparison with the larger number of biochemical studies that have been published on the mechanism of carotenoid biosynthesis in plants using the distal precursors, i.e. CO_2, acetate and mevalonic acid, relatively few papers have appeared that deal at an enzymological level with the question of how isopentenyl pyrophosphate is formed in plastids and channeled towards the carotenoid pathway. In previous experiments the incorporation of CO_2 into various isoprenoids by isolated plastids has been demonstrated (Rogers *et al.*, 1966). This plastid autonomy was also observed for the first set of enzymes which catalyze the first committed steps in the biosynthesis of isoprenoid compounds (hydroxymethylglutary1 CoA reductase, mevalonic acid kinase, mevalonate phosphate kinase, and pyrophosphate mevalonate anhydrodecarboxylase) (Camara and Monéger, 1982; Bach, 1987; and Schulze-Siebert and Schultz, 1987). Taken together, the results of all these investigations suggest the existence of at least two separate sites of isoprenoid synthesis in the plant cell, inside and outside the plastids. A critical point that deserves mention is that, in general, the activities detected in isolated plastids for the above enzymes were always low. As a consequence of this situation and of its implication, it is assumed but not yet proved that isopentenyl pyrophosphate is synthesized solely in the cytoplasm and transported into plastids (Kreuz and Kleinig 1982; Gray, 1987). This view is to be modified to some extent (Kleinig, 1989) in the light of more recent results confirming (Moore and Shepard, 1978) that in chloroplasts isolated from *Acetabularia* the capacity to synthesize isoprenoid derivatives from CO_2 is preserved (Bauerle *et al.*, 1990). The significance of these apparently contradictory results is not yet clear. Some light has been shed on the basis of studies performed recently with developing barley leaves (Schultz, 1990). With this system it has been shown that the capacity of isolated plastids to efficiently incorporate CO_2 into isoprenoid compounds is subject to developmental control, i.e. the activity progressively decreases from the basal region of the leaves containing young chloroplasts towards the tip of the leaves containing mature chloroplasts. A very similar phenomenon can be deduced from developing tomato fruits (Narita and Gruissem, 1989). In spite of this evidence many problems are far from being resolved. For example the use of heterologous probes to identify the gene(s) encoding hydroxymethylglutaryl CoA reductase has led to suggestions for the presence of several genes in tomato (Narita and Gruissem, 1989) and one (Learned and Fink, 1989) or several (Caelles *et al.*, 1989) genes for *Arabidopsis.*

Biosynthesis of the Central Substrate Geranylgeranyl Pyrophosphate

Geranylgeranyl pyrophosphate, the basic skeleton of most plastid isoprenoids, arises from two basic reactions which involve the isomerization of isopentenyl pyrophosphate to yield the first allylic pyrophosphate partner dimethylallyl pyrophosphate followed by a series of prenyl transfers. The isomerization of isopentenyl pyrophosphate to dimethylallyl pyrophosphate is catalyzed by the enzyme isopentenyl pyrophosphate isomerase. The reaction catalyzed by this enzyme is reversible but the activity is directed mainly towards the synthesis of dimethylallyl pyrophosphate. This enzyme has been characterized recently as a monomeric protein having a molecular weight of 34 kD from tomato (Spurgeon *et al.*, 1984) and pepper (Dogbo and Camara, 1987) fruits. The enzyme is sensitive to thiol reagents and the only cofactor required is divalent cation Mg^{2+} or Mn^{2+}. Subsequent reactions catalyzed by a prenyltransferase affords geranyl pyrophosphate the C_{10} allylic pyrophosphate. According to the same mechanism (for further details see Poulter and Rilling, 1981a), the latter undergoes further condensation with a second molecule of isopentenyl pyrophosphate. Ultimately, farnesyl pyrophosphate condenses with a third molecule of isopentenyl pyrophosphate to yield geranylgeranyl pyrophosphate. From this sequence of reactions (Figure 2) one could invoke the existence of several prenyltransferases whose specificity is directed towards the synthesis of C_{10}, C_{15} and C_{20} prenyl pyrophosphates. From presently available data, only one prenyltransferase termed geranylgeranyl pyrophosphate synthase catalyzes the three prenyl transfer reactions. This enzyme has been purified from *Capsicum* chromoplasts (Dogbo and Camara, 1987) and from *Sinapis* etioplasts (Laferrière, 1989). Based on sodium dodecylsulfate polyacrylamide gel electrophoresis, the monomeric molecular weight of geranylgeranyl pyrophosphate is 37 kD and by gel filtration its molecular weight is 74 kD, demonstrating that this enzyme exists as dimer. The activity of the enzyme is strictly dependent upon the presence of divalent cations Mg^{2+} or Mn^{2+}, the latter being preferred at low concentrations. In the presence of

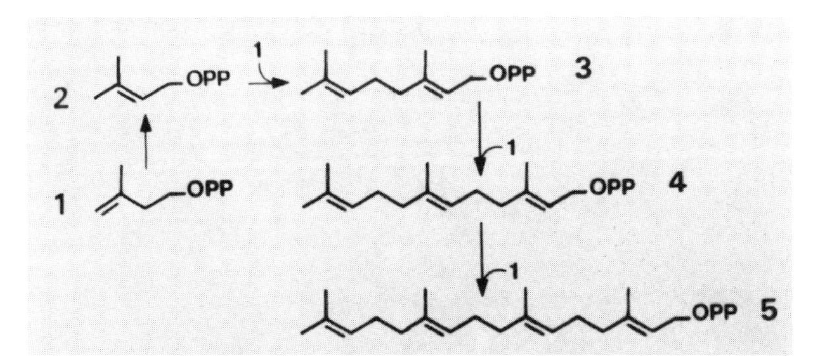

Figure 2. Sequential synthesis of geranylgeranyl pyrophosphate in plastid stroma. The isomerization of isopentenyl pyrophosphate (1) into dimethylallyl pyrophosphate (2) is followed by a series of prenyl transfers which afford geranyl pyrophosphate (3), farnesyl pyrophosphate (4) and finally geranylgeranyl pyrophosphate (5) which leaves the active site of the enzyme. The whole prenyl transfer reaction is catalyzed by one enzyme termed geranylgeranyl pyrophosphate synthase.

isopentenyl pyrophosphate, geranylgeranyl pyrophosphate accepts dimethylallyl pyrophosphate, geranyl pyrophosphate or farnesyl pyrophosphate as allylic pyrophosphate primers. In this context it is worth noting that though the cytosolic (farnesyl pyrophosphate synthase) and the plastidial (geranylgeranyl pyrophosphate synthase) prenyltranferases are formally analogous in their mechanism of catalysis or in the identity of their C_5 (dimethylallyl pyrophosphate) or C_{10} (geranyl pyrophosphate) allylic pyrophosphate substrates, they are immunologically distinct (Hugueney and Camara, 1990). Furthermore, using antibodies against the purified geranylgeranyl pyrophosphate synthase, we have demonstrated that this enzyme is exclusively localized in the plastid compartment in *Capsicum* cells (Camara *et al.*, 1989; and manuscript in preparation), in good agreement with previous data (Camara and Monèger, 1982; Dogbo *et al.*, 1987).

Biosynthesis of Phytoene, the First C_{40} Carotenoid

Following the formation of geranylgeranyl pyrophosphate, further reactions of carotene biosynthesis involve the dimerization of two molecules of geranylgeranyl pyrophosphate which affords phytoene through the intermediacy of prephytoene pyrophosphate (Figure 3). An analogous reaction has been observed earlier during the synthesis of squalene via presqualene pyrophosphate (Poulter

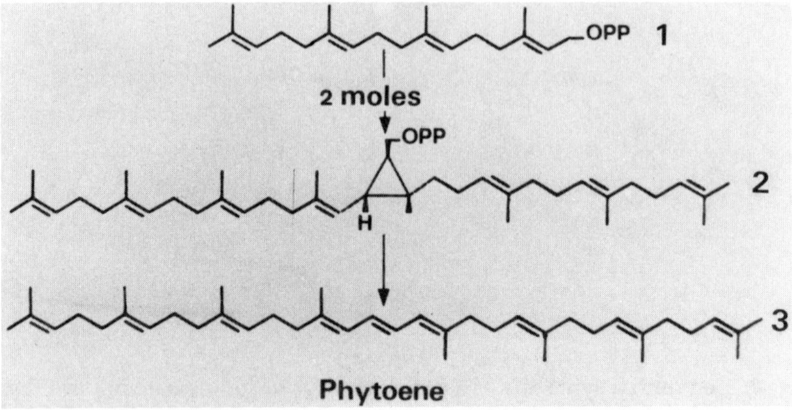

Figure 3. Synthesis of phytoene from geranylgeranyl pyrophosphate (1) through the intermediacy of prephytoene pyrophosphate (2). The different steps are catalyzed by a single enzyme phytoene synthase.

and Rilling, 1981b). Data gained from these studies has allowed us to define to some extent the mechanism involved. Accordingly, if the mechanism were to proceed essentially like that involved during the formation of squalene, NADPH would be required and lycopersene would be formed. Studies with purified chromoplasts devoid of endoplasmic reticulum and cytosolic contaminations clearly showed that lycopersene is not involved in the biosynthesis of carotenoids (Camara and Monèger 1980; Camara *et al.*, 1983), nor is NADPH required. Indeed, during the head-to-head joining of the two units of geranylgeranyl pyrophosphate the sterochemistry of the hydrogen loss affords *cis*-or *trans*-phytoene. In higher plants *cis* phytoene is largely predominant (Camara *et al.*, 1983). The enzyme phytoene synthase which catalyzes the synthesis of phytoene from geranylgeranyl pyrophosphate has been purified recently from *Capsicum* chromoplasts (Dogbo *et al.*, 1988). The condensing activity enzyme and the subsequent rearrangement of prephytoene to phytoene are both catalyzed by the same protein which is a monomer of 48 kD. The activity of this enzyme is strictly dependent upon Mn^{2+}. This requirement could not be fulfilled by any other divalent cation tested so far. This feature allows a selective regulation of the synthesis of phytoene synthesis compared to that of geranylgeranyl pyrophosphate for which either Mg^{2+} or Mn^{2+} are effective. Furthermore this specificty for Mn^{2+} offers a key distinction between phytoene synthase and squalene synthase which requires specifically Mg^{2+} for activity. Finally, antibodies raised against phytoene synthase reveal that *in vivo* phytoene synthase is strictly located in the plastid compartment (Camara *et al.*, 1989; and manuscript in preparation) as demonstrated from *in vitro* studies performed with different plastid preparations (Camara and Monèger, 1982; Mayfield *et al.*, 1986; Dogbo *et al.*, 1987).

Later Reactions of Carotenoid Biosynthesis

A survey of the present state of our knowledge concerning these reactions reveals that after 30 years of investigation and despite accumulation of numerous and significant pieces of information, we are still far from understanding the nature and the number of enzymes involved. Data gained from *in vitro* studies point to the fact that, in contrast to the formation of phytoene where the different steps are catalyzed by operationally soluble enzymes, the desaturation of phytoene is mediated by membrane-bound enzymes. These enzymes catalyze the sequential reactions depicted in the "consensus" pathway (Figure 1). Several herbicides have shown to interfere directly with the desaturation process; however, since most of them have pleiotropic effects (Ridley, 1982), the specificity of these compounds as real phytoene desaturase inhibitors must still be considered an open question. The enzyme cyclization of lycopene to B-carotene is very sensitive to thiol reagents (Camera *et al.*, 1985b) and is selectively blocked *in vitro* by several amine derivatives (Camara *et al.*, 1985a). Deviations to the normal pathyway shown in Figure 1 have been reported. Recently, it has been proposed that a pathway of phytoene desaturation very similar to that previously observed in Tangerine tomato operates in daffodil flowers (Beyer *et al.*, 1989), i.e., the desaturation of phytoene affords *cis*-carotene which is converted into *trans*-B-carotene through the intermediacy of proneurosporene and prolycopene. In considering the role of proneurosporene and prolycopene to participate in the biosynthesis of *trans*-carotene, it is worth noting that their formation in Tangerine tomato is induced by a recessive mutation linked to a single gene (Kirk, 1978). The existence of this recessive gene in plants is also suggested by the fact that several onium compounds which induce the accumulation of *cis*-carotenes probably act at the gene level (Yokoyama *et al.*, 1982). A rather different deviation to the normal pathway has been suggested using *Dunaliella bardawill* exposed to high light intensity (Ben-Amotz *et al.*, 1988). Under these conditions they suggested that 9-*cis*-B-carotene, which accumulates as a major product, is synthesized by the sequential desaturation of 9-*cis*-phytoene. Obviously the above observations leave the enzymology of phytoene desaturation largely unresolved. In this context it is worth noting that antibodies to a putative phytoene desaturase from *Rhodobacter capsulatus* (Bartley and Scolnick, 1989) reacts with plant phytoene desaturase (Schmidt *et al.*, 1989). Though informative, this approach must be complemented with the identification of the individual plant enzymes involved in these reactions.

Regulation and Molecular Biology of Carotenogenic Enzymes

With the availability of antibodies to isopentenyl pyrophosphate isomerase, geranylgeranyl pyrophosphate synthase and phytoene synthase, it has been possible to begin understanding the regulation of carotenoid biosynthesis at the molecular level. When Poly(A+) RNA isolated from ripening *Capsicum* fruits are translated *in vitro* using a cell free system, the polypeptide product of immune precipitation with antiserum against isopentenyl pyrophosphate isomerase, geranylgeranyl pyrophosphate synthase and phytoene synthase had a molecular weight approximatively 5 kD higher than the corresponding mature enzymes (Camara *et al.*, 1989; and manuscript in preparation). These data show that the basic enzymes of carotenoid biosynthesis are synthesized as high molecular weight precursors in good agreement with plastid destined polypeptides encoded by the nuclear genome.

In line with these studies an unanswered question was to identify the mechansim by which the massive synthesis of carotenoids is regulated during chromoplast differentiation. As a preliminary exploration of this feature we have started studying the mechanism of the induction of carotene synthesis in *Capsicum annuum* cell cultures. To this end, chlorophenylthiotriethylamine (CPTA), which triggers massive accumulation of the acyclic carotenoid lycopene in several plants (Yokoyama *et al.*, 1982), has been used. To elucidate the biochemical basis for the overproduction of lycopene, two cell lines were made tolerant to CPTA and designated CAR+ and CAR++. Compared to the wild types which contained 0.3 μg of carotenoids per g of fresh weight, CAR+ and CAR++ contained 7 and 17 μg of carotenoids per g of fresh weight. Biochemical analysis revealed that the response of *Capsicum* cells to CPTA is mediated at least by enhanced induction of geranylgeranyl pyrophosphate synthase and phytoene synthase activities as shown by the data displayed in Table 1.

Table 1. Geranylgeranyl pyrophosphate synthase and phytoene synthase and phytoene synthase activities in wild and CPTA adapted cells of *Capsicum annuum*.

Cell types	Geranylgeranyl pyrophosphate synthase %	Phytoene synthase %
Wild	1	2
CAR +	55	40
CAR + +	44	58

A more detailed analysis is presented in Figure 4A which shows that geranylgeranyl pyrophosphate synthase and phytoene synthase polypeptides practically undetectable in wild cells are clearly visualized from extract derived from CAR + and CAR + + cells. To enable the study of the rate of synthesis of the two enzymes, the different cell lines were incubated with ^{35}S methionine for 12h. At the end of the incubation, an equal quantity of cells was lysed and separated into an insoluble fraction and a soluble fraction. The resulting soluble fraction was used to immunoprecipitate geranylgeranyl pyrophosphate and phytoene synthase polypeptides. The results (Figure 4B and C) obtained show that CPTA induces the synthesis of geranylgeranyl pyrophosphate synthase and phytoene synthase polypeptide. This phenomenon is severely inhibited by cycloheximide (Figure 4C), thus demonstrating the prevalence of the nuclear control of plastid enzymes of carotenoid synthesis. Finally from this model system one may conclude that the enhanced accumulation of carotenoid which occurs during chromoplast differentiation is mediated at least by new synthesis of carotenogenic enzymes. Though less amenable to this kind of experimentation we have observed the same trend during the ripening of pepper and tomato fruits (Walter *et al.*, 1990; and manuscript in preparation). For a better understanding, cDNA encoding geranylgeranyl pyrophosphate synthase and phytoene synthase isolated from ripening fruits are now being used to further delineate the different mechanisms involved in the induction and the control of carotenoid biosynthesis in plants.

Finally, the study of carotenoid metabolism is now an expanding field. In this context it is worth noting that with the aid of a chromoplast model, many aspects of the metabolism of carotenoid in plants have been established. However, the available data raise

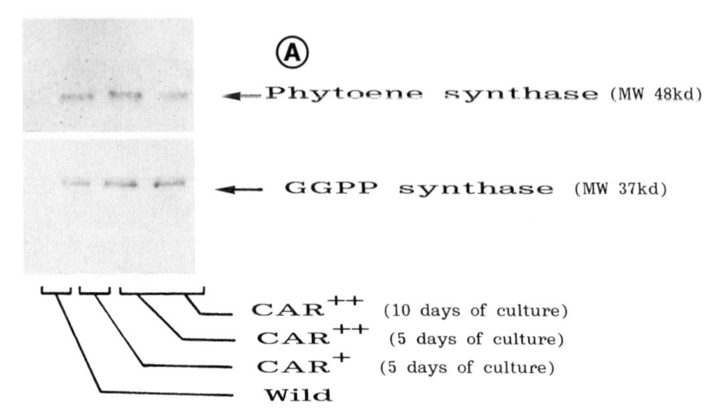

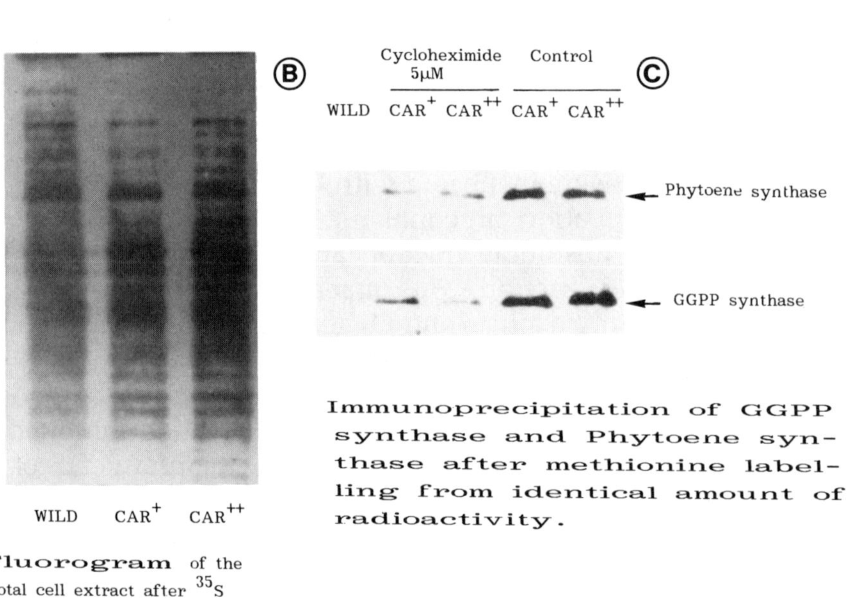

Figure 4. Analysis of the synthesis of geranylgeranyl pyrophosphate synthase and phytoene synthase in *Capsicum* cells treated with chlorohenylthiotriethylamine to elicit the accumulation of carotenoids.

a series of questions concerning the role of plastids in the elaboration of distal precursors of carotenoid biosynthesis, the nature of polypeptides and the mechanism involved in the conversion of phytoene into different carotenes and xanthophylls. Answers to some of these facets promise to give new insights into the molecular regulation of carotenoid biosynthesis in plants.

References

Bach, T.J. 1987. Synthesis and metabolism of mevalonic acid in plants. Plant Physiol. Biochem. 25: 163.

Bartley, G.E. and Scolnik, P.A. 1989. Carotenoid biosynthesis in photosynthetic bacteria. Genetic characterization of the *Rhodobacter capsulatus* Crt protein. J. Biol. Chem. 264: 13109.

Bauerle, R.E., Lutke-Brinkhaus, F., Ortmann, B., Berger, S. and Kleinig, H. 1990. Prenyl and fatty-acid synthesis in isolated *Acetabularia* chloroplasts. Planta 181: 229.

Ben-Amotz, A., Lers, A. and Avron, M. 1988. Stereoisomers of B-carotene and phytoene in the Alga *Dunaliella bardawil*. Plant Physiol. 86: 1286.

Beyer, P., Mayer, M. and Kleinig, H. 1989. Molecular oxygen and the state of geometric isomerism of intermediates are essential in the carotene desaturation and cyclization reactions in daffodil chromoplasts. Eur. J. Biochem. 184: 141.

Bramley, P.M. and Mackenzie, A. 1988. Regulation of carotenoid biosynthesis. Curr. Top. Cell Regul. 29: 291.

Britton, G. 1988. Biosynthesis of carotenoids. In "Plant Pigments," p. 133, Academic Press, London.

Caellas, C. Ferrer, A., Balcells, L., Hegardt, F.G. and Boronat, A. 1989. Isolation and structural characterization of cDNA encoding *Arabidopsis thaliana* 3- hydroxy-3-methylglutaryl coenzyme A reductase. Plant Molec. Biol. 13: 627.

Camara, B., Bardat, F., Dogbo, O., Brangeon, J. and Monèger, R. 1983. Terpenoid metabolism in plastids. Isolation and biochemical characteristics of *Capsicum annuum* chromoplasts. Plant Physiol. 74: 94.

Camara, B., Bardat, F. and Monèger, R. 1988. Sites of biogenesis of carotenoids in *Capsicum annuum*. Eur. J. Biochem. 127: 255.

Camara, B., Dogbo, O., d'Harlingue, A. and Bardat, F. 1985a. Inhibition of lycopene cyclization from *Capsicum* chromoplast membranes by 2-aza-2,3 dihydrosqualene. Phytochemistry 24: 2751.

Camara, B., Dogbo, O., d'Harlingue, A., Kleinig, H. and Monèger, R. 1985b. Metabolism of plastid terpenoids: lycopene cyclization by *Capsicum* chromoplast membranes. Biochim. Biophys. Acta 836: 262.

Camara, B., Bousquet J., Cheniclet, C., Carde, J.P., Kuntz, M. and Weil J.H. 1989. Enzymology of isoprenoid biosynthesis and expression of plastid and nuclear genes during chromoplast differentiation in pepper fruits (*Capsicum annuum).* In "Physiology, Biochemistry, and Genetics of Nongreen Plastids," p. 141, ASPP, Rockville, Maryland.

Camara, B. and Monèger, R. 1980. Carotenoid biosynthesis: Biogenesis of capsanthin and capsorubin in pepper fruits (*Capsicum annuum).* In "Biogenesis and Function of Plant Lipids," p. 363, Elsevier Biomed. Press, Amsterdam.

Camara, B. and Monèger, R. 1982. Biosynthetic capabilities and localization of enzymatic activities in carotenoid metabolism of *Capsicum annum* isolated chromoplasts. Physiol. Vèg. 2O: 757.

Dogbo, O. and Camara, B. 1987. Purification of isopentenyl pyrophosphate isomerase and geranylgeranyl pyrophosphate synthase from Capsicum chromoplasts by affinity chromatography. Biochim. Biophys. Acta 920: 140.

Dogbo, O., Bardat, F., Laferrière, A., Quennemet, J., Brangeon, J. and Camara, B. 1987. Metabolism of plastid terpenoids. I. Biosynthesis of phytoene in plastid stroma isolated from higher plants. Plant Sci. 49: 89.

Dogbo, A., Laferrière A., d'Harlingue A. and Camara B. 1988. Carotenoid biosynthesis: isolation and characterization of a bifunctional enzyme catalyzing the synthesis of phytoene. Proc. Natl. Acad Sci. USA 85: 7054.

Gray, J.C. 1987. Control of isoprenoid biosynthesis in higher plants. Adv. Bot Res 14: 25.

Gray, J.C. and Kekwick, RG.O. 1973. Mevalonate kinase in green leaves and etiolated cotyledons of the french beans *Phaseolus vulgaris*. Biochem J. 133: 335.

Hugueney, P. and Camara, B. 1990. Purification and characterization of farnesyl pyrophosphate synthase from Capsicum *annuum.* FEBS Letters. (In press.)

Jones, B.L. and Porter, J.W. 1986. Biosynthesis of carotenes in higher plants. CRC Critical Rev. Plant Sci. 3: 295.

Kirk, J.T.O. 1987. The biochemical basis of plastid autonomy. In "The Plastids," p. 525, Elsvier North Holland, Biomed Press, Amsterdam.

Kleinig, H. 1989. The role of plastids in isoprenoid biosynthesis. Annu Rev. Plant Physiol Mol. Biol. 4O: 39.

Kreuz, K. and Kleinig, H. 1982. On the compartmentation of isopentenyl disphosphate systhesis and utilization in plant cells. Planta 153: 578.

Laferrière, A. 1989. Enzyme de biosynthèse des isoprenoîdes: Purification et caractèrisation de la gèranylgèranyl pyrophosphate synthase des ètioplastes de moutarde (*Sinapis* alba L..). Ph. D. Thesis Universitè Bordeaux I, Bordeaux.

Learned, R.M. and Fink, G.R. 1989. 3-hydroxy-3-methylglutaryl-coenzyzme A reductase from *Arabidopsis thaliana* is structurally distinct from the yeast and animal enzymes. Proc. Natl Acad Sci, USA. 86: 2779.

Mayfield, S.P., Nelson, T., Taylor, W.C. and Malkin, R. 1986. Carotenoid synthesis and pleiotropic effects in carotenoid deficient seedlings of maize. Planta. 169: 23.

Moore, F.D. and Shepard, D.C. 1978. Chloroplast autonomy in pigment synthesis. Protoplasma 94: 1.

Narita, J.O. and Gruissem, W. 1989. Tomato hydroxymethyl-CoA reductase is required early in fruit development but not during ripening. Plant Cell. 1: 181.

Porter, J.W. and Spurgeon, S.L. 1979. Enzymatic synthesis of carotenes. Pure Appl. Chem. 51: 609.

Porter, J.W. and Spurgeon, S.L. 1983. Biosynthesis of carotenoids. In "Biosynthesis of Isoprenoid Compounds," Vol. 2, p.1, Wiley-Inerscience Publication, New York.

Porter, J.W. and Lincoln, R.E. 1950. *Lycopersicon* selections containing high content of carotenes and colorless polyenes. II. The mechanism of carotene biosynthesis. Arch. Biochem. Biophys. 27: 39O.

Poulter, C.D. and Rilling H.C. 1981a. Prenyltransferases and isomerases. In "Biosynthesis of Isoprenoid Compounds," Vol. 1, p. 161, Wiley-Interscience Publication, New York.

Poulter, C.D. and Rilling, H.C. 1981b. Conversion of farnesyl pyrophosphate to squalene. In "Biosynthesis of Isoprenoid Compounds," Vol. 1, p. 413, Wiley-Interscience Publication, New York.

Ridley, S.M. 1982. Carotenoids and herbicide action. In "Carotenoid Chemistry and Biochemistry," p. 353, Pergamon Press, Oxford.

Rogers, L.J., Shah, S.P.J. and Goodwin, T.W. 1966. Intracellular localization of mevalonate-activating enzymes in plant cells. Biochem. J. 99: 381.

Schmidt, A., Sandmann, G., Armstrong, G.A., Hearst, J.E. and Böger, P. 1989. Immunological detection of phytoene desaturase in algae and higher plants using an antiserum raised against a bacterial fusion-gene construct. Eur J. Biochem. 184: 375.

Schultz, G. 1990. Biosynthesis of tocopherol in chloroplasts of higher plants. Fat Sci. Technol. 92: 86.

Schulze-Siebert, D. and Schultz, G. 1987. Full autonomy in isoprenoid synthesis in spinach chloroplast. Plant Physiol. Biochem. 25: 145.

Spurgeon S.L., Sathamoorthy, N. and Porter, J.W. 1984. Isopentenyl pyrophosphate isomerase and prenylatranferase from tomato fruit plastids. Arch. Biochem. Biophys. 23O: 446.

Walter, J., Aguejouf, O., Launay, J. and Camara, B. 1990. Induction and control of carotenoid synthesis in chlorophenylthiotriethylamine tolerant cell of *Capsicum annuum.* Presented at the VII International Congress on Plant Tissue and Cell Culture, June 24–29, Amsterdam.

Yokoyama, H., Hsu, W.J., Poling, S.M. and Hayman, E. 1982. Chemical regulation of carotenoid biosynthesis. In "Carotenoid Chemistry and Biochemistry," p. 371, Pergamon Press, Oxford.

Cloning of a Gene Related to the Missing Key Enzyme for Biosynthesis of Ascorbic Acid in Humans

Morimitsu Nishikimi, Takuya Koshizaka and Kunio Yagi
Institute of Applied Biochemistry
Yagi Memorial Park, Mitake
Gifu, Japan

Humans and other primates are lacking in L-gulono-γ-lactone oxidase, a key enzyme in the biosynthesis of L-ascorbic acid in animals, and depend on dietary intake of vitamin C to prevent scurvy. To understand the genetic basis for the enzyme deficiency as well as to produce L-ascorbic acid by genetic engineering, we isolated a cDNA clone encoding rat liver L-gulono-γ-lactone oxidase. Enzymatically active protein was expressed by transfection of monkey cells with the cDNA inserted into a eukaryotic expression vector. By use of the cDNA as a hybridization probe, L-gulono-γ-lactone oxidase genes were isolated from rat and human genomic DNA libraries in phage vectors. Comparison of the nucleotide sequences of several exons between the two species revealed that the human gene has rapidly accumulated mutations under no selective pressure once it ceased to be active, and now exists as a pseudogene in the human genome.

Introduction

Most phylogenetically higher animals are able to synthesize L-ascorbic acid from D-glucose through the pathway shown in Figure 1 (Burns, 1959). The conversion of D-glucose to D-glucuronic acid is effected by several enzymes participating in the uronic acid cycle, and L-ascorbic acid is formed from D-glucuronic acid by three enzymatic reactions, viz, reduction, lactonization, and oxidation. Humans, other primates, and guinea pigs are incapable of synthesizing L-ascorbic acid, and consequently depend on a dietary source

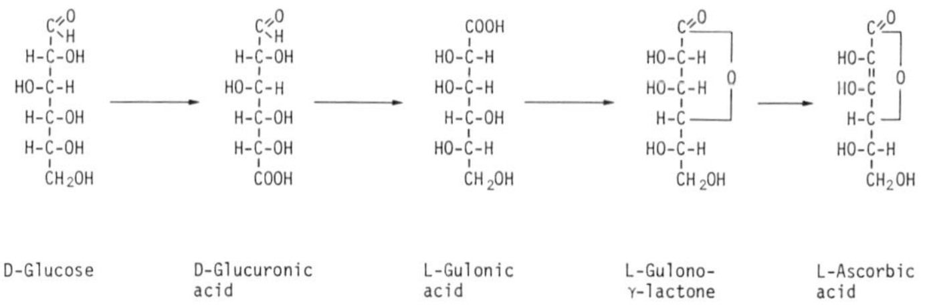

D-Glucose → D-Glucuronic acid → L-Gulonic acid → L-Gulono-γ-lactone → L-Ascorbic acid

Figure 1. The metabolic pathway of L-ascorbic acid biosynthesis in animals.

of vitamin C to prevent scurvy. This metabolic defect arose during evolution of these animals and is carried in all individuals of them. In this sense, the inability to synthesize L-ascorbic acid in such scurvy-prone animals may be regarded as an unusual type of inborn error of metabolism. As early as in the late 1950's, the metabolic defect was found to be caused by the loss of L-gulono-γ-lactone oxidase (GLO) (Burns, 1957), the enzyme that catalyzes the last step of L-ascorbic acid biosynthesis in most animals.

Elucidation at the gene level of the molecular mechanism underlying this enzyme deficiency should provide a definite answer to the interesting question of what has become of the GLO gene that was once active in the ancestors of the scurvy-prone animals. This knowledge should also help us understand how this deficiency could be normalized by application of biotechnological methods such as gene transfer. With these views in mind, we have been investigating the genetic defect in the GLO deficiencies in the scurvy-prone animals mentioned above. In this article, we will summarize our study, focusing on the GLO deficiency in humans, and also describe the expression of a minigene for GLO in monkey cells.

Cloning of a cDNA for Rat GLO

In an attempt to elucidate the genetic defect in the GLO deficiency at the gene level, it is essential to isolate a cDNA for GLO of an L-ascorbic acid-synthesizing animal. In so doing, screening of cDNA libraries constructed with prokaryotic expression vectors is the method of choice among screening methods, provided that antiserum against the enzyme in question is available. In our previous study on GLO (Nishikimi *et al.*, 1976; Kiuchi *et al.*, 1982), the methods for purification of this enzyme from several animals were established and rabbit antiserum directed against it was obtained. By using the antiserum against rat GLO as a probe to

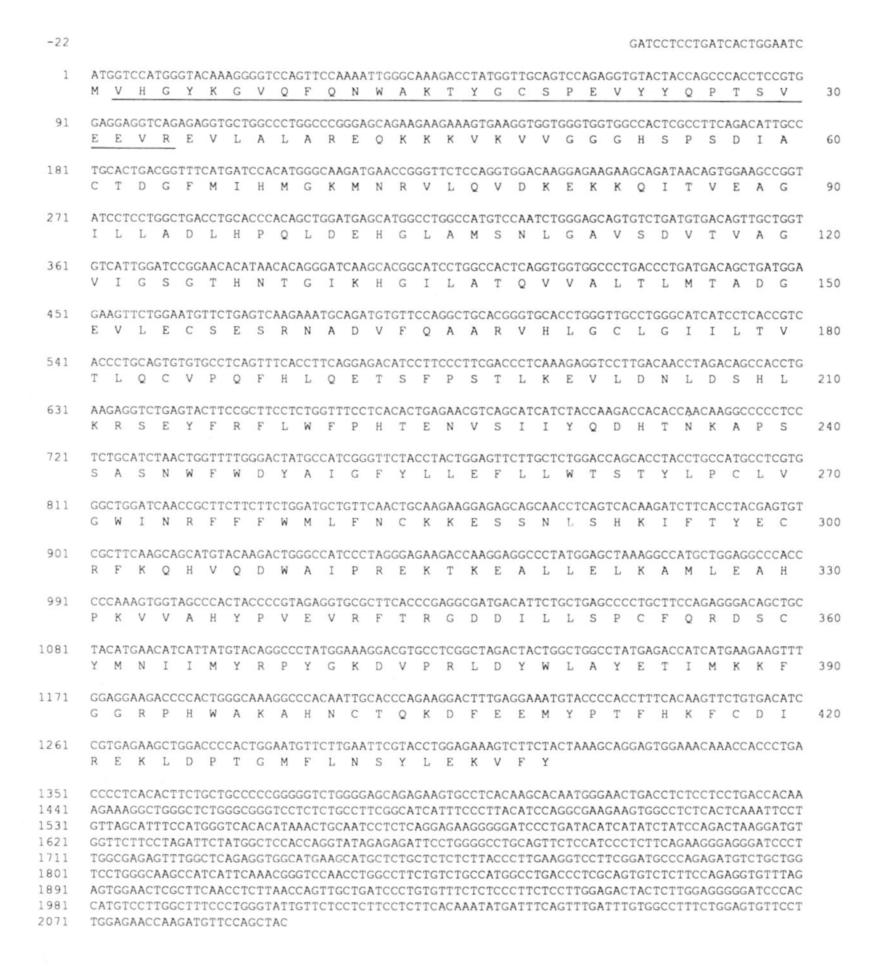

Figure 2. The nucleotide sequence of a cDNA encoding rat liver GLO and its deduced amino acid sequence. The data are taken from our previous paper (Koshizaka *et al.*, 1988) with a few revisions. The revisions made are as follows: a change from G to C at nucleotide position 567 accompanying a change from glutamine to histidine at amino acid residue 189; and deletion of the G at nucleotide position 1 and the C's at positions 1460 and 2097 of the previous sequence. The amino-terminal amino acid sequence determined by the sequence analysis of purified rat liver GLO is underlined. Adapted from Koshizaka *et al.* (1988). (Reproduced with permission.)

screen a rat liver cDNA library in an expression vector, λgt11, we succeeded in isolating a cDNA clone encoding the entire amino acid sequence of rat GLO (Koshizaka *et al.*, 1988). The nucleotide sequence of the cDNA and its deduced amino acid sequence are shown in Figure 2. The authenticity of the cloned cDNA was confirmed by the following: (1) complete matching of the amino-terminal 33-amino acid sequence of rat liver GLO with that of the deduced amino acid sequence, (2) reasonable agreement of the amino acid composition of rat GLO with that calculated from the deduced sequence, and (3) expression of enzymatically active GLO in COS-1 cells by transfection with a eukaryotic expression vector containing the cDNA. We will discuss the study of expression of this minigene in some detail below.

In the nucleotide sequence of the cDNA, there is an open reading frame of 1320 nucleotides, which starts from the first ATG codon and ends with a TAA termination codon. The sequence ATC*ATG*G surrounding the ATG codon agrees well with the consensus initiation sequence (PCC*ATG*G) described by Kozak (1984). Thus, it was concluded that the cDNA encodes a 440-amino acid polypeptide. As the amino-terminal methionine is cleaved off to form the mature GLO, the mature enzyme protein consists of 439 amino acids with a molecular weight of 50,483. Since this enzyme possesses a flavin adenine dinucleotide in the covalently bound form (Kenney *et al.*, 1976), the molecular weight of the mature enzyme is presumed to be 51,267. This value is comparable to that estimated for rat GLO by sodium dodecyl sulfate-polyacrylamide gel electrophoresis (Nishikimi *et al.*, 1976). In addition to the coding sequence, the cDNA has a fairly long 3′-noncoding region (776 nucleotides) and a short 5′-noncoding region (22 nucleotides). There is neither a poly(A) tail nor a polyadenylation signal, and thus the cDNA is not of full length. The presence of several strongly hydrophobic regions was indicated by the hydropathy analysis of Kyte and Doolittle (1982). Since GLO is tightly associated with the microsomal membrane (Nakagawa *et al.*, 1970), these regions may anchor the protein into the membrane; however, they appear not to form the typical transmembrane α-helical structure, as deduced by the method of Chou and Fassman (1974).

Expression of a Minigene for GLO

The GLO cDNA has the potential to be used to produce L-ascorbic acid by genetic engineering techniques. Since humans and other primates are lacking in GLO, their cells provide a convenient system to investigate the expression of the enzyme. We constructed a minigene for GLO by inserting a truncated form of the rat cDNA into the eukaryotic expression vector pSVL (Templeton and Eckhart, 1984) (Figure 3), and introduced it into COS-1 cells, a cell line derived from the kidney of the African green monkey (Yagi *et al.*, 1990). The results indicated that the cDNA placed under the control of the SV40 late promoter could function to produce enzymatically active GLO as shown in Figure 4. The size of the enzyme produced was

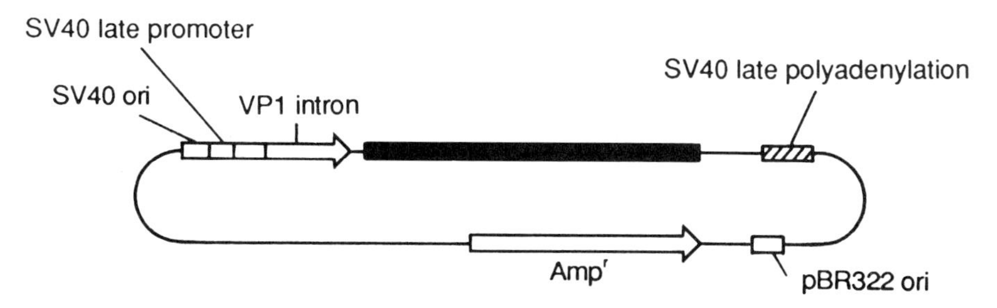

Figure 3. **The construct of a minigene for GLO. A truncated form of the rat GLO cDNA (nucleotides 9 to 1610 in Figure 2) was placed downstream of the SV40 late promoter in the eukaryotic expression vector pSVL (Templeton *et al.*, 1984. Reproduced with permission.) The part of the cDNA is represented by the solid box.**

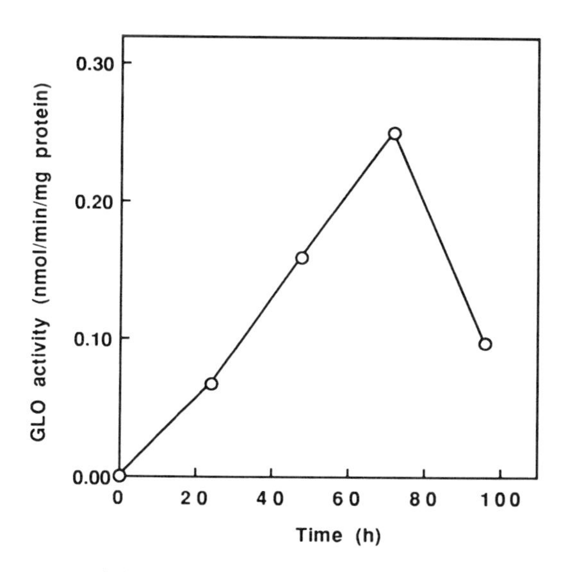

Figure 4. **Time course of the expression of a minigene for rat GLO in COS-I cells, a monkey cell line. The construct of the rat GLO minigene shown in Figure 3 was transferred into COS-1 cells by the calcium phosphate coprecipitation method. The cells were collected at the indicated times, and GLO activity in the cells disrupted in phosphate-buffered saline containing 0.4 % sodium deoxycholate was determined as described (Nishikimi *et al.*, 1989. Reproduced with permission.)**

indistinguishable from the rat GLO, as demonstrated by Western blot analysis. Furthermore, cell fractionation study showed that the enzyme was present in the microsomal fraction, indicating that it is localized in the endoplasmic reticulum of the cell, the site of GLO localization in rat liver cells.

Genetic Defect of GLO Deficiency in Humans

Humans are deficient in GLO activity as mentioned above. Then, what is the molecular basis for this deficiency? The first question addressed at the protein level is whether humans have an aberrant form of GLO. Radioimmuno-assay using anti-rat GLO rabbit antibody showed that the amount of cross-reacting material in human liver was below the limit of detection (Sato and Udenfriend, 1978), a situation similar to the GLO deficiency in guinea pigs (Nishikimi and Udenfriend, 1976).

With the rat GLO cDNA in hand, it became possible for us to see whether or not the human genome contains any sequence that is related to GLO (Nishikimi *et al.*, 1988). Southern blot analysis of genomic DNAs from various species of animals showed that the human genome does have a nucleotide sequence that is hybridizable with the rat GLO cDNA, as do the genomes of several species of animals (mouse, dogs, cow, chicken) that are capable of synthesizing L-ascorbic acid (Figure 5). However, the human sequence was less homologous to the rat GLO cDNA than were the sequences of the L-ascorbic acid-synthesizing mammals and even the chicken sequence, as judged by the intensities of the hybridization signals. The latter finding indicates that the human sequence has altered to a large extent during the course of human evolution.

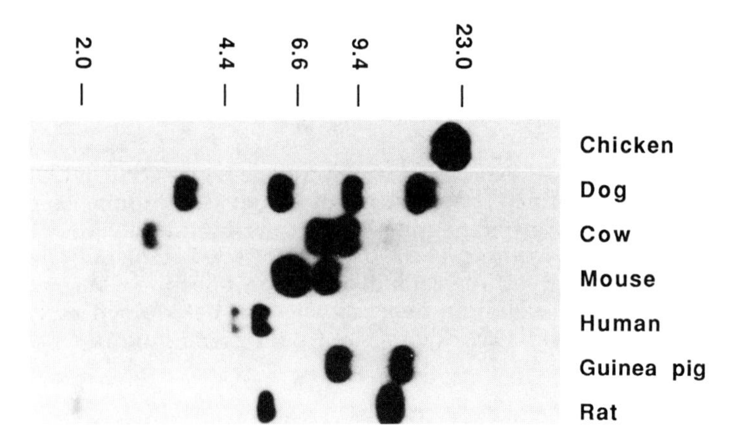

Figure 5. Southern blot analysis of genomic DNAs from various animals. Genomic DNAs from the indicated animals were digested with *Eco*RI and analyzed by Southern blot hybridization using, as a probe, the rat GLO cDNA fragment covering the 5′-noncoding region and most of the coding region (nucleotides-22-1293). Size markers are shown on the left. Adapted from Nishikimi *et al.* (1988). (Reproduced with permission.)

To gain quantitative information regarding the molecular evolution of GLO genes, it is essential to isolate the rat GLO gene and also to isolate the GLO-related sequences in the human genome. Screening of partial *Eco*RI- and *Hae*III-cut rat genomic DNA libraries in Charon 4A led to isolation of three overlapping clones encoding the entire amino acid sequence of rat GLO (Nishikimi and Yagi, 1990, unpublished data). The rat GLO gene was found to be over 20 kbp in length. Intron-exon junctions were located by comparison with the rat cDNA sequence. The coding sequence is interupted by 11 introns. All of the splice donor and acceptor sites follow the "GT/AG" rule (Mount, 1982). The determined nucleotide sequence of the exons of the cloned genomic DNA was identical with the cDNA sequence, with the exception of three differences at nucleotide positions 252, 432, and 1211 of the cDNA. The first two differences (G to A at position 252 and C to T at position 432) occur in the third position of the codons for glutamine and threonine, respectively, and most probably are due to allelic polymorphism. The third difference (A to G at position 1211) gives a change of amino acid from glutamine to arginine. This may be a polymorphism too, or may be an artifact that arose during the cloning procedures.

By screening a human genomic DNA library in EMBL3, we isolated a clone that contained nucleotide sequences homologous to at least four exons of the rat GLO gene (Nishikimi and Yagi, 1990, unpublished data). This is a partial clone for the human GLO gene, since the human genome seems to contain the nucleotide sequence covering the entire coding region (Nishikimi *et al.*, 1988). The determined human nucleotide sequences are shown in Figure 6. Although the sequences corresponding to the two relatively short rat exons are expected to occur in the cloned genomic DNA, they were not detected by Southern blot analysis using the rat GLO cDNA as a probe. This may be due to too weak hybridization resulting from a large number of substitutions in the human exon sequences. There were two single-base and one three-base deletions and one single-base insertion in the four exon-related regions of the human sequences. It was also noted that two out of the eight intron-exon junctions sequenced in the human GLO gene did not follow the "GT/AG" rule. When the sequences of the exon-related regions were compared with those of the corresponding four rat exons, the overall homology between the two species was approximately 80 percent. When compared at the amino acid level, they showed still lower homology (approximately 70 percent) with many drastic changes of

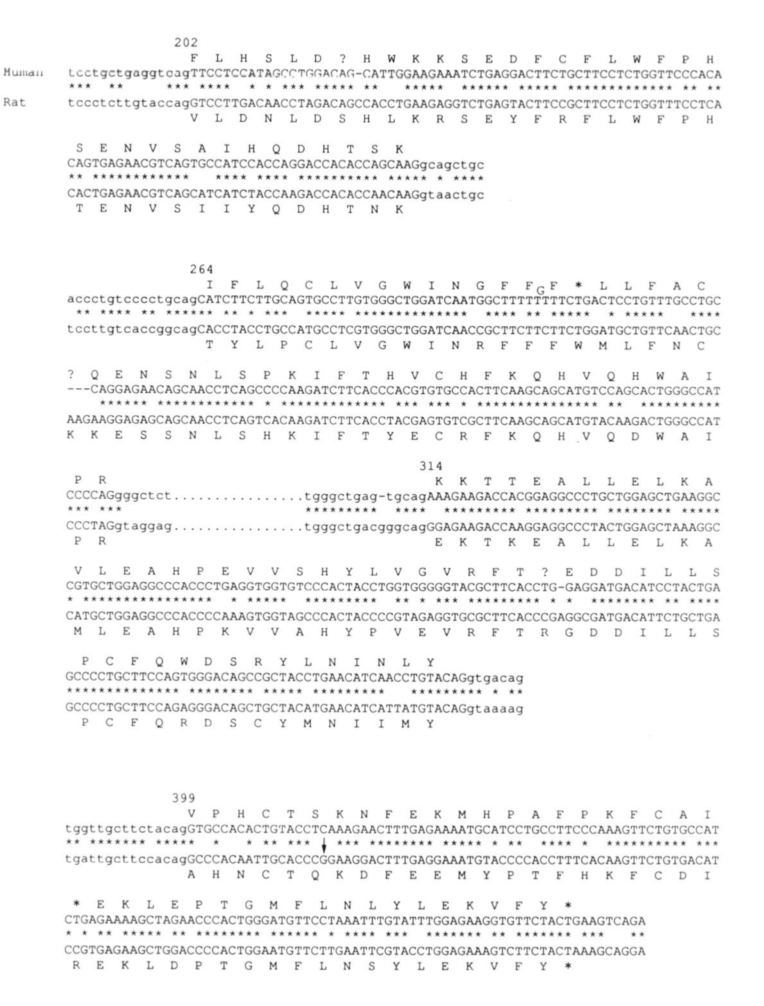

Figure 6. Comparison of four exons of the rat GLO gene with the human nucleotide sequences related to them. Exon sequences are shown by uppercase letters, and intron sequences by lowercase letters. Identical residues in the two sequences are indicated by asterisks. The deduced amino acid sequences for humans and rats are shown above and below the respective sequences. The numbers above the human amino acid sequences indicate the residue numbers of rat GLO shown in Figure 2. Question marks in the human amino acid sequences represent the positions where there is a deletion of nucleotide(s) in the human sequence; and *, stop codon. In the genomic sequence of the rat GLO gene, the A at nucleotide position 1211 of the cDNA shown in Figure 2 is G (arrow).

amino acids and appearance of a stop codon in the human sequence. These results clearly indicate that the human GLO gene has rapidly accumulated mutations under no selective pressure once it ceased to be active, and now exists as a pseudogene in the human genome.

References

Burns, J.J. 1957. Missing step in man, monkey and guinea pig required for the biosynthesis of L-ascorbic acid. Nature 180: 553.

Burns, J.J. 1959. Biosynthesis of L-Ascorbic acid: Basic defect in scurvy. Am. J. Med. 26: 740.

Chou, P.Y. and Fassman, G.D. 1974. Prediction of protein conformation. Biochemistry 13: 222.

Kenney, W.C. Edmondson, D.E., Singer, T.P., Nakagawa, H., Asano, A. and Sato, R. 1976. Identification of the covalently bound flavin of L-gulono-γ-lactone oxidase. Biochem. Biophys. Res. Commun. 71: 1194.

Kiuchi, K., Nishikimi, M. and Yagi, K. 1982. Purification and characterization of L-gulonolactone oxidase from chicken kidney microsomes. Biochemistry 21: 5076.

Koshizaka, T., Nishikimi, M., Ozawa, T. and Yagi, K. 1988. Isolation and sequence analysis of a complementary DNA encoding rat liver L-gulono-γ-lactone oxidase, a key enzyme for L-ascorbic acid biosynthesis. J. Biol. Chem. 263: 1619.

Kozak, M. 1984. Compilation and analysis of sequences upstream from the translational start site in eukaryotic mRNA. Nucleic Acids Res. 12: 875.

Kyte, J. and Doolittle, R.F. 1982. A simple method for displaying the hydropathic character of protein. J. Mol. Biol. 157: 105.

Mount, S.M. 1982. A catalogue of splice junction sequences. Nucleic Acids Res. 10: 459.

Nakagawa, H. and Asano, A. 1970. Ascorbate synthesizing systems. I. Gulonolactone-reducible pigment as a prosthetic group of gulonolactone oxidase. J. Biochem. (Tokyo) 68: 737.

Nishikimi, M., Tolbert, B. and Udenfriend, S. 1976. Purification and characterization of L-gulono-γ-lactone oxidase from rat and goat liver. Arch. Biochem. Biophys. 175: 427.

Nishikimi, M. and Udenfriend, S. 1976. Immunologic evidence that the gene for L-gulono-γ-lactone oxidase is not expressed in animals subjected to scurvy. Proc. Natl. Acad. Sci. U.S.A. 73: 2066.

Nishikimi, M., Koshizaka, T., Ozawa, T. and Yagi, K. 1988. Occurrence in human and guinea pigs of the gene related to their missing enzyme L-gulono-γ-lactone oxidase. Arch. Biochem. Biophys. 267: 842.

Nishikimi, M., Koshizaka, T., Ozawa, T. and Yagi, K. 1989. Expression of the mutant gene for L-gulono-γ-lactone oxidase in scurvy-prone rats. Experientia (Basel) 45: 126.

Sato, P. and Udenfriend, S. 1978. Scurvy-prone animals, including man, monkey, and guinea pig, do not express the gene for gulonolactone oxidase. Arch. Biochem. Biophys. 187: 158.

Templeton, D. and Eckhart, W. 1984. N-terminal amino acid sequences of the polyoma middle-size T antigen are important for protein kinase activity and cell transformation. Mol. Cell. Biol. 4: 817.

Regulation of Iron Accumulation In Food Crops: Studies Using Single Gene Pea Mutants

---

Ross M. Welch
Leon V. Kochian
U.S. Department of Agriculture
Agricultural Research Service
US Plant, Soil and Nutrition Laboratory
Tower Road
Ithaca, NY 14853

Principal plant foods (including legume seeds and cereal grains) are poor sources of dietary iron because they don't contain adequate quantities of available iron. Also, they can contain compounds that inhibit iron bioavailability. Supplying more iron to seed and grain crops during growth does not dramatically increase the iron concentration in these plant foods, because homeostatic mechanisms highly regulate and control iron absorption, translocation, redistribution, and deposition in plant organs. We have been using single gene pea mutants to characterize the regulation of iron uptake and translocation in plants. Once the control mechanisms are understood, it may be possible to develop new varieties of seed and grain crops which accumulate adequate available quantities of dietary iron in their edible parts.

Introduction

Iron is the nutrient most commonly deficient in people of the United States (U.S. Department of Health and Human Services and U.S. Department of Agriculture, 1989). Iron deficiency occurs most frequently in adolescents, women of menstrual age, pregnant women, and young children. The consequences of iron deficiency include: poor work performance, impaired body temperature regulation, diminished intellectual performance, decreased resistance to infection, and increased sensitivity to heavy metal poisoning (U.S. Department of Health and Human Services, 1988).

Currently, both edible legume seeds and cereal grains are considered to be poor dietary sources of iron because they have relatively low iron concentrations and are reported to contain "antinutritive" substances, which interfere with the absorption and/or utilization of non-heme iron by humans (Welch and House, 1984). Increasing the available iron content of edible portions of major food crops would result in important improvements in the nutritional quality of plant foods and possibly would decrease the incidence of iron deficiency in humans throughout the world.

Unfortunately, current knowledge of the mechanisms that control and regulate the accumulation of iron in plants is meager and insufficient to allow scientists to use modern genetic engineering techniques to improve the iron nutritional quality of legume seeds and cereal grains. Here, we discuss current concepts concerning iron uptake and translocation in plants and present some future research goals that need to be addressed before significant improvements can be made in the iron nutritional quality of these crops.

Additionally, we present some original research findings concerning the regulation of iron transport by root cells using our newly discovered, single-gene, pea mutant, E107 [*Pisum sativum*, L., cv *Sparkle* E107 (*brz brz)*] (Kneen *et al.*, 1990). This mutant, E107, was produced by treating pea seeds with the chemical mutagen, ethylmethane sulfonic acid. E107 accumulates toxic levels of iron in its older leaves when grown in either acid or neutral pH soils, or on nutrient media containing standard concentrations of synthetic iron chelates (Welch and LaRue, 1990). The mutant acts functionally as an iron-deficient plant unable to regulate the amount of iron it accumulates (Grusak *et al.*, 1989, 1990a-c). The availability of this type of mineral transport mutant as a research tool should enable us to better understand the regulation of iron absorption and translocation in plants at the molecular level (i.e., gene expression). A more complete understanding of the molecular mechanisms of iron acquisition are required before rapid progress can be made in improving the iron nutritional quality of seed and grain crops using genetic engineering techniques.

Regulation of Iron Absorption in Higher Plants

At least two distinct types of iron acquisition strategies have evolved in higher plants, one scheme for dicotyledons and non-gramineous monocotyledons, and another for members of the Gramineae family of monocotyledons (i.e., the grasses). Römheld and Marschner (1986a) have assigned the names Strategy I (see Figure 1) and Strategy II (see Figure 2) for the systems proposed for the non-grass and grass species, respectively.

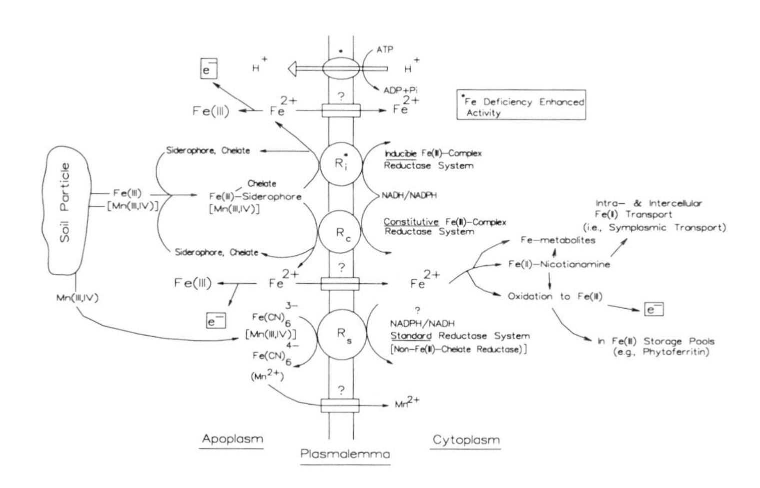

Figure 1. Proposed iron transport model (Strategy I) for non-grasses. Ferric reductases are: R_i, inducible reductase (i.e., "Turbo" reductase); R_c, constitutive reductase; R_s, standard reductase. Other symbols are: e-, electron; ?, possible transport channel; *, indicates enhanced activity or stimulation when iron deficiency stress conditions exists. Modified from Kochian, 1990. (Reproduced with permission.)

Non-grasses. In Strategy I plants (Figure 1), iron is absorbed as the Fe^{2+} cation (i.e., in the Fe(II) oxidation state). However, in most well-aerated soils, iron is not present in any significant amounts in this oxidation state; the Fe(III) oxidation state predominates. The Fe(III) oxidation state forms very insoluble iron oxide and hydroxide precipitates on soil particles and epidermal root-cell surfaces at H^+ ion activities generally found in most soils (i.e., pH values between 5 and 8) . Thus, roots of non-grasses must first solubilize Fe(III) from soil particle surfaces and then reduce Fe(III) to Fe(II) before the iron can be absorbed as the Fe^{2+} ion across the plasma membrane of the root cells (Römheld and Marschner, 1986a).

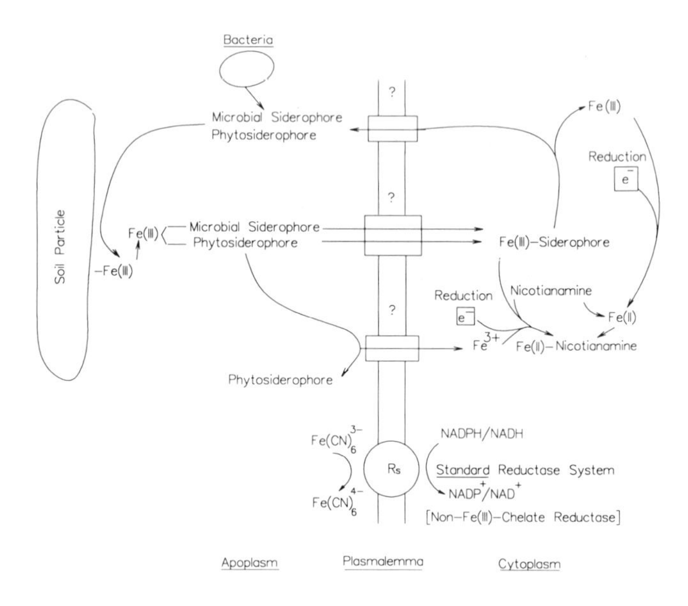

Figure 2. Proposed iron transport model (Strategy II) for grasses. Symbols are: e-, electron; ?, possible membrane transport protein induced under iron deficiency conditions. Modified from Kochian, 1990. (Reproduced with permission.)

Strategy I plants regulate iron absorption by regulating: 1) the rate of reduction of Fe(III) to Fe(II) via a root-cell plasma membrane ferric reductase enzyme system; and 2) the rhizosphere pH, via control of H^+ efflux (Bienfait, 1988a). By extruding protons from the symplasm of root epidermal cells into the soil solution (primarily via the plasma membrane H^+-ATPase), the rhizosphere pH is decreased, which in turn increases the activity of Fe^{3+} ions in soil solution and solubilizes more iron. Reduction of Fe(III) to Fe(II) also increases the solubility of iron and de-stabilizes Fe(III) complexes [e.g., various Fe(III)-siderophores] that form in soil solution via the action of various soil-borne organisms (Chaney, 1987; Römheld and Marschner, 1986a). Other secondary iron deficiency stress responses have been reported for various non-grass species but will not be discussed here (see Longnecker and Welch, 1986 and Kochian, 1990, for a discussion of these responses).

Interestingly, some researchers have suggested that non-grasses may also induce (in addition to the plasma membrane ferric reductase) the synthesis of a Fe(II) transport protein in the plasma membrane of root epidermal cells in response to iron deficiency (Kochian, 1990). However, this hypothesis is still highly speculative even though several ubiquitous soil-borne bacteria can induce the synthesis of iron transport proteins in response to iron deficiency stress (Neilands, 1981).

Grasses. As with non-grasses, grass species also face the problem of acquiring iron from soils containing highly insoluble precipitates of Fe(III). In contrast to non-grasses, current evidence suggests that epidermal and cortical cells of grass roots do not rely on Fe(III) reduction to absorb iron (Bienfait, 1988a; Chaney, 1987; Römheld and Marschner, 1986a). Root cells of Strategy II plant species appear to release phytosiderophores [i.e., naturally occurring Fe(III) binding compounds] into the rhizosphere to mobilize iron to transport sites at the outer surface of the root-cell plasma membrane (Figure 2). At this time it is not clear whether the Fe(III)-phytosiderophore complex is absorbed *in toto,* or whether the iron is split from the chelate prior to absorption, possibly via reduction of Fe^{3+} to the Fe^{2+} ion (Kochian, 1990).

The regulation of iron transport processes in grasses is not as well understood as the regulation of iron uptake by non-grasses. It appears that the plasma membrane of grass root cells does not contain a Fe(III)-chelate reductase system capable of reducing Fe(III) to Fe(II), when the iron is supplied as a highly stable synthetic or naturally occurring Fe(III)-chelate [e.g., Fe(III)-EDTA or Fe(III)-aerobactin]. Nevertheless, a reductase capable of reducing ferricyanide is present in the plasma membrane of grass root cells (Kochian, 1990; Kochian and Lucas, 1990). Recently, Brüggemann *et al.* (1990) have suggested that both non-grasses and grasses contain a similar constitutive Fe(III) reductase system but that, in grasses, this reductase is not very active, functioning only at a very reduced level.

Grasses might regulate iron absorption by controlling both phytosiderophore efflux and the subsequent influx of Fe(III)-phytosiderophore complexes into root cells (Crowley *et al.*, 1988; Kochian, 1990; Römheld and Marschner, 1986b). If true, then there should be specific Fe(III)-phytosiderophore binding sites on the outer surface of the plasma membrane of root cells. This speculation has not been proven although some evidence exists in support of this concept (see references cited above in this paragraph).

Plasma membrane reductases. Over the past decade, considerable evidence has been accumulated indicating that several different electron transport systems function at or across the plant cell plasmalemma that could be involved in physiological processes associated with growth and development. Both intact and excised plant organs such as roots and leaves, as well as protoplasts and plasmalemma vesicles derived from these organs can mediate oxidation-reduction reactions at the plasmalemma. Plasmalemma electron transport systems have been implicated in a number of plant processes, including hormonal control, regulation of cell division and expansive growth, peroxidative defense mechanisms, and ion transport. However, this is an area filled with confusion and controversy, and the involvement and relevance of most of the observed reductase activities to physiological processes are still unclear (for a review, see Kochian and Lucas, 1990). The only physiological process where plasma membrane reductase activity is clearly involved is in Fe(III) reduction and subsequent Fe^{2+} uptake at the root-cell plasma membrane of dicots and non-gramineous monocots.

There is a large body of evidence that strongly supports the existence of a plasma membrane-bound Fe(III) reductase system that facilitates Fe(III) reduction in response to iron deficiency. This was first suggested in the work of Chaney *et al.* (1972), which demonstrated that Fe(III)-chelates must be reduced prior to uptake. The authors speculated that the reduction was mediated by a membrane-bound reducing system that transferred electrons from the cytosol to the external face of the plasma membrane. Subsequently, Bienfait and his coworkers demonstrated that Fe(III) reduction by roots of Fe-deficient bean (*Phaseolus vulgaris* L.) plants exhibited kinetics consistent with that of matrix-bound enzymes (Bienfait *et al.*, 1983). Subsequently, they showed that exposure of Fe-deficient roots to Fe(III)-EDTA elicited a rapid depolarization of the root-cell membrane potential (Sijmons *et al.*, 1984). More recently, we have conducted a more detailed characterization of the electrophysiology of roots of a single-gene mutant of pea (*Pisum sativum* L.), named E107, that accumulates toxic levels of iron in its leaves and will be discussed in detail later. We have demonstrated that in the regions of the root that were involved in Fe(III) reduction, exposure to Fe(III)-EDTA elicited a depolarization of the membrane potential that was not due to secondary alterations of other electrogenic (H^+ and K^+) transport processes (Figure 3 and Grusak *et al.*, 1989). Using the Fe(II) chelate, BPDS, it was possible to separate the depolarization into a reduction component (transfer of electrons from the cytosol to the outer face of the plasma membrane), and a depolarizing Fe(II) ion uptake component. These results are strong evidence for a membrane-mediated reduction.

Wergin *et al.* (1988) conducted an electron microscope examination of Prussian blue staining [a specific stain for Fe(III) reduction] of root hairs from Fe-deficient tomato plants. They showed that the Prussian blue accumulated between the plasma membrane and cell wall. These observations also support the existance of a plama membrane bound Fe(III) reductase at the surface of root epidermal cells.

The strongest evidence for an iron reductase localized in the plasma membrane of root cells comes from the recent work of Buckhout *et al.* (1989), utilizing plasma membrane vesicles isolated from Fe-deficient and Fe-replete tomato (*Lycopersicum esculentum,* Mill.) roots via aqueous two-phase partitioning techniques. They reported that plasma membranes isolated from the roots of Fe-deficient plants exhibited a 2-fold increase in NADH-dependent Fe(III)-citrate reduction over plasma membranes from roots grown under Fe-sufficient conditions. This is a conclusive demonstration of a plasma membrane-bound reductase in plant cells. Based on the above evidence, the model illustrating iron uptake via Fe(III) reduction presented in Figure 1 includes an inducible Fe(III) reductase (R_i) as an integral component.

Bienfait (1985; 1988a) has suggested that roots contain two reductase activities in dicots and nongramineous monocots; one that can reduce Fe(III) chelates and ferricyanide and another capable of reducing only ferricyanide (called the standard system - R_s in Figure 1). The system that could reduce both Fe(III) chelates and ferricyanide would be the inducible reductase (R_i in Figure 1) described above. Bienfait has hypothesized that the ferricyanide reducing system is a separate reductase, and there is evidence for a reductase that is not involved in iron uptake which appears to be expressed in all root cells, including those of grasses, and may be involved in other physiological functions (Rubinstein *et al.*, 1984). Quite possibly, one of these functions might involve the reduction of Mn(III) and/or Mn(IV) to Mn(II) (see Figure 1). Buckhout *et al.* (1989) have challenged this hypothesis, based on their observation that both ferricyanide reduction and Fe(III) chelate reduction increased 2-fold in plasma membranes isolated from Fe-deficient plants over those isolated from Fe-sufficient plants. However, in our recent work on iron absorp tion and Fe-dependent electrical properties in pea roots, it was found that in regions of the root involved in ferric reduction, both Fe(III)EDTA and ferricyanide elicited a depolarization, while in root regions not reducing Fe, only ferricyanide caused a depolarization of the membrane potential (see Figure 3). This is strong evidence in support of two distinct reductase systems.

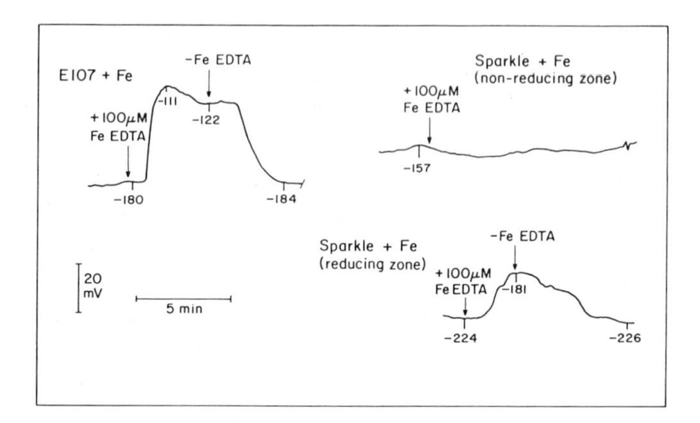

Figure 3. Effect of 100 μM Fe(III)-EDTA on the root-cell membrane potential in +Fe-grown E-107 and *Sparkle* plants. For E107, which exhibits Fe(III) reductase activity along all of the lateral roots except for the terminal 2 cm, the response shown is typical for impalements made in the reducing zone. If a root cortical cell was impaled near the apex [non Fe(III) reducing zone], no electrical response to Fe(III)-EDTA exposure is seen. For roots of +Fe-grown *Sparkle,* Fe(III) reducting activity is observed only in patchy zones. When impalements are made in these zones, a small Fe(III) EDTA-induced depolarization is observed (lower trace). Otherwise, no response to Fe(III)EDTA is observed (upper trace). Exposure to 100 μM ferricyanide, however, causes a significant depolarization both in reducing and non reducing zones of roots of both E107 and *Sparkle.*

Higher Plant Iron Transport Mutants

There are at least three reported single-gene mutations of dicotyledonous plants having defects in their ability to regulate iron transport processes. One, a tomato mutant (*Lycopersicon esculentum* Mill., cv T3820fer, genotype *fer*) cannot induce the synthesis of the root-cell plasma membrane Fe(III)-chelate reductase which regulates Fe(III) reduction in the parent plant, *FER*, cv *Floradel* (Brown *et al.*, 1971). Bienfait (1988b) has proposed that the normal *FER* gene in tomato codes for a protein which controls the transcription of genes that mediate iron efficiency responses in plants, including the synthesis of an "activator" protein responsible for the induction of the synthesis of the "Turbo" or inducible Fe(III)-chelate reductase protein. Because of this single-gene defect, the *fer* genotype develops iron deficiency early after germination unless it is supplied with an adequate level of Fe(II).

Another single-gene tomato mutant, cv *Bonner Beste,* genotype *chloronerva (chln),* also suffers from intractable iron transport problems (Böhme and Scholz, 1960; Rudolph and Scholz, 1972; Scholz,

1983). The *chloronerva* tomato mutant exhibits reduced growth and leaf chlorosis, and expresses typical iron deficiency stress responses [e.g., increased proton extrusion, thickened root tips and root hair zones and elevated root-cell plasma membrane Fe(III)-reductase activity] even while accumulating significant amounts of iron within its roots and shoots (Stephan and Grün, 1989). The chloronerva mutant cannot synthesize nicotianamine (see Figure 4), an amino acid which is capable of forming stable Fe(II)-complexes within physiological pH and E_H ranges. Nicotianamine has been suggested to be an intracellular transporter of Fe(II) (Scholz, 1989; Stephan and Grün, 1989). If true, it could play a key role in the regulation of biosynthetic processes involved in iron uptake and translocation in higher plants.

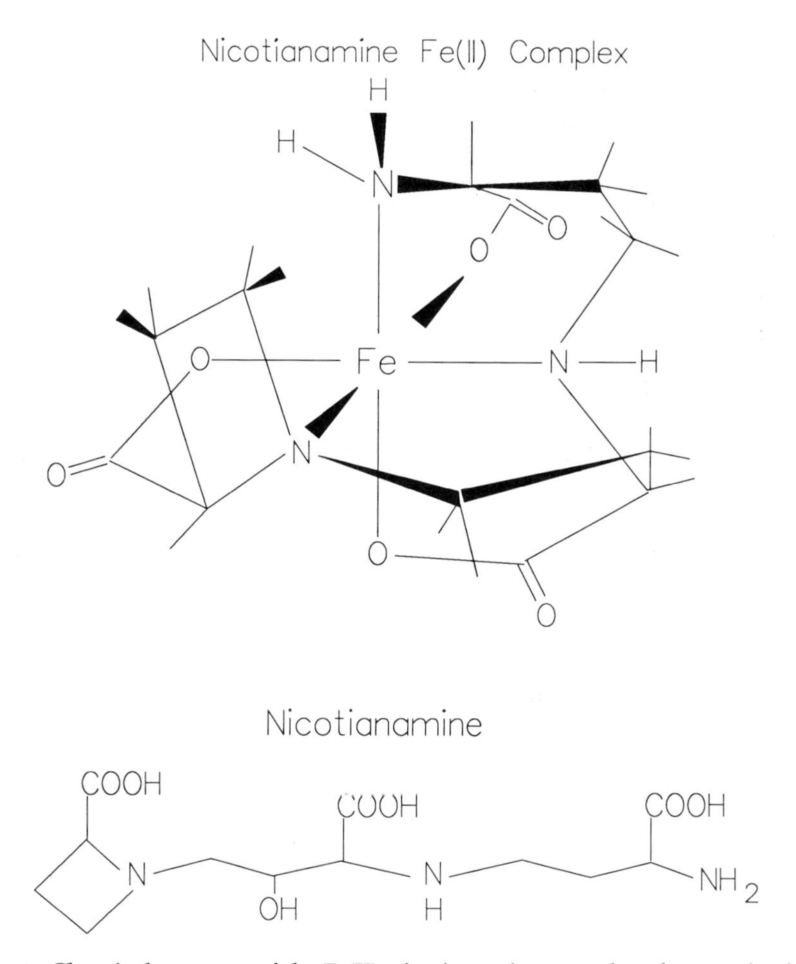

Figure 4. Chemical structure of the Fe(II)-nicotianamine complex, the putative intra- and inter-cellular Fe(II) transporter in plant cells and organs.

In addition to manifesting iron deficiency symptoms and developing iron deficiency stress responses, *chloronerva* also accumulates unusually high concentrations of other divalent cations (e.g., Mn^{2+} and Zn^{2+}) in its roots and tops when compared to the parent genotype *Bonner Beste* (Stephan and Grün, 1989).

The E107 pea genotype is the most recently discovered single-gene higher plant mutant discovered which has abnormal iron transport characteristics (Kneen *et al.*, 1990). It cannot regulate the amount of iron it absorbs, accumulating toxic amounts of iron in its older leaves when grown on conventional nutrient solutions or in soils containing adequate levels of iron for normal plant growth (Welch and LaRue, 1990). We have been studying this pea mutant to determine why it cannot regulate the amount of iron it accumulates (Welch and LaRue, 1990; Grusak *et al.*, 1989, 1990a-c).

Studies Using E107, The Hyper Iron Accumulating Pea Genotype

Ferric reductase activity. Recently, we have studied the effect of iron deficiency on the ability of the roots of the mutant (E107) and parent (*Sparkle*) genotypes to reduce Fe(III)-chelates (Grusak *et al.*, 1989, 1990a, 1990b). We wanted to determine if E107 was capable of regulating the activity of its root-cell Fe(III)-chelate reductase system in a manner similar to the parent genotype, *Sparkle*. Figure 5 shows the effects of growth on nutrient solution containing +/- Fe(III)-EDDHA on the activity of the root Fe(III)-chelate reductase system in E107 and Sparkle, as the plants either maintain iron sufficiency or develop iron deficiency. Daily measurements revealed consistently enhanced Fe(III) reduction rates in +Fe-treated E107 over that of +Fe-treated *Sparkle*, -Fe-treated *Sparkle*, or -Fe-treated E107. The reduction rates in the -Fe-treated *Sparkle* plants did achieve the level of that found for the -Fe-treated E107 plants but never did reach the level of the +Fe-treated E107 plants. From this data we concluded that E107 was not capable of regulating its Fe(III)-chelate reductase system. Apparently, it lacks the genetic capability to synthesize a certain component(s) required for the regulation of this reductase.

Relationship between Fe^{2+} influx and Fe(III) reduction. As stated previously, non-grass species absorb iron in the reduced, Fe^{2+} form. However, in normal well-aerated soil and nutrient media, iron is predominately in the Fe(III) oxidation state and root cells of these species must reduce Fe(III) to Fe(II) in order to absorb iron and meet

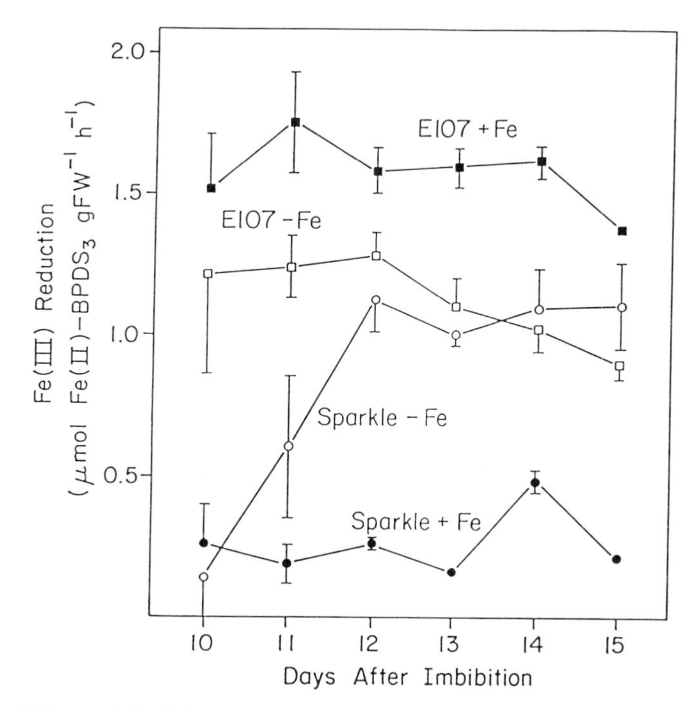

Figure 5. Effects of Fe(III) supply on Fe(III) reduction rates determined in roots of E107 or *Sparkle* pea plants. Iron treatments were (as Fe(III)-EDDHA): +Fe= 1, μM until day 12, then 2 μM for remainder of the growth period; -Fe = no iron added for entire growth period. Error bars represent SEM (n=3). (Reproduced with permission from Grusak *et al.,* 1990b.)

their iron requirements. Thus, these plant species could augment the amount of iron they absorb by either: 1) increasing the activity of the root-cell membrane Fe(III) reductase system, or 2) by enhancing the activity of a Fe^{2+} transport protein (e.g., possibly a transmembrane channel or an ion "carrier" protein), or 3) by modifying both systems. We have studied the effects of iron deficiency on Fe(III)-EDTA reduction and short-term Fe^{2+} absorption rates by E107 and Sparkle pea genotypes in order to determine which process(es) govern iron absorption across the root-cell plasma membrane (Grusak *et al.*, 1990c).

Figure 6 shows the correlation between Fe^{2+} influx and Fe(III) reduction rates determined in excised primary roots from Fe-deficient and Fe-adequate E107 and *Sparkle* plants (Grusak *et al.*, 1990c). The data depict a linear relationship between Fe^{2+} influx and Fe(III) reduction until absorption saturates at higher Fe(III)-EDTA concentrations in the absorption solution. The correlation between Fe^{2+}

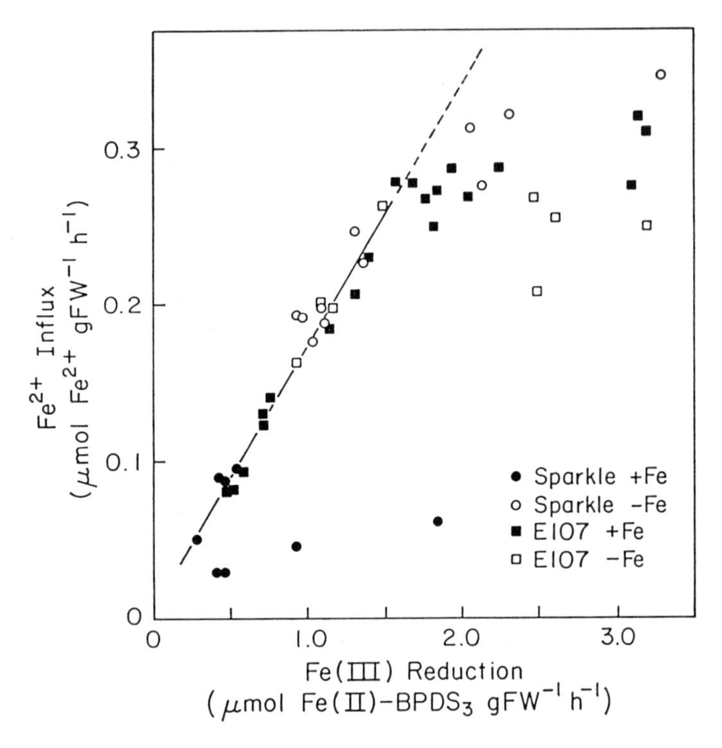

Figure 6. Correlation between Fe(II) influx and Fe (III) reduction rates determined in excised primary lateral roots of Fe-suficient (+Fe) and Fe-deficient (-Fe) Sparkle and E107 pea plants. Values plotted represent data obtained via sequential Fe(III) reduction and Fe^{2+} influx determinations using various concentrations of Fe(III)-EDTA. (Reproduced with permission from Grusak *et al.,* 1990c.)

influx and Fe(III) reduction is independent of iron status of the plants or of the plant genotype. The ratio between iron influx and reduction remains constant, revealing no enhanced Fe^{2+} absorption in roots from either Fe-deficient *Sparkle* or E107 relative to Fe-adequate *Sparkle*. We concluded that the physiological control of iron uptake by plants grown in well-aerated nutrient media is the regulation of Fe(III) reduction, and not Fe^{2+} influx. Apparently, plants do not induce the synthesis of an Fe^{2+} transport protein in response to iron deficiency even though certain microorganisms are known to do so (Neilands, 1981).

Effects of magnesium and manganese supply on iron accumulation. Besides accumulating toxic levels of iron in its older leaves, E107 also accumulates high levels of other divalent cations (e.g., Mg^{2+}, Zn^{2+} and Mn^{2+}), and to a much lesser degree, under some conditions,

Ca^{2+} and the monovalent cation K^+. Furthermore, we observed that growing E107 on full-concentration, complete, nutrient solution greatly reduced the total amount of iron accumulated when compared to E107 plants grown on nutrient solutions containing one-fifth the concentration of nutrients supplied in full-concentration nutrient solutions (unpublished data). *Sparkle*, however, was not greatly affected by the supply of nutrients within this range. Thus, increasing the supply of other macronutrients and micronutrients to E107 was shown to dramatically affect the amount of iron and other cations accumulating by E107.

Initially, we hypothesized that the increased amount of magnesium and/or manganese supplied to E107 plants grown in the full-concentration nutrient solution might be responsible for the dramatic decrease in iron concentrations found in E107 plants, when compared to E107 grown in one-fifth concentration nutrient solutions. We reasoned that Fe^{2+} has many chemical characteristics in common with Mg^{2+} and Mn^{2+} ions and is known to interact with these ions in a number of ways. For example, all have similar hydrated ionic radii in aqueous solution, and they interact competitively in a number of biochemical processes. Additionally, supplying excess manganese levels to plants is known to induce iron deficiency in various plant species and conversely, excessive quantities of iron can lead to manganese deficiency in some plant species (Olsen, 1972). Therefore, we hypothesized that increases in the manganese and magnesium supply in nutrient solution could be responsible for the reduction of iron content observed in E107 plants grown under high nutrient supply conditions, possibly because of the mass action of these ions on reducing Fe^{2+} absorption. We designed experiments to test the hypothesis that increasing the magnesium and manganese supply to E107 would reduce iron uptake [supplied as 2 or 5 μM Fe(III)-EDDHA].

Figure 7 illustrates the effects of low (0.2 mM magnesium and 0.4 μM manganese) and high (1.0 mM magnesium and 2.0 μM manganese) magnesium and manganese supply in the nutrient solution on the total plant iron content (normalized for root dry weights) in 14-day-old E107 and *Sparkle* seedlings. Antithetic to our hypothesis, the results show that neither high levels of magnesium nor high levels of manganese reduced the amount of iron accumulated in plants provided either full-concentration nutrient solution or fifth-concentration nutrient solution. Indeed, E107 plants receiving the

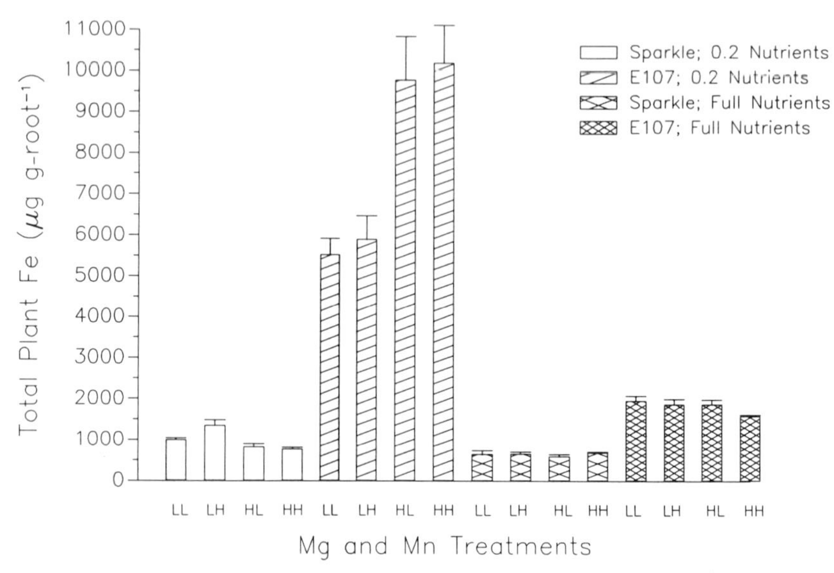

Figure 7. Effects of magnesium and manganese supply on total iron content (roots plus shoots) of 14-day-old pea plants normalized per g dry weight or roots; L represents low (0.2 Mm magnesium or 0.4 μM manganese) and H represents high (1.0 mM magnesium or 2.0 μM manganese) magnesium or manganese treatments. Error bars represent SEM (n=3)

fifth-concentration nutrient solutions and high levels of magnesium actually accumulated significantly more iron than did E107 plants supplied the same nutrient solution treatments but with low magnesium supplies. Varying the manganese supply had little effect on the amount of iron accumulated by either E107 or *Sparkle* plants regardless of the concentration of the nutrient culture media used. Interestingly, providing full-concentration nutrient solutions to E107 plants dramatically reduced their total plant iron content as compared to those supplied one-fifth concentration nutrient solutions while the total content of iron in *Sparkle* plants was much less affected by increasing their nutrient media supplies from one-fifth to full-concentration nutrient solutions.

Apparently, some other nutrient (or combination of nutrients), besides magnesium and manganese, was responsible for the marked reduction of iron in E107 plants grown on full-concentration nutrient solutions as compared to those supplied fifth-concentration nutrient solutions. Currently, we speculate that possibly increasing the calcium supply may have been responsible for the reduction of iron in E107 plants grown with full-concentration nutrient solution. We suggest this for several reasons.

First, the concentration of calcium in mutant E107 plants is within ordinary limits, i.e., calcium concentrations in E107 organs have always been determined to be within the normal range found for the parent genotype *Sparkle* grown under identical conditions while most other cations are not. Therefore, E107 appears to function normally in regulating its calcium content but not when regulating its content of other cations. Interestingly, the regulation of Ca^{2+} absorption by plant cells involves both calcium influx and efflux mechanisms (Hanson, 1984; Marmé, 1983; Poovaiah and Reddy, 1987). These mechanisms are tightly controlled by the plant cell because free cytosolic Ca^{2+} concentrations have to be maintained within narrow limits [i.e., less than 1 M Ca^{2+} for normal cell metabolism and plant growth (Hanson, 1984; Poovaiah and Reddy, 1987)]. Second, in our experiments, providing higher calcium levels to E107 are associated with lower total levels of iron in the whole plant. Third, in aqueous solutions, free calcium exists as a divalent cation (Ca^{2+}) as does the ionic form of iron (Fe^{2+}) absorbed by plant root cells; some non-competitive interaction between these divalent cations in their uptake by root cells is reasonable. Therefore, calcium supply may affect the accumulation of iron in E107 plants via some unrecognized process(es). Could these processes involve the Ca efflux pump proposed to operate at the plasma membrane of plant cells (Briars *et al.*, 1989; Gräf and Weiler, 1989)? Is it possible that Fe^{2+} can be pumped out of cells along with Ca^{2+} via the Ca^{2+} efflux pump if the concentration of free Ca^{2+} and Fe^{2+} ions exceed a certain concentration limit in the cytosol? Further research is in progress to test this possibility.

Future Research Needs

Much more research is needed before scientists can take advantage of modern genetic engineering techniques to significantly increase the amount of iron accumulated in edible portions of major food crops. Among these crops, seeds and grains contribute the greatest amount of dry matter to most human diets while leafy vegetables contribute the least because of their high moisture content. Therefore, in order to have a significant impact on the iron nutritional status of humans, plant scientists need to develop new varieties of major seed and grain crops that contain significantly more bioavailable iron than current commercial varieties.

Unfortunately, only a few of the processes that potentially control the accumulation of iron in seeds and grains have been studied in detail. The processes that have been studied in depth have been limited, almost entirely, to the mechanisms of absorption of iron by root-cells. Almost nothing is known about the transport and deposition of iron within the plant. Much more research is needed on these processes, including: 1) the radial symplasmic movement of iron across the root to vascular elements in the root stele; 2) iron loading into xylem vessel in the stele; 3) the iron species transported in xylem vessels during translocation of iron to plant tops; 4) iron unloading in the apoplasmic spaces of the mesophyll cells of leaves, and iron absorption processes by these leaf cells; 5) re-mobilization and movement processes between leaf cells; 6) phloem loading of iron into vascular elements of the leaf; 7) the movement of iron within the phloem sap and its form in the phloem sap; 8) unloading of iron out of the phloem sap and into sink sites within cells forming reproductive organs; 9) the major forms of iron deposited in seed and grain tissues via the phloem; and 10) the bioavailability of these forms of iron to people. Furthermore, the presence of promotor or inhibitor substances that enhance or inhibit iron bioavailability to humans, that may occur in the edible portions of seeds and grains, have received only modest attention by nutritional scientists (for a more detailed discussion of these research topics see Welch, 1986; Welch and House, 1984). Without much more knowledge concerning these processes, it will be extremely difficult to improve the nutritional quality of these major food crops with respect to iron. Additionally, there may be important evolutionary reasons for the relatively low concentrations of iron found in reproductive organs. Dramatically changing the iron content of seeds and grains may have unforseen negative effects on seed and grain quality and vigor. Thus, this possibility also needs to be studied. Clearly, much more research needs to be carried out in order to answer the questions posed by this brief review. Hopefully, this paper will stimulate new research interest among plant scientists in this important but largely neglected area of crop nutritional quality.

References

Bienfait, H.F. 1985. Regulated redox processes at the plasmalemma of plant root cells and their function in iron uptake. J. Bioenerg. Biomembr. 17: 73-83.

Bienfait, H.F. 1988a. Mechanisms in Fe-efficiency reactions of higher plants. J. Plant Nutr.11: 605.

Bienfait, H.F. 1988b. Proteins under the control of the gene for Fe efficiency in tomato. Plant Physiol. 88: 785.

Bienfait, H.F., Bino, R.J., VanDer Bliek, A.M., Duivenvoorden, J.F., and Fontaine, J.M.1983. Characterization of ferric reducing activity in roots of Fe-deficient *Phaseolus vulgaris*. Physiol. Plant. 59: 196.

Bhöme, H. and Scholz, G. 1960. Versuche zur Normalisierung des Phänotyps der Mutante *chloronerva vön Lycopersicon esculentum* Mill. Kulturpflanze 8: 93.

Briars, S-A, Kessler, F. and Evans, D.E. 1989. The calmodulin-stimulated ATPase of maize coleoptiles is a 140000-MR polypeptide. Planta 176: 283.

Brown, J.C., Chaney, R.L. and Ambler, J.E. 1971. A new tomato mutant is efficient in the transport of iron. Physiol. Plant. 25: 48.

Brüggemann, W., Moog, P.R., Nakagawa, H., Janiesch, P. and Kuiper, P.J.C. 1990. Plasma membrane-bound NADH: Fe^{3+}-EDTA reductase and iron deficiency in tomato (*Lycopersicon esculentum*). Is there a turbo reductase? Physiol. Plant. 79: 339.

Buckhout, T.J., Bell, P.F., Luster, D.G. and Chaney, R.L. 1989. Iron-stress induced redox activity in tomato (*Lycopersicum esculentum* Mill.) is localized on the plasma membrane. Plant Physiol. 90: 151.

Chaney, R.L. 1987. Complexity of iron nutrition: lessons for plant-soil interaction research. J. Plant Nutr. 10: 963.

Chaney, R.L., Brown, R.C. and Tiffin, L.O. 1972. Obligatory reduction of ferric chelates in iron uptake by soybeans. Plant Physiol. 50: 208.

Crowley, D.E., Reid, C.P. and Szaniszlo, P.J. 1988. Utilization of microbial siderophores in iron acquisition by oat. Plant Physiol. 87: 680.

Gräf, P. and Weiler, E.W. 1989. ATP-driven Ca^{2+} transport in sealed plasma membrane vesicles prepared by aqueous two-phase partitioning from leaves of *Commelina communis*. Physiol. Plant. 75: 469.

Grusak, M.A., Kochian, L.V. and Welch, R.M. 1990a. A transport mutant of pea (*Pisum sativum*) for the study of iron absorption in higher plant roots. In "Plant nutrition—physiology and applications," p. 219, Kluwer. Academic Publishers, Boston, MA.

Grusak, M.A., Welch, R.M. and Kochian, L.V. 1989. A transport mutant for the study of plant root iron absorption. In "Plant membrane transport: The current position," p. 61. Elsevier Press, Amsterdam, The Netherlands.

Grusak, M.A., Welch, R.M. and Kochian, L.V. 1990b. Physiological characterization of a single-gene mutant of *Pisum sativum* exhibiting excess iron accumulation. I. Root iron reduction and iron uptake. Plant Physiol. 93: 976.

Grusak, M.A., Welch, R.M. and Kochian, L.V. 1990c. Does iron deficiency in *Pisum sativum* enhance the activity of the root plasmalemma iron transport protein? Plant Physiol. In press.

Hanson, J.B. 1984. The functions of calcium in plant nutrition. Adv. Plant Nutr. 1: 149.

Kneen, B.E., LaRue, T.A., Welch, R.M. and Weeden, N.F. 1990. A mutation in *Pisum sativum* (L.) cv "Sparkle" conditioning decreased nodulation and increased iron uptake and leaf necrosis. Plant Physiol. 93: 717.

Kochian, L.V. 1990. Mechanisms of micronutrient uptake and translocation in plants. In "Micronutrients in Agriculture," 2nd ed., Soil Science Society of America, Madison, WI. In press.

Kochian, L.V and Lucas, W.J. 1990. Do plasmalemma oxidoreductases play a role in plant mineral ion transport? In "Oxidoreductases at the plasma membrane: Relation to growth and development," Vol. Plants. CRC Press, Boca Raton, FL. In press.

Longnecker, N. and Welch, R.M. 1986. The relationships among iron-stress response, iron-efficiency and iron uptake of plants. J. Plant Nutr. 9: 715.

Marmè, D. 1983. Calcium transport and function. In "Inorganic Plant Nutrition," p. 599. Springer-Verlag, New York.

Neilands, J.B. 1981. Iron absorption and transport in microorganisms. Ann. Rev. Nutr. 1: 27.

Olsen, S.R. 1972. Micronutrient interactions. In "Micronutrients in Agriculture," 1st ed. p.243. American Society of Agronomy, Madison, WI.

Poovaiah, B.W. and Reddy, A.S.N. 1987. Calcium messenger system in plants. In "CRC Critical Reviews in Plant Science," Vol. 6, p. 47, CRC Press, Inc., Boca Raton, FL.

Römheld, V. and Marschner, H. 1986a. Mobilization of iron in the rhizosphere of different plant species. Adv. Plant Nutr. 2: 155.

Römheld, V. and Marschner, H. 1986b. Evidence for a specific uptake system for iron phytosiderophores in roots of grasses. Plant Physiol. 80: 175.

Rubinstein, B., Stern, A.I. and Stout, R.G. 1984. Redox activity at the surface of oat root cells. Plant Physiol. 76: 386.
Rudolph, A. and Scholz, G. 1972. Physiologische Untersuchungen an der Mutante *chloronerva* von *Lycopersicon esculentum* Mill. IV. Über eine Methode zur quantitativen Bestimmung des "Normalisierungsfaktors" sowie über dessen Vorkommen im Pflanzenreich. Biochem. Physiol. Pflanzen 163: 156.
Scholz, G. 1983. The amino acid nicotianamine, an effector of root elongation for the tomato mutant chloronerva (Lycopersicon esculentum Mill.). Plant. Sci. Lett. 32: 327.
Scholz, G. 1989. Effect of nicotianamine on iron re-mobilization in de-rooted tomato seedlings. Biol. Metals 2: 89.
Sijmons, P.C., Lanfermeijer, F.C., DeBoer, A.H., Prins, H.B.A. and Bienfait, H.F. 1984. Depolarization of cell membrane potential during trans-plasma membrane electron transfer to extracellular electron acceptors in iron-deficient roots of *Phaseolus vulgaris* L. Plant Physiol. 76: 943.
Stephan, U.W. and Grün, M. 1989. Physiological disorders of the nicotianamine-auxotroph tomato mutant *chloronerva* at different levels of iron nutrition II. Iron deficiency response and heavy metal metabolism. Biochem. Physiol. Pflanzen 185: 189-200.
U.S. Department of Health and Human Services and U.S. Department of Agriculture. 1989. "Nutrition Monitoring in the United States," DHHS Publication No. (PHS) 89-1255, U.S. Government Printing Office, Washington, D.C.
U.S. Department of Health and Human Services. 1988. "The Surgeon General's Report on Nutrition and Health," DHHS Publication No. (PHS) 88-50210, U.S. Government Printing Office, Washington, D.C.
Wergin, W.P., Bell, P.F. and Chaney, R.L. 1988. Use of SEM, X-ray analysis and TEM to localize Fe^{3+} reduction in roots. In "Proceedings 46th Microscopy Society of America," p. 392. San Francisco Press, San Francisco, CA.
Welch, R.M. 1986. Effects of nutrient deficiencies on seed production and quality. Adv. Plant Nutr. 2: 205.
Welch, R.M. and House, W.A. 1984. Factors affecting the bioavailability of mineral nutrients in plant foods. In "Crops as Sources of Nutrients for Humans," ASA Special Publication No. 48, p. 37. American Society of Agronomy, Madison, WI.

Welch, R.M. and LaRue, T.A. 1990. Physiological characteristics of Fe accumulation in the "Bronze" mutant of *Pisum sativum* L., cv "Sparkle" E107 *(brz brz)*. Plant Physiol.93:723.

Acknowledgements

We thank Dr. Michael A. Grusak for his helpful discussions during the preparation of this manuscript.

Ellagic Acid Enhancement in Strawberries

John L. Maas, Gene J. Galletta and Shiow Y. Wang
Fruit Laboratory
Agricultural Research Service
United States Department of Agriculture
Beltsville, MD 20705

Ellagic acid is a naturally occurring phenolic constituent of plants, including many that are important in our diet. Interest in ellagic acid has increased greatly during the past decade due to its effectiveness as an antimutagen and its potential as an inhibitor of chemically induced cancer. Much has been learned concerning the diverse clinical attributes of ellagic acid, but relatively little is known about its physiological, genetic, and ecological aspects and its ability to form derivatives (ellagitannins) in the plant.

Manipulation of the gene(s) responsible for ellagic acid synthesis to develop ellagic acid-enhanced cultivars will depend on a better understanding of the constitutive and inducible enzyme systems in plant cells that control the phenylpropanoid end-product pathways to primarily ellagitannins, other hydrolyzable and/or condensed tannins, or flavonoids, including anthocyanins. Phenylalanine ammonia-lyase, the key enzyme of phenylpropanoid metabolism, may or may not be directly responsible for end-product synthesis. We review here the biosynthesis of ellagic acid in general and of strawberry in particular, the genetics of ellagic acid biosynthesis in strawberry, and the possibility to produce improved strawberries with increased ellagic acid content.

Introduction

Ellagic acid is a naturally occurring phenolic constituent of many plants that are important in the human diet, especially fruits and nuts. Interest in ellagic acid has increased greatly during the last decade due to its effectiveness as an antimutagen and anticarcinogen and its potential as an inhibitor of chemically induced cancer. Much has been learned since 1980 concerning the clinical attributes of ellagic acid, but relatively little is known about the physiological, genetic, and ecological aspects of ellagic acid and its naturally occurring derivatives in the plant. A large amount of research is also being conducted, especially in Japan, on ellagitannins as active constituents of medicinal plants (Okuda *et al.*, 1989a).

In nature, ellagic acid may occur in free form, but more commonly in the form of ellagitannins as esters of the diphenic acid analog on glucose. These ellagitannins differ in solubility, mobility, and reactivity in plant as well as in animal systems. Research has progressed at a rapid rate among medical research programs studying ellagic acid as a potential anticarcinogen, as seen in reviews by De Flora and Ramel (1988), Hayatsu *et al.* (1988), Stoner (1989), and Maas *et al.* (1991a), as an inhibitor of the human immunodeficiency virus (Take *et al.*, 1989; Asanaka *et al.*, 1988), and in blood clotting research (Bock *et al.*, 1981).

Significantly, ellagic acid may mediate cancer inhibition through several mechanisms. (1) Ellagic acid may inhibit the metabolic activation of carcinogens. For example, ellagic acid may inhibit the conversion of polycyclic aromatic hydrocarbons [e.g., benzo(*a*)pyrene, 7,12- dimethylbenz(*a*)anthracene, and 3-methylcholanthrene], nitroso compounds (e.g., *N*-nitrosobenzylmethylamine and *N*-methyl-*N*-nitrosourea), and aflatoxin B_1 into forms that induce genetic damage (Dixit *et al.*, 1985; Teel *et al.*, 1985; Mandal *et al.*, 1987; Mandal *et al.*, 1988). (2) Carcinogens may be detoxified by stimulation of enzyme (e.g., glutathione s-transferase) activity. Such increase in enzyme activity towards benzo(a)pyrene-4,5-oxide and 1-chloro-2,4-dinitrobenzene substrates has been demonstrated in mice (Das *et al.*, 1985). (3) Ellagic acid may bind to reactive metabolic forms of the carcinogen to form a harmless complex that is incapable of reacting with cellular DNA, that is, it acts as a scavenger. For example, ellagic acid reacts with benzo(*a*)pyrene diol epoxide by taking a sterically favorable position to form a covalently linked product in which the reactive epoxide ring of the pyrene is opened, rendering the carcinogen harmless (Sayer *et al.*, 1982). (4) Ellagic acid may occupy sites in DNA that might otherwise react with carcinogens or their metabolites. In this manner, ellagic acid inhibited the binding of *N*-methyl-*N*-nitrosourea to salmon sperm DNA by reacting with the O^6 position in quanine and preventing methylation at that site (Dixit and Gold, 1986; Teel, 1986).

Although less is known about the function and attributes of ellagic acid and ellagitannins in plants, recent studies indicate these compounds may participate in plant hormone regulatory systems (Runkova *et al.*, 1972), allelopathy (Rice, 1984), fungistasis (Hart and

Hillis, 1972), auto-inhibition (Vieitez and Ballester, 1988; Hildebrandt and Harney, 1989), insect growth inhibition (Klocke *et al.*, 1986), and insect feeding deterrency (Jones and Klocke, 1987). Recent commercial interest in ellagic acid in plant systems has been largely for fruit juice processing (Boyle and Hsu, 1990) and wine applications (Singleton *et al.*, 1966) and in wood pulping (Press and Hardcastle, 1969).

Biosynthesis of Phenylpropanoids

In flavonoid (especially anthocyanin pigments) biosynthesis (Fig. 1), three major metabolic pathways apparently are active; these are the shikimate (producing phenylalanine and other aromatic amino acids), phenylpropanoid (producing cinnamic acid derivatives), and the flavonoid pathways (Hrazdina and Jensen, 1990). These pathways, and the factors affecting them, may have a large influence on other pathways including ellagitannin biosynthesis.

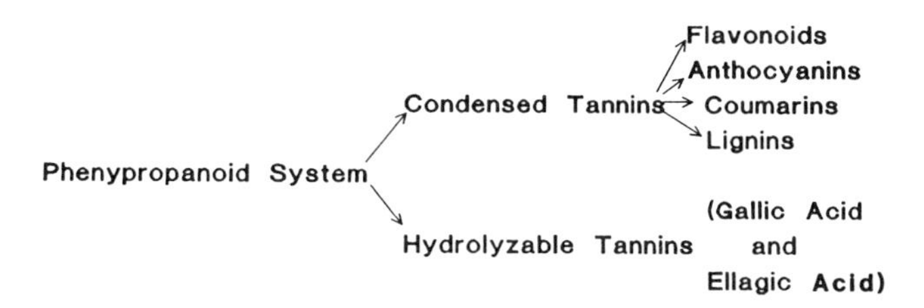

Figure 1. Polyphenol derivatives of the phenylpropanoid pathway.

Ellagic acid is believed to be synthesized primarily by way of the shikimic acid pathway to phenylpropanoid metabolism. Phenylalanine is deaminated by phenylalanine ammonia lyase (PAL) to cinnamic acid (Fig. 2), which, in turn, is converted to 4-coumaric acid by cinnamate 4-hydroxylase and to 4-coumaroyl-CoA by 4-coumarate: CoA ligase (Halbrock and Scheel, 1989). These core reactions in the phenylpropanoid metabolism have been variously studied but the deamination of phenylalanine by PAL has received the most attention (Camm and Towers, 1973). PAL activity is known to be affected by changes in light intensity, wounding and ethylene production, hormones and growth modifiers, carbohydrate levels, tissue and organ type, age and development stage, and genotype (Camm and Towers, 1973).

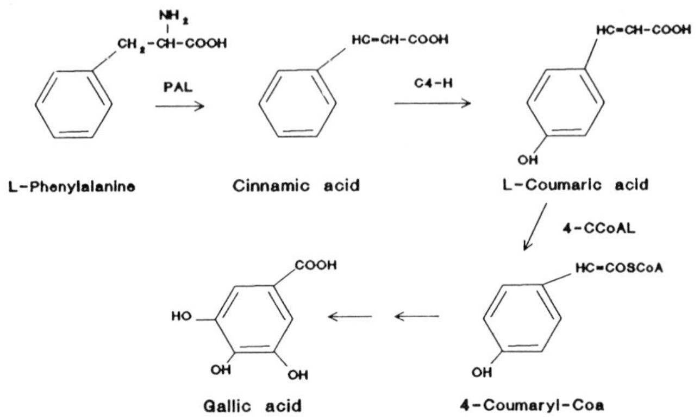

NH₂
CH₂-CH-COOH
PAL
HC=CH-COOH
C4-H
HC=CH-COOH
OH
L-Phenylalanine
Cinnamic acid
L-Coumaric acid
4-CCoAL
COOH
HC=COSCoA
HO
OH
OH
OH
Gallic acid
4-Coumaryl-Coa

Figure 2. Biosynthesis of gallic acid from L-phenylalanine. PAL=phenylalanine ammonia-lyase; C4-H=cinnamate 4-hydroxylase; 4-CCoA1= 4-coumarate CoA ligase.

The key branch point enzymes of the shikimic and phenylpropanoid pathways have been shown to exist as isoenzymes. Hrazdina and Jensen (1990) suggest that the existence of isoenzymes is in the duplication of pathways or pathway segments in the subcellular environment. Isozymes involved in the shikimic and phenylpropanoid segments may be localized to function, on one hand, in protein biosynthesis in chloroplasts from aromatic amino acids (phenylalanine, tyrosine, and tryptophan) and on the other hand, in polyphenol biosynthesis in the cytoplasm.

Increasing evidence indicates that enzymes functioning in phenylpropanoid metabolism and flavonoid biosynthesis exist in an organized manner associated with cytoplasmic membrane structures and not disassociated in the cytosol (Hrazdina and Jensen, 1990). Data are lacking to indicate whether or not enzymes involved in hydrolyzable tannin biosynthesis (e.g., ellagitannins) are membrane associated. However, in view of the association of other major polyphenol biosynthesis pathways with cytoplasmic membranes, it may be logical to assume that components of ellagitannin synthesis are also membrane associated.

Chalcone synthase, the first committed enzyme of the flavonoid biosynthesis section, is present as a series of isoenzymes, each of which may be involved in specialized pathways of flavoid compound synthesis (Hrazdina and Jensen, 1990). According to Hrazdina and Jensen (1990) each of these isoenzymes may be under the control of specific genes. Six to eight chalcone synthase genes, some tightly clustered, have been determined in *Phaseolus vulgaris* (Ryder *et al.*, 1987).

PAL isolated from bean (*Phaseolus vulgaris*) cell suspension cultures is a mixture of four isozymes (Bolwell *et al.*, 1985), although studies with other species showed fewer or no PAL isozymes (Camm and Towers, 1973). The genetics of PAL may be extremely complicated. For example, in potato, as many as 20 or more cloned genes have been demonstrated (Halbrock and Sheel, 1989). In addition, PAL genes may be regulated differently in different plant parts (root, stem, leaf and flower) as well as induction by wounding, infection or illumination (Halbrock and Sheel, 1989).

The existence of isozymes in the biosynthesis of ellagic acid and ellagitannins has not been determined. However, the presence or absence of ellagic acid has been considered to be important cytotaxonomically (Bate-Smith, 1956; 1961a), indicating that major genes for ellagitannin pathways are absent or repressed in major taxons. Species of related genera may also differ in their biosynthesis of either gallotannins or ellagitannins, where gallic acid is synthesized but the dimerization to ellagic acid does not occur. In the genus *Vitis* (grapes), muscadine grapes (*V. rotundifolia* Michx.) produce such quantities of ellagic acid that it becomes troublesome in juice processing (Boyle and Hsu, 1990). *Vitis vinifera* L. (European, or wine, grape), on the other hand, synthesizes gallic acid and gallotannins but not ellagic acid (Singleton and Esau, 1969; Singleton and Trousdale, 1983). In the rose family, Roseaceae, Bate-Smith (1961b) showed that ellagic acid is produced in the genera *Geum, Fragaria* (strawberry), *Potentilla, Rosa* (rose), and *Rubus* (blackberry and raspberries) but not in *Malus* or *Crataegus*. Their paper chromatographic methods may have not been sensitive enough to detect small amounts of ellagic acid. Using high-performance liquid chromatography, we (Wang *et al.*, 1990) have found ellagic acid in fruit samples of *Malus domestica* Borkh. (common apple), *Crataegus aestivalis* (Walter) Torrey and Gray and *Crataegus rufula* Sarg. (Mayhaws), both genera found in earlier reports to lack ellagic acid (Bate-Smith, 1961a).

Ellagic Acid Biosynthesis in the Strawberry

Ellagic acid is formed by oxidation and dimerization of gallic acid, forming the biologically active hexahydroxydiphenoyl (HHDP) group (Fig. 3). In general, oxidation of gallic acid is hastened by alkaline conditions, whereas hydrolysis and lactonization is favored by acidic conditions (Tulyathan *et al.*, 1989). Working with *Rhus* and *Acer*, Ishikura *et al.* (1984) postulated the existence of two pathways for gallic acid formation: through β-oxidation of phenylpropanoid and through dehydrogenation of shikimic acid. The preferential pathway, according to Ishikura *et al.* (1984), may be determined by leaf age. Dehydrogenation of shikimic acid appeared to be the major pathway in young leaves, while in older leaves the phenylpropanoid pathway predominated.

The pathway from phenylalanine to cinnamic acid via phenylacetic acid is probably the major pathway in strawberries (Creasy, 1971). Little is known concerning the terminal enzyme systems involved directly in ellagic acid and ellagitannin biosynthesis in strawberry. Since PAL is a key enzyme in polyphenol synthesis, including ellagitannins, we will discuss growth and environmental factors that affect PAL activity.

Activity of the deaminating enzyme PAL corresponds to the accumulation of anthocyanin (red color) in strawberry fruit (Given *et al.*, 1988a; Hyodo, 1971) and of flavonoids and cinnamic acids in strawberry leaf disks (Creasy, 1971); however, the relationship has not been determined between PAL activity and ellagic acid production.

Hyodo (1971) suggested that increases in PAL activity in ripening strawberry fruits may be required for high rates of anthocyanin synthesis. He found that PAL activity reached a maximum level of activity but anthocyanin production continued to increase as fruits ripen. Anthocyanin synthesis is generally stimulated by light exposure, however, white unripe fruits incubated in darkness at 20°C for 24 hr. showed a rise in anthocyanin production and PAL activity. Hyodo (1971) suggests that ethylene production in harvested, unripe fruits may have stimulated PAL activity. More direct evidence that increased PAL activity is necessary for antho-cyanin accumulation is that PAL-specific inhibitors, such as L-α-aminooxy-β-phenylpropionic acid, also inhibit anthocyanin synthesis *in vitro* (Given *et al.*, 1988a). Similar associations between increased PAL activity and flavonoid synthesis in strawberry leaf disks have been determined (Creasy, 1968a; 1971). Factors influencing increased PAL activity and flavonoid synthesis include carbon dioxide supply in illuminated leaf disks but not in darkness (Creasy, 1968b). Sucrose supplied externally substituted for the carbon dioxide dependency and 3(3,4 dichlorophenyl)-1,1-dimethylurea (DCMU); conversely, an inhibitor of carbon dioxide fixation and sucrose uptake inhibited PAL activity and flavonoid synthesis. Creasy (1968b) concluded that

Table 1. Ellagic acid content of tissues of selected strawberry cultivars and clones (mg/g dry weight).

Fruit pulp - green, unripe (mean for 32 clones = 3.36; SE = 1.82)			
Dana	8.51	MDUS-4588	1.32
Blakemore	8.43	MDUS-5524	1.59
Arking	8.00	Lester	1.63
Cesena	5.26	Earlibelle	1.89
Vesper	4.86	MDUS-5288	2.10
Fruit pulp - red, ripe (mean for 35 clones 1.55; SE = 0.77)			
Arking	4.64	Honeoye	0.43
Micmac	2.65	Marlate	0.51
Midland	2.46	Scott	0.75
Vesper	2.38	Earliglow	0.83
Blakemore	2.33	Allstar	0.83
Achenes from green, unripe fruit (mean for 32 clones 7.24; = SE = 2.84)			
Redchief	20.73	MDUS-4588	1.37
MDUS-5146	16.33	Delite	1.58
MDUS-5524	12.82	Earliglow	2.69
MDUS-5393	12.16	Scott	2.84
Fairfax	12.16	Tangi	2.92
Achenes from red, ripe fruit (mean for 34 clones = 8.46; SE = 5.03)			
Kent	21.65	Tangi	1.37
Earliglow	19.54	Sunrise	2.04
MDUS-5517	18.96	Blakemore	2.09
MDUS-5279	16.56	MDUS-5456	2.34
Lateglow	12.71	MDUS-4588	3.56
Leaves, after fruit harvest (mean for 13 clones = 14.71; SE = 6.39)			
Tribute	32.30	Earliglow	8.08
Delite	20.05	Midway	8.36
Cesena	17.90	Lester	8.89
MDUS-4588	16.01	MDUS-5455	11.73
Blakemore	15.41	Redchief	12.35

Adapted from Maas *et al.*, 1990c.

photosynthesis and carbohydrate metabolism were of extreme importance in controlling synthesis of flavonoids in strawberry leaf disks. Temperature is important in anthocyanin synthesis. Creasy *et al.* (1965) found the maximum rate of anthocyanin production occurred at 30°C in light and at 25°C in darkness. Work of Given *et al.* (1988b) suggests that the increase in PAL activity in ripening strawberry fruit is due to *de novo* synthesis of the enzyme. Ellagic acid, however, is usually less in red strawberries compared with green fruit (Table 1). Thus, we cannot fully equate conditions that produce anthocyanins with those that produce ellagitannins.

Purification of PAL extracts from strawberry fruit show that the native molecular weight of the enzyme is a tetramer of approximately 266,000 with subunit molecular weight of 72,000 and is a single isozyme and that the increase in PAL activity in ripening strawberry fruit is due to *de novo* synthesis (Given *et al.*, 1988b).

Ellagic Acid Genetics in the Strawberry

The manner of ellagic acid inheritance is not known. Strawberry (Maas *et al.*, 1991b), cranberry (Wang *et al.*, 1990), blackberry and apple (Maas and Wang *et al.*, unpublished), and muscadine grape (Boyle and Hsu, 1990) cultivars differ in amounts of ellagic acid in tissues and among genotypes. We presume by analogy with other metabolites that cultivar differences are characteristic of each cultivar and that inheritance of relative ellagic acid content has either a high general or high specific combining ability (Galletta and Maas, 1990). Since the major portion of ellagic acid is presumably derived from a single metabolic pathway through gallic acid, we may find in strawberry that specific combining ability is of greater importance than general combining ability in ellagic acid inheritance. The same may be true for inheritance of specific polyphenols, such as gallic acid and geraniin, but probably not for massed derivatives of either constituent (e.g., ellagitannins as a group). If this hypothesis is correct, breeding for high (or low) ellagic acid content of fruit, or any other structure, should be highly successful, although not nearly as easy to achieve as the above statement may indicate.

According to Given *et al.* (1988b) the PAL enzyme exists in strawberry as a single isozyme, suggesting single gene control. However, the activity of PAL is highly influenced by external environmental factors, such as light and temperature, and by internal stimuli, such as carbon dioxide concentration and carbohydrate metabolism. In our work with strawberry we found not only a wide variation among cultivars, but that tissues also varied considerably in their ellagic acid contents (Table 1). This is not totally unexpected since products of the general phenylpropanoid pathway generally are not distributed uniformly throughout the plant. Anthocyanins, absent in roots but abundant in ripening fruit, also occur in leaves but are generally masked by their green chlorophyll pigment. Anthocyanins in strawberry leaves become conspicuous under certain stress conditions, such as those induced by drought, cold, nitrogen deprivation and fungal infection. Appearance of anthocyanins under these conditions may be due to *de novo* synthesis of PAL, or it may reflect changes in flavonoid biosynthesis beginning with chalcone synthase, the first dedicated enzyme of flavonoid synthesis. We have noted that the flavonoid spectra of strawberry leaves do change in response to seasonal decreases in photoperiod and temperature (Maas *et al.*, 1990) and that ellagic acid contents show a seasonal response in leaves (Maas *et al.*, unpublished). Preliminary results suggest that leaf ellagic acid decreases as daylight and temperature decrease.

In attempting to genetically enhance the ellagic acid content of strawberry cultivars, we must ask whether or not sufficient variation in ellagic acid production exists among genotypes to be useful in breeding for greater concentrations; if selection for enhanced quantities of ellagic acid should be done on a whole plant basis or for specific tissue type alone; and if screening progenies for ellagic acid content is adequate, or evaluations should be limited to one or more ellagitannin based on their biological activities in mammalian systems.

We believe that sufficient genotype variation among the octoploid *Fragaria* cultivars exists and can be exploited for enhancement of ellagic acid content. Greater variability may be present in cultivars not examined and in other *Fragaria* spp., especially those that are also octoploids. To answer whether or not whole plant or tissue content should be used in screening for enhanced ellagic acid levels, we have begun a series of experiments designed to determine the heritability of ellagic acid synthesis in various plant parts. Several reciprocal crosses were made, using parental material whose selection was based on earlier results (Maas *et al.*, 1991b), e.g., high fruit x high leaf, high fruit x high fruit, high fruit x low leaf, etc. Maternal cytoplasmic contributions were not taken into account at this time. Initial screening will be for leaf ellagic acid content, and their fruit sampled, after plants have gone through flower bud initiation and produced fruit. The latter generally requires that plants overwinter and fruit the following spring. The process of making a cross and following the seedling progenies through the first year's growth and then flowering and fruiting requires 2.5–3 years for most cultivars. Investment in time, labor, and field space is considerable. It is quite possible that some of this investment can be lessened by restricting the number of progenies to a critical number that may be considered to represent the recombination limits of the parents and remain statistically significant for all comparisons. At this point, assuming such a number would be arbitrary, it could conceivably be between 20 and 100 seedlings per parental combination. Assuming that cultivars with enhanced ellagic acid contents must also satisfy the general requirements for productivity, disease resistance, etc., the number of hybrid seedlings generated for doubling ellagic acid content may be as high as 80,000 to 100,000.

If selection for enhanced ellagic acid content is successful and if selection can be specific for tissue type, then the reverse situation may be advantageous in certain crops. For example, moderation of ellagic acid synthesis may be desirable in fruits of loganberry (Singleton *et al.*, 1966) and in muscadine grapes (Boyle and Hsu, 1990), where ellagic acid is one component of the troublesome phenolics extracted with juice for winemaking or other commercial processing.

Further knowledge of the metabolic pathways involved in biosynthesis of ellagic acid from phenylpropanoids may be essential in fully understanding the role of ellagitannins in strawberry. Research is in progress to determine if ellagic acid synthesis is dependent entirely on PAL activity to synthesize cinnamic acid or if an alternate pathway involving the dehydrogenation of shikimic acid is functional (Ishikura *et al.*, 1984). We have found that strawberry fruit cell suspension cultures can be manipulated to produce either predominantly ellagic acid or predominantly flavonoids, especially anthocyanins (Harlander, Maas and Wang, unpublished). By using feedback inhibition of PAL with *trans*-cinnamic acid and direct PAL inhibition with L-α-aminooxy-β-phenylpropionic acid and substrate addition, we should learn a great deal about the distinct but fundamentally related basic processes involved in these biosynthetic pathways.

One important aspect of ellagic acid and its potential as an anticarcinogen to be considered is that ellagic acid seldom is present naturally in plants as ellagic acid. Ellagic acid as such is relatively insoluble. The reactive intermediary form is the hexahydroxydiphenoyl (HHDP) group which presumably is produced in the plant by forming a C-C bond between two galloyl groups (Fig. 3). With the simpler ellagitannins, the HHDP group condenses with a gluco-pyranose ring to form a monomeric ellagitannin (Fig. 4). Further condensation may occur with HHDP or galloyl groups on the glucopyranose ring. The HHDP groups may be metabolized to other aromatic groups through oxidation, reduction, ring cleavage, and C-O(C) oxidative coupling (Okuda *et al.*, 1989a). Ellagitannins may form condensates with ascorbic acid, catechins, or form oligomers with other ellagitannins. The property of ellagitannins to form stable free radicals is due to the presence of several phenolic hydroxyl groups in a molecule. This is the basis of their radical scavenging activities that inhibit peroxidation of lipids and other substances (Okuda *et al.*, 1989b). The ellagitannins are biologically soluble and each may be isolated from naturally occurring mixtures as a comparatively stable compound. Several ellagitannins have been shown to have antitumor activity, e.g., camellin B and nobotanin I (Yoshida *et al.*, 1989), while most of the earlier biomedical research involved ellagic acid (De Flora and Ramel, 1988; Hayatsu *et al.*, 1988; Stoner, 1989; Maas *et al.*, 1991b).

Nothing is known about the bioavailability of ellagic acid in strawberries; however, some progress is being made on characterizing the ellagitannins that occur in strawberry fruit. At least five

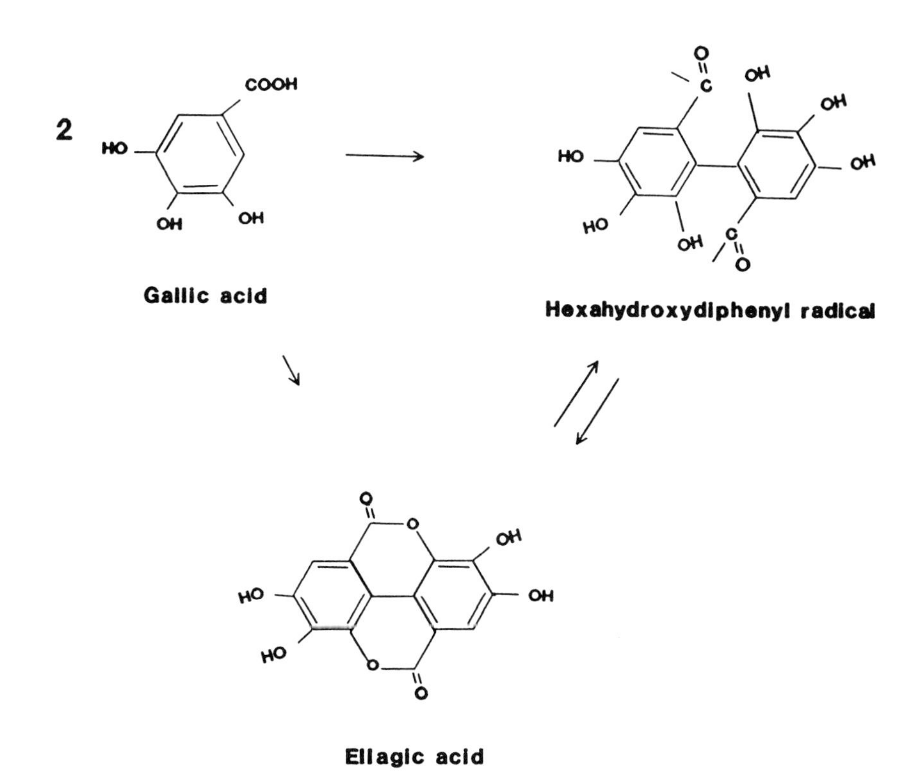

Figure 3. Ellagic acid and the hexahydroxydiphenyl (HHDP) radical.

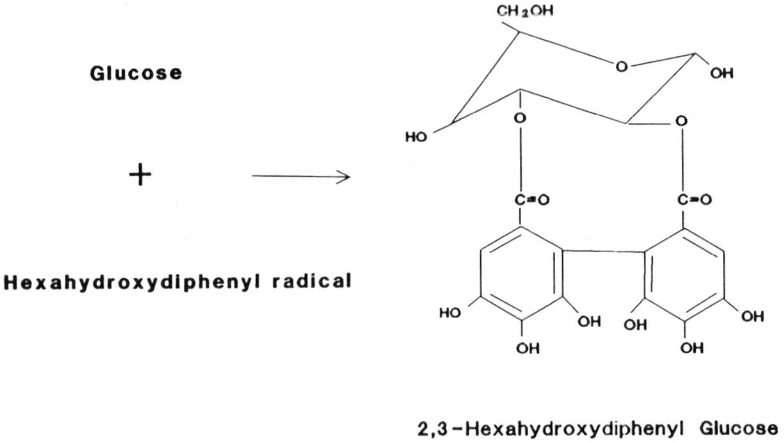

Glucose

+

Hexahydroxydiphenyl radical

2,3-Hexahydroxydiphenyl Glucose

Figure 4. Biosynthesis of a simple ellagitannin, 2,3-hexahydroxydiphenyl glucose.

ellagitannins are produced and three have been identified. The individual ellagitannins will be synthesized and their antitumor and anticarcinogenic properties determined (G.D. Stoner, pers. comm.). These results will aid in determining the genetics of ellagitannin inheritance and their physiology in the strawberry. It is conceivable that among certain cultivar and/or species genotypes some ellagitannins will predominate in the mixture and that these differences may be used to further genetic and enzymological knowledge in strawberry.

Eventually, genetic transformation at the molecular level of ellagitannin enzymes may be feasible. Once specific ellagitannins are identified and their enzyme synthesis genes determined, genetic transformation of strawberry may be possible. Considerable progress has been made in developing an efficient and reliable transformation procedure in strawberries (Jelenkovic *et al.*, 1990; Nehra *et al.*, 1990). Genetic transformation may alleviate problems associated with parasexual manipulation of the commercial strawberry due to its highly heterozygous, octoploid nature and interspecific origin. Transferring of qualitative traits in these genotypes by traditional methods produces a tremendous amount of variability among the progenies, necessitating the evaluation of large seedling populations to select the desired trait. Strawberry cultivars are vegetatively propagated, virtually eliminating the possible unintentional loss or modification of a desired trait once it is acquired.

Quality and Consumer Acceptance

Enhanced production of ellagic acid in strawberry fruits may have negligible effects on flavor, color, or other quality factors. Ellagitannins in general are very mild, being weakly astringent on the tongue and weakly irritating to mucous membranes. Extracts of medicinal plants containing high concentrations of ellagitannins have been administered orally without marked unfavorable responses (Okuda *et al.*, 1989a). In large doses, ellagic acid and ellagitannins were found to be nontoxic in various human and animal feeding studies. Rats fed ellagic acid at doses of 50 mg/kg/day for up to 45 days (Blumenberg *et al.*, 1966) and humans administered ellagic acid at 0.2 mg/kg intravenously (Girolami and Cliffton, 1967) showed no toxological effects due to ellagic acid.

Conclusions

Our study was undertaken to determine the ellagic acid content of strawberry fruit and other plant parts in response to increased interest in this naturally occurring plant phenol as a potential anticarcinogenic constituent of strawberries. A large number of cultivars and advanced selections were tested so that parents may be chosen for ellagic acid heritability studies.

To date, our results with strawberry indicate that:

1) Variation occurs among cultivars in ellagic acid content;
2) Variation occurs in different plant organs (fruit, leaves, etc.);
3) Ellagic acid content changes as the physiology of the tissue changes, e.g., fruit ripening and seasonal environmental effects;
4) There is an apparent lack of correlation in ellagic acid content among tissue types among newer and older tissues, and among related and nonrelated cultivars;
5) This apparent lack of correlation suggests that ellagic acid content is not closely linked with other fruit or plant characteristics that have been selected for; and
6) Variation in ellagic acid expression suggests that ellagic acid content of cultivars and tissue types can be manipulated genetically.

We will continue these studies to further characterize strawberry cultivars, certain advanced selections, and *Fragaria* species for ellagic acid content of fruit, leaf, crown, and root tissues. With this information, parental material will be chosen to test the inheritance of ellagic acid and specific ellagitannins produced by the strawberry.

These experiments should result in new fundamental knowledge on ellagic acid inheritance and physiology in plants generally, and in strawberry specifically. This information may provide a foundation for further clinical work with ellagic acid as an anticarcinogenic constituent of plants, and the information will be of great value in understanding the role of ellagic acid in the strawberry, including possible contributions to disease and insect resistance.

Intangible benefits of this research include the possibility that the strawberry may be a source of one more beneficial dietary factor, along with vitamin C and fiber.

References

Asanaka, M., Kurimura, T., Koshiura, R., Okuda, T., Mori, M. and Yokoi, H. 1988. Tannins as candidates for anti-HIV drug. Fourth International Conference on Immunopharmacology. May, 1988. Osaka, Japan. Abstract No. WS25-4.

Bate-Smith, E.C. 1956. Chromatography and systematic distribution of ellagic acid. Chem. Ind. B.I.F. Review. April, 1956. pp. 32-33.

Bate-Smith, E.C. 1961a. The phenolic constituents of plants and their taxonomic significance. I. Dicotyledons. J. Linn. Soc. (Bot.) 58: 95-173.

Bate-Smith, E.C. 1961b. Chromatography and taxonomy in the Rosaceae, with special reference to *Potentilla* and *Prunus*. J. Linn. Soc. (Bot.) 58: 39-54.

Blumenberg, F.W., Enneker, C. and Kessler, F.-J. 1966. On the question of the hepatic toxic effect of orally administered tannins and their galloyl component. Arzneim-Forsch. 10: 223-226.

Bock, P.E., Srinivasan, K.R. and Shore, J.D. 1981. Activation of intrinsic blood coagulation by ellagic acid: insoluble ellagic acid-metal ion complexes are the activating species. Biochemistry 20: 7258.

Bolwell, G.P., Bell, J.N., Cramer, C.L., Schuch, W., Lamb, C.J. and Dixon, R.A. 1985. L-Phenylalanine ammonia-lyase from *Phaseolus vulgaris*. Characterization and differential induction of multiple forms from elicitor-treated cell suspension cultures. Eur. J. Biochem. 149: 411-419.

Boyle, J.A. and Hsu, L. 1990. Identification and quantification of ellagic acid in muscadine grape juice. Amer. J. Enol. Vitic. 41: 43-47.

Camm, E.L. and Towers, G.H.N. 1973. Phenylalanine ammonia lyase (Review article). Phytochemistry 12: 961-973.

Creasy, L.L., Maxie, E.C. and Chichester, C.O. 1965. Anthocyanin production in strawberry leaf disks. Phytochemistry. 4: 517-521.
Creasy, L.L. 1968a. The increase in phenylalanine ammonia-lyase activity in strawberry leaf disks and its correlation with flavonoid synthesis. Phytochemistry 7: 441-446.
Creasy, L.L. 1968b. The significance of carbohydrate metabolism in flavonoid synthesis in strawberry leaf disks. Phytochemistry 7: 1743-1749.
Creasy, L.L. 1971. Role of phenylalanine in the biosynthesis of flavonoids and cinnamic acids in strawberry leaf disks. Phytochemistry 10: 2705-2711.
Das, M., Bickers, D.R. and Mukhtar, H. 1985. Effect of ellagic acid on hepatic and pulmonary xenobiotic metabolism in mice: studies on the mechanism of its anticarcinogenic action. Carcinogenesis 6: 1409-1413.
De Flora, S. and Ramel, C. 1988. Mechanisms of inhibitions of mutagenesis carcinogenesis. Classification and overview. Mutation Res. 202: 285-306.
Dixit, R. and Gold, B. 1986. Inhibition of *N*-methyl-*N*-nitrosourea-induced mutagenicity and DNA methylation by ellagic acid. Proc. Nat. Acad. Sci. (U.S.A.) 83: 8039-8043.
Dixit, R., Teel, R.W., Daniel, F.B. and Stoner, G.D. 1985. Inhibition of benzo(*a*)pyrene and benzo(*a*)pyrene-*trans*-7,8-diol metabolism and DNA binding in mouse lung explants by ellagic acid. Cancer Res. 45: 2951-2956.
Galletta, C.J. and Maas, J.L. 1990. Strawberry genetics. HortScience 25: 871-879.
Girolami, A. and Cliffton, E.E. 1967. Hypercoagulable state induced in humans by intravenous administration of purified ellagic acid. Throm. Diath. Haemorrh. 17: 165-175.
Given, N.K., Venis, M.A. and Grierson, D. 1988a. Phenylalanine ammonia-lyase activity and anthocyanin synthesis in ripening strawberry fruit. J. Plant Physiol. 133: 25-30.
Given, N.K., Venis, M.A. and Grierson, D. 1988b. Purification and properties of phenylalanine ammonia-lyase from strawberry fruit and its synthesis during ripening. J. Plant Physiol. 133: 31-37.
Halbrock, K. and Sheel, D. 1989. Physiology and molecular biology of phenylpropanoid metabolism. Annu. Rev. Plant Physiol. Plant Mol. Biol. 40: 347-369.

Hart, J.H. and Hillis, W.E. 1972. Inhibition of wood-rotting fungi by ellagitannins in the heartwood of *Quercus alba*. Phytopathology 62: 620-626.

Hayatsu, H., Arimoto, S. and Negishi, T. 1988. Dietary inhibitors of mutagenesis and carcinogenesis. Mutation Res. 202: 429-446.

Hildebrandt, V. and Harney, P.M. 1989. Identity of a phenolic exudate inhibitor from geranium plants. Can. J. Plant Sci. 69: 569-575.

Hrazdina, G. and Jensen, R.A. 1990. Multiple parallel pathways in plant aromatic metabolism? In "Structural and Organizational Aspects of Metabolic Regulation," pp. 27-41. Alan R. Liss, Inc.

Hyodo, H. 1971. Phenylalanine ammonia-lyase in strawberry fruits. Plant Cell Physiol. 12:989-991.

Ishikura, N., Hayashida, S. and Tazaki, K. 1984. Biosynthesis of gallic and ellagic acids with ^{14}C-labeled compounds in *Acer* and *Rhus* leaves. Bot. Mag. Tokyo 97: 355-367.

Jelenkovic, G., Chin, C., Billings, S. and Eberhardt, J. 1990. Transformation studies in the cultivated strawberry, *Fragaria* x *ananassa* Duch. Proc. 2nd North American Strawberry Conference, Houston, Texas. (In press).

Jones, K.C. and Klocke, J.A. 1987. Aphid feeding deterrency of ellagitannins, their phenolic hydrolysis products, and related phenolic derivatives. Entomol. Exp. Appl. 44: 229-234.

Klocke, J.A., Van Wagenen, B. and Balandrin, M.F. 1986. The ellagitannin geraniin and its hydrolysis products isolated as insect growth inhibitors from semi-arid land plants. Phytochemistry 25: 85-91.

Maas, J.L., Galletta, G.J. and Stoner, G.D. 1991a. Ellagic acid, an anticarcinogen in fruits, especially in strawberries: A review. HortScience 26: 10-14.

Maas, J.L., Griesbach, R.J. and Galletta, G.J. 1990. Changes in strawberry leaf flavonoid pigment composition: An indicator of plant dormancy status? Adv. Strawberry Prod. 9: 28-30.

Maas, J.L., Wang, S.Y. and Galletta, G.J. 1991b. Evaluation of strawberry cultivars for ellagic acid content. HortScience 26: 66-68.

Mandal, S., Ahuja, A., Shivapurkar, N.M., Cheng, S.-J., Groopman, J.D. and Stoner, G.D. 1987. Inhibition of aflatoxin B1 mutagenesis in *Salmonella typhimurium* and DNA damage in cultured rat and human tracheobronchial tissues by ellagic acid. Carcinogenesis 8: 1651-1656.

Mandal, S., Shivapurkar, N.M., Galati, A.J. and Stoner, G.D. 1988. Inhibition of *N*-nitrosobenzylmethylamine metabolism and DNA binding in cultured rat esophagus by ellagic acid. Carcinogenesis 9: 1313-1316.

Nehra, N.S., Chibbar, R.N., Kartha, K.K., Datla, R.S.S., Crosby, W.L. and Stushnoff, C. 1990. Agrobacterium-mediated transformation of strawberry calli and recovery of transgenic plants. Plant Cell Repts. 9: 10-14.

Okuda, T., Yoshida, T. and Hatano, T. 1989a. Ellagitannins as active constituents of medicinal plants. Planta Med. 55: 117-122.

Okuda, T., Yoshida, T. and Hatano, T. 1989b. New methods of analyzing tannins. J. Nat. Prod. 52: 1-31.

Press, R.E. and Hardcastle, D. 1969. Some physio-chemical properties of ellagic acid. J. Appl. Chem. 19: 247-251.

Rice, E.L. 1984. Allelopathy. 2nd ed. Academic Press, New York, New York.

Runkova, K.V., Lis, E.K., Tomaszewski, M. and Antoszewski, R. 1972. Function of phenolic substances in the degradation system of indole-3-acetic acid in strawberries. Biol. Plant. (Praha) 14: 71-81.

Ryder, T.B., Hedrich, S.A., Bell, J.N., Liang, X., Clouse, S.D. and Lamb, C.J. 1987. Organization and differential activation of a gene family encoding the plant disease enzyme chalcone synthase in *Phaseolus vulgaris*. Mol. Gen. Genet. 210: 219-233.

Sayer, J.M., Yagi, H., Wood, A.W., Conney, A.H. and Jerina, D.M. 1982. Extremely facile reaction between the ultimate carcinogen benzo(*a*)pyrene-7,8,-diol 9,10-epoxide and ellagic acid. J. Amer. Chem. Soc. 104: 5562-5564.

Singleton, V.L. and Esau, P. 1969. Phenolic substances in grapes and wine, and their significance. Academic Press, New York, New York, Adv. Food Res. Suppl. 1: 1-282.

Singleton, V.L., Marsh, G.L. and Coven, M. 1966. Identification of ellagic acid as a precipitate from loganberry wine. J. Agr. Food Chem. 14: 5-8.

Singleton, V.L. and Trousdale, E. 1983. White wine phenolics: varietal and processing differences as shown by HPLC. Am. J. Enol. Vitic. 34: 27-34.

Stoner, G.D. 1989. Ellagic acid: A naturally occurring inhibitor of chemically-induced cancer. Proc. 1989 Ann. Mtg. N. Amer. Strawb. Grow. Assn., Grand Rapids, Michigan. pp. 20-34.

Take, Y., Inouye, Y., Nakamura, S., Allaudeen, H.S. and Kubo, A. 1989. Comparative studies of the inhibitory properties of antibiotics on human immunodeficiency virus and avian myeloblastosis virus reverse transcriptases and cellular DNA polymerases. J. Antibiotics 42: 107-115.

Teel, R.W. 1986. Ellagic acid binding to DNA as a possible mechanism for its antimutagenic and anticarcinogenic action. Cancer Lett. 30: 329-336.

Teel, R.W., Dixit, R. and Stoner, G.D. 1985. The effect of ellagic acid on the uptake, persistence, metabolism and DNA-binding of benzo[*a*]pyrene in cultured explants of strain A/J mouse lung. Carcinogenesis 6: 391-395.

Tulyathan, V., Boulton, R.B. and Singleton, V.L. 1989. Oxygen uptake by gallic acid as a model for similar reactions in wines. J. Agric. Food Chem. 37: 844-849.

Vieitez, F. J. and Ballester, A. 1988. Effect of etiolation and shading on the formation of rooting inhibitors in chestnut trees. Phyton (B. Aries) 48: 13-20.

Wang, S.Y., Maas, J.L., Daniel, E.M. and Galletta, G.J. 1990. Improved HPLC resolution and quantification of ellagic acid from strawberry, blackberry and cranberry. ASHS 1990 Annu. Mtg., Tucson, Arizona, Prog. and Abstr. (In press).

Yoshida, T., Chou, T., Haba, K., Okano, Y., Shingu, T., Miyamoto, K.-I., Koshiura, R. and Okuda, T. 1989. Camellin B and nobotanin I, macrocyclic ellagitannin dimers and related dimers, and their antitumor activity. Chem. Pharm. Bull. 37: 3174-3176.

MOLECULAR APPROACHES IN THE MODIFICATION AND PRODUCTION OF EDIBLE OILS

Plant Fatty Acid Biosynthesis and Its Potential For Manipulation

John Ohlrogge, Dusty Post-Beittenmiller, Alenka Hloušek-Radojčić, Katherine Schmid
Department of Botany and Plant Pathology
Michigan State University
East Lansing, MI 48824
Jan Jaworski
Department of Chemistry
Miami University
Oxford, OH 45056

In the American diet, plant oils have gradually replaced animal fats and now account for 15–20 percent of total calories consumed. In addition, plant oils are employed in the manufacture of a wide range of specialty products including lubricants, plastics and detergents. Success in altering seed oils through mutation breeding has indicated that relatively large alterations can be made in triacylglycerol fatty acid composition without exerting any obvious deleterious effect on plant growth and development. It, therefore, may be possible to use molecular genetic methods to provide more directed and extensive modifications of the plant fatty acid biosynthetic pathway. Alterations in the amount and type of unsaturated fatty acids may have particular importance in obtaining optimum nutritional value from plant oils. Our current understanding of the biochemistry and molecular biology of plant fatty acid production will be reviewed, and the potential for its modification will be discussed.

Introduction

The diets of America and other developed nations are characterized by their high and increasing content of fat (Rizek *et al.*, 1983). The contribution of calories derived from fat to the U.S. food supply increased one-third, from 32 percent in 1909 to 42 percent in 1980 (Figure 1). The current average daily consumption of 170 grams is the equivalent of three-fourths cup of salad oil. In addition to this general increase in fat content, the source of fat has gradually changed from predominantly animal in origin (83 percent in 1909) to the current situation where plant and animal fats contribute approximately equal quantities (Figure 2). In large part, the transition to more plant products has reflected the declining use of butter and lard and their replacement by margarine and vegetable shortening (Figure 3).

The largest source of vegetable oil in American diets is soybean oil which accounts for roughly 70 percent of U.S. consumption. The predominant role of this oil in a wide variety of food products has resulted in it becoming a major single source of calories in our diets. (Table 1).

Concomitant with the shift to vegetable oils has been a change in the fatty acid composition of the diet. Saturated fatty acids and oleic acid each contributed ca. 40 percent of the fat in the 1909 food supply. However, by 1980 the proportion from saturated fatty acids had dropped to 34 percent. In contrast, the share of fatty acids provided by linoleic acid doubled from 7 to 15 percent during this period.

The nutritional consequences of the above changes are complex and controversial. On the one hand, the increase in unsaturated fatty acids has been considered beneficial for the reduction of serum cholesterol. However, the overall increase in fat consumption may be associated with increased risks of cancer and coronary heart disease. In addition, the rise in linoleic acid consumption, a precursor of several prostaglandins and leukotrienes may contribute to overproduction of these hormone-like fatty acid derivatives which in turn is associated with a variety of health disorders such as atherosclerosis, thrombosis and asthma (Lands, 1986).

Clearly, a complete understanding of the role of the quantity and quality of dietary fatty acids in health and nutrition is not yet available. Nevertheless, evidence has accumulated that certain changes in the fatty acid composition of our foods could lead to improved health. The following are considered desirable objectives:

1. Decrease the total fat intake.
2. Decrease the proportion of saturated fatty acids.
3. Among the saturates, decrease the proportion of palmitic acid.
4. Decrease the content of "n-6" unsaturated fatty acids, in particular linoleic.
5. Increase the content of "n-3" unsaturated fatty acids.

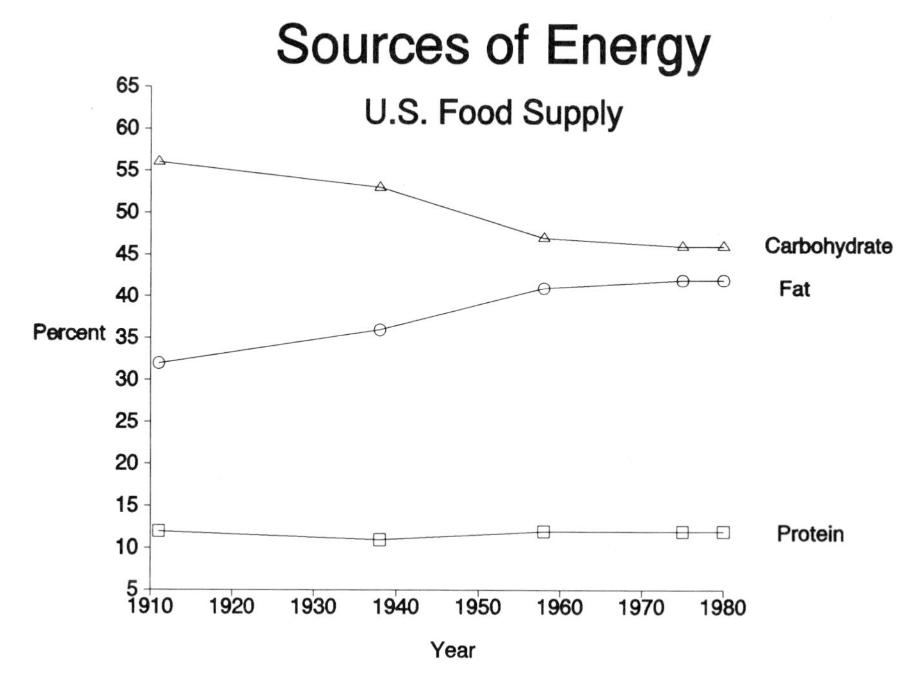

Figure 1. Sources of energy in U.S. food supply from 1909 to 1980

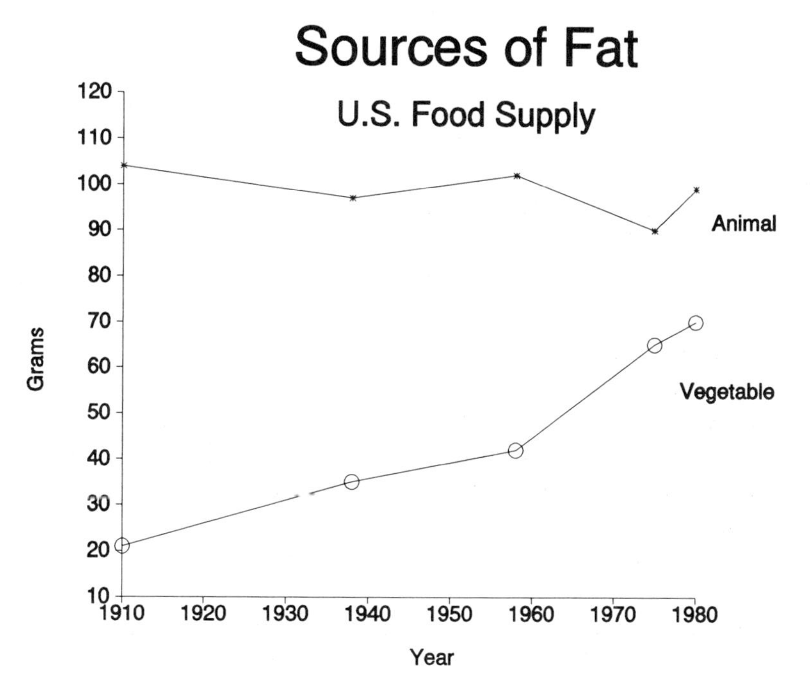

Figure 2. Contributions of vegetable and animal fat to U.S. food supply

Table 1. Major sources of calories in American diets

	Percent of Total Calories
Soybean Oil	12–14
Beef	12–14
Wheat	10
Milk	5

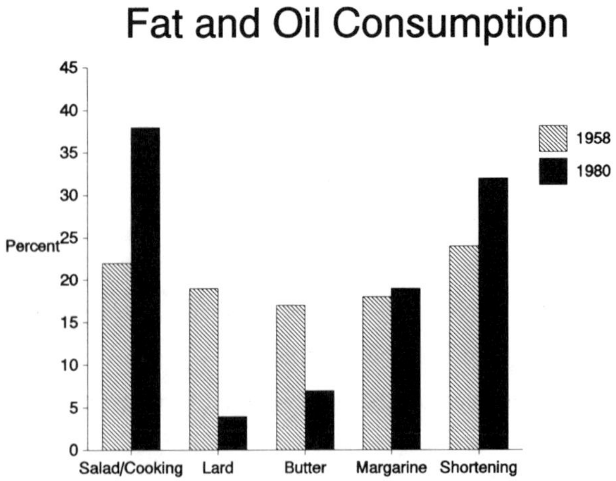

Fat and Oil Consumption

Figure 3. Types of fat consumed in U.S. food supply

Because almost 50 percent of dietary fat is derived from oilseeds, it may be possible to achieve at least some of the above goals through a combination of conventional breeding and through genetic engineering of oilseeds. Breeding of oilseeds has already proven particularly successful in altering the fatty acid composition available for edible oils. Important examples include the elimination of erucic acid from rapeseed oil which has led to the development of canola oil as a major commodity. More recently, high oleic acid sunflower oil has become available. In addition to its greater freedom from undesirable oxidation during frying, the high oleic acid content may in part meet objective 4 above. Although breeding of oilseeds will continue to provide a major avenue toward altered oilseed fatty acid compositions, in the future it can be expected that molecular genetic approaches will add substantially to the repertoire of available vegetable oil varieties.

The viability of current gene transfer technology for oilseeds has been demonstrated by the transformation of rapeseed, flax, cotton, sunflower and safflower with foreign DNA (Weising *et al.*, 1988).

Seed specific promotors have been used to target introduced proteins to this organ (Weising, *et al.* 1988), increasing the probability that modifications of seed oil may be undertaken with disruption of whole plant physiology. However, the synthesis of oil by plants requires the interaction of multiple gene products. To what degree must the oil synthetic apparatus be modified to bring about industrially applicable changes and what are the prospects for achieving significant modification of plant oils through genetic engineering?

Oil Biosynthesis

Plants store oil in the form of triacylglycerol (TAG), a molecule which has three fatty acids esterified to the three hydroxyl positions of glycerol. As shown in Figure 4, the machinery responsible for synthesis of the fatty acids and their incorporation into TAG is distributed between several compartments within a plant cell. The backbones of fatty acid species are produced in the plastids, where a collection of at least six individual enzymes, termed the fatty acyl synthase, assembles 2-carbon units into chains up to 18 carbons long (Figure 5) (Harwood, 1988). Three condensing enzymes with different chain length specificity and sensitivity to inhibitors have been identified. The enzyme, 3-keto-acyl-ACP synthase-I accepts acyl-ACP primers from 2 to 16 carbons long. A second condensing enzyme appears to be specific for elongation of palmitoyl- (C16) to stearoyl-ACP (C18). Recently, a third condensing enzyme has been described which is insensitive to the antibiotic cerulenin and which catalyzes the first condensation between acetyl-CoA and malonyl-ACP (Jaworski *et al.*, 1989). The existence of a separate condensing enzyme for the C16 to C 18 elongation suggests that manipulation of this enzyme's activity through genetic engineering might allow objective 3 above to be addressed.

The primary products of plastid fatty acid synthesis are palmitate and stearate, the 16 and 18 carbon saturated fatty acids, linked to acyl carrier protein (ACP) (Ohlrogge, 1987). Oleate, the major monounsaturated fatty acid of plant oils, is also a plastid product formed by insertion of double bond into stearoyl ACP (McKeon and Stumpf, 1982). Recently a cDNA clone for this enzyme was isolated (Shanklin and Somerville, in press). Overexpression of this desaturase in transgeneic plants may address objective 2. In addition, reduction in the activity of the stearoyl-ACP desaturase by

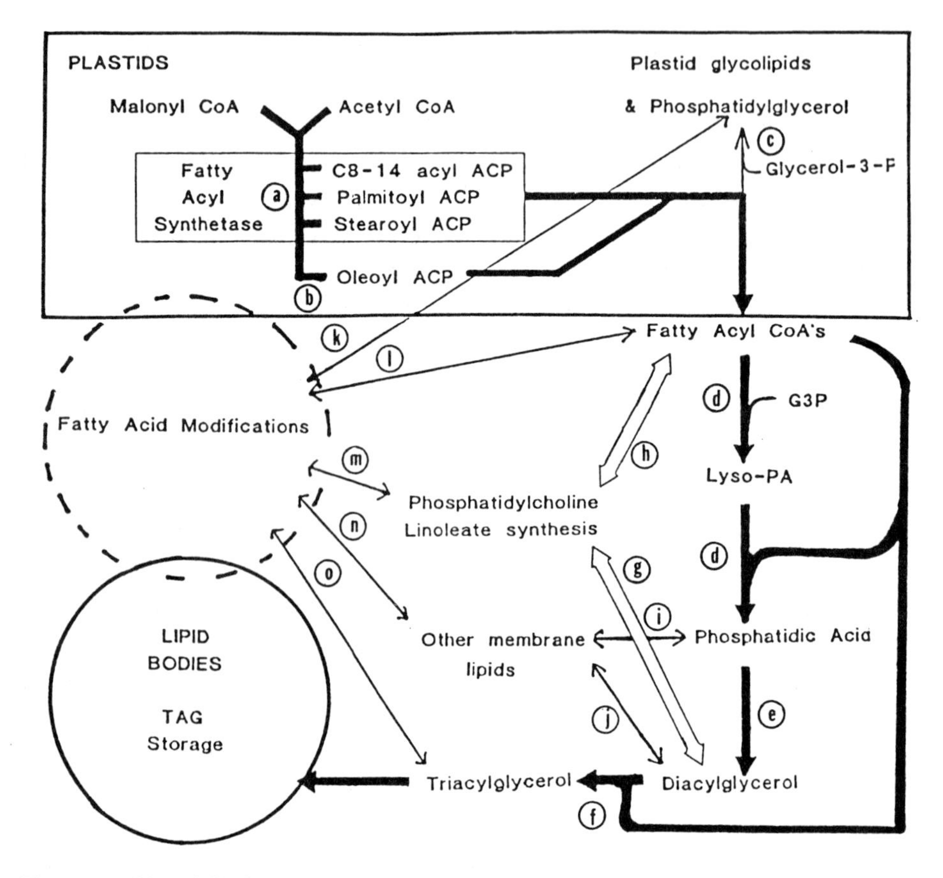

Figure 4. Simplified scheme for triacylglycerol synthesis in plants.

▬▬▬ Direct route of fatty acid incorporation into TAG. A. Fatty acyl synthase. B. Stearoyl ACP desaturase. D. Attachment of the first two fatty acids to the glycerol backbone by glycerol-3-phosphate (G3P) acyltransferase and lysophosphatidic acid (lyso PA) acyltransferase. E. Phosphatidic acid phosphatase removes phosphate. F. Diacylglycerol acyltransferase adds final fatty acid.

▭ Entry of linoleate into TAG. G. Linoleate is formed on phosphatidylcholine synthesized from diacylglycerol. Diacylglycerol released by reverse reaction carries linoleate back to direct TAG synthesis route. H. Linoleoyl CoA released from phosphatidylcholine enters direct route at reactions D or F.

——— Reactions affecting TAG composition. C,G,I and J compete for TAG precursors. Products of these reactions may also serve as substrates for modification of fatty acids. K-O. Modification of fatty acids. Compartmentation and substrates for synthesis of many unusual fatty acids are unknown. Modificaton of fatty acids has been shown to occur on substrates of K-N. The transfer of fatty acids modified by K to TAG has not been demonstrated.

antisense expression may yield plants which produce fats similar in properties to cocoa butter. This very high value product is characterized by a fatty acid composition of approximately 30 percent stearate, 30 percent oleate and 30 percent palmitate.

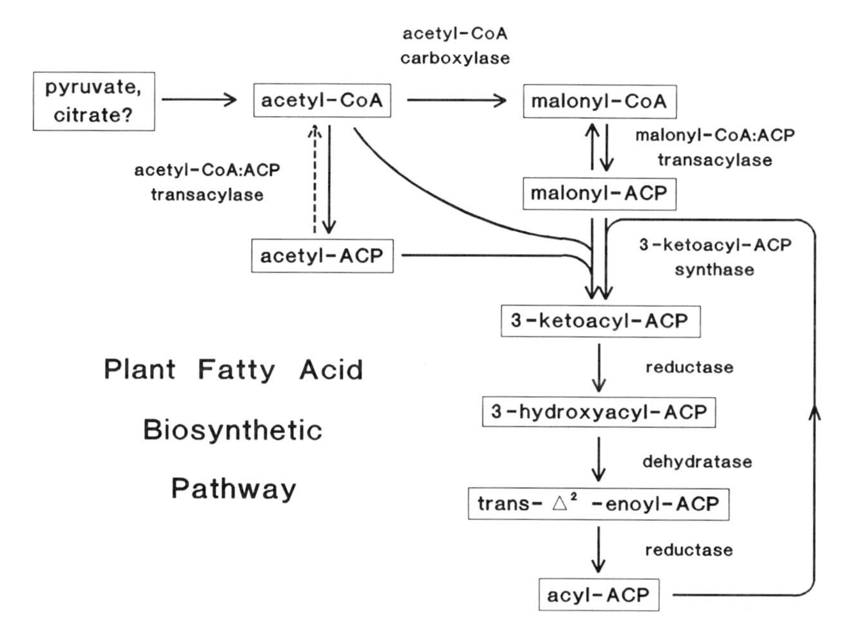

Figure 5. Reactions of plant fatty acid biosynthetic pathway. The path of carbon for acetyl-CoA production is not yet established.

Plastids are capable of transferring fatty acids directly from ACP to glycerol phosphate (Frentzen, 1986). However, evidence to date indicates that fatty acids destined for storage oils are cleaved from ACP, exported into the cytoplasm, and reesterified to coenzyme A (Roughan and Slack, 1982). Only then are two acyl units transferred to glycerol-3-phosphate, forming phosphatidic acid at the endoplasmic reticulum. TAG may be synthesized from this phospholipid in two steps: removal of phosphate to form diacylglycerol, and addition of a final fatty acid. The enzymes required for these steps are believed to be present in the plant endoplasmic reticulum, although it has been suggested that the final acylation can occur at the lipid bodies, ultimate sites of TAG deposition (Gurr, Blades and Appleby, 1974). However, both phosphatidic acid and diacylglycerol also serve as intermediates in a range of lipid synthesis reactions, some of which are readily reversible. In fact, many fatty acids bound for TAG probably spend time associated with lipids other than those on the "direct" route to TAG. For example, linoleate, the chief diunsaturated fatty acid of plants, is produced by insertion of a double bond into oleate bound to phosphatidylcholine (Jaworski, 1987), a lipid derived from diacylglycerol. Linoleate-containing diacylglycerol may then be released from phosphatidylcholine and used for TAG synthesis. Alternatively, acyl exchange is known to occur between phosphatidylcholine and the acyl coenzyme A pool, making linoleoyl CoA available for TAG synthesis (Stymne and Stobart, 1987).

Organization and Regulation of FAS Genes

This section is adapted from a more compreshensive review on "The Genetics of Plant Lipids" (Ohlrogge *et al.*, in press). At present, clones are available for only a few of the many components required for plant TAG biosynthesis. ACP was the first protein of plant lipid metabolism purified to homogeneity (Simoni, Criddle and Stumpf, 1967), and it has also been the first to yield to cloning efforts. The availability of amino acid sequence data for barley, spinach and *E. coli* ACPs allowed identification of regions highly conserved between these species. These regions have been used to design oligonucleotide probes which proved successful in identifying ACP cDNAs from *Brassica napus* embryos (Safford *et al.*, 1988), spinach leaf (Scherer and Knauf, 1987) and barley leaf (Hansen, 1987). Antibodies to spinach ACP-I have also been used successfully in the isolation of a clone of ACP-II from spinach root (Schmid and Ohlrogge, 1989) and an ACP from an *Arabidopsis thaliana* cDNA library (Hloušek-Radojčić, Post-Beittenmiller and Ohlrogge, 1989). In addition to the several cDNA clones, a genomic clone of ACP from *Arabidopsis thaliana* has been isolated using a brassica ACP cDNA as a heterologous probe (Post-Beittenmiller, Hloušek-Radojčić and Ohlrogge, 1989) and recently, two genomic clones of ACP have been characterized from *Brassica napus* (De Silva *et al.*, 1990).

The initial cDNA clones of ACP made it possible to confirm earlier *in vitro* translation evidence (Ohlrogge and Kuo, 1984b) that at least some ACP genes are nuclear encoded. This conclusion can now be extended to include all ACP genes because examination of the complete sequence of the tobacco plastid genome has revealed no ACP-like sequences.

Tissue specific expression: All higher plants examined express multiple ACP isoforms in their photosynthetic tissues. Although almost all leaf ACP can be attributed to the chloroplasts (Ohlrogge, Kuhn and Stumpf, 1979), recent evidence indicates that mitochondria from pea leaves and potato tubers also contain distinct forms of ACP (Chuman and Brody, 1989). The trait of multiple forms of ACP appears to have evolved early in plant speciation because even lower vascular plants, non-vascular liverworts and mosses, and multicellular algae contain electrophoretically distinguishable ACP isoforms (Battey and Ohlrogge, 1990). An ancient gene duplication is consistent with the observation that the degree of identity of ACP amino acid sequences within a species is similar to the sequence identity of ACPs between species.

Protein immunoblot analysis of spinach, castor, *Arabidopsis thaliana* and other species has indicated a common pattern of organ specific expression of ACP isoforms (Ohlrogge, 1987). In all cases there appear to be two or more ACPs expressed in leaves whereas in most species only a single ACP band is detected in western blots of SDS-PAGE of seeds or roots. In spinach, separation of mesophyll and epidermal cells has revealed an enrichment of ACP-I in the former and ACP-II in the latter indicating cell type as well as tissue specificity (Battey and Ohlrogge, 1990). The determination of distinct protein and cDNA sequences for the isoforms confirms that at least some of the ACP polymorphism results from expression of distinct genes rather than from post-translational modifications of the proteins.

In spinach and *Arabidopsis thaliana,* the predominant seed and root ACP isoform co-migrates on both SDS and native PAGE with a minor form in leaf. This pattern of expression is also reflected in RNA blot analysis using cDNA probes which indicate the ACP-II-like sequences are expressed in spinach roots, leaves and seeds. These data might result from the expression in spinach of a family of closely related but organ-specific ACP-II messages rather than expression of a single gene in all three organs. However, RNase protection assays which distinguish even very closely related sequences have indicated that a 490 bp region of ACP-II is expressed in roots, seeds, and leaves (Schmid and Ohlrogge, 1990). Similar results are observed with *Arabidopsis thaliana.* Thus, at least in spinach and *Arabidopsis thaliana,* there appears to be an ACP gene that may be considered "constitutive" based on its expression in all tissues so far examined.

In addition to evidence for leaf specific forms of ACP, *Brassica napus* appears to express some ACP sequences predominantly in seeds. Both Northern blot analysis (Safford *et al.*, 1988) and quantitative slot blots (De Silva *et al.*, 1990) indicate that sequences corresponding to an embryo ACP cDNA are at least 25-fold more prevalent in developing seeds than in leaves.

Regulation of ACP during seed development: Major oilseeds such as rapeseed, safflower, or sunflower contain approximately 40 percent oil by weight. These contents are much higher than found in most other plant organs, such as spinach leaves which contain approximately 10 percent lipid by weight. This high oil content of seeds has often led to the assumption that seeds have high specific activity for lipid biosynthetic enzymes and that these enzymes are "induced" or turned on during seed development. For example, during the development of the soybean seed, fatty acid synthesis increases markedly, reaches a peak at 40–50 days after flowering (DAF) and then declines as the seed reaches maturity (Ohlrogge and Kuo, 1984a). The major increase in lipid synthesis occurs after cell division stops at about 20 DAF. When the level of ACP is measured either enzymatically, or immunologically, it is found that there is a close coordination between increases in ACP level per seed and the increase in fatty acid synthesis (Ohlrogge and Kuo, 1984a). Similar results have been observed in rapeseed (Slabas *et al.*, 1987). Futhermore, the level of ACP continues to increase after cell division stops, leading to an increase in ACP content per cell. The close correlation between the level of active ACP and *in vivo* lipid biosynthesis suggests that the quantity of FAS proteins present in the cell may be a rate-determining component of the cell's overall lipid biosynthetic capacity. Other FAS enzymes such as acetyl-CoA carboxylase (Turnham and Northcote, 1983) and enoyl-ACP reductase (Slabas *et al.*, 1986) undergo similar patterns of increasing activity during seed development, suggesting that the complete complement of FAS enzymes may be under a common coordinant control.

The increase in ACP per seed can now be accounted for by *de novo* synthesis from increased ACP mRNA expression. Examination of ACP mRNA levels during soybean seed development indicate that ACP mRNA per seed reached a peak at about 20 DAF, slightly before both the peak of ACP activity and the major accumulation of storage protein mRNA (Hannapel and Ohlrogge, 1988).

ACP gene structure: The first available genomic sequence data for fatty acid biosynthetic protein is that for an *Arabidopsis thaliana* ACP gene (AD4) (Post-Beittenmiller, Hloŭsek-Radojčić and Ohlrogge, 1989) and two *Brassica napus* clones (De Silva *et al.*, 1990). The *Arabidopsis thaliana* gene contains three introns: one (445 bp) near the middle of the transit peptide, the second (80 bp) at the transit peptide cleavage point, and a third (76 bp) in the highly conserved sequence surrounding the prosthetic group attachment site. Both *Brassica napus* clones have introns in analogous locations. The transcription start site of the *Brassica napus* ACP05 gene has been mapped by RNase protection and found to occur 69 nucleotides 5′ to the coding region. Examination of the cDNA sequences of 10 other ACP clones near the prosthetic group attachment site suggests that other ACP genes also contain introns in this region. Although the overall sequence identity between ACP clones is 60–70 percent, there is one segment of 14 bases which is 90–100 percent conserved. In all but one case, cDNA clones from spinach, barley, *Brassica napus, Brassica campestris* and *Arabidopsis thaliana* contain the same 5 base sequence 5′ and 8 base sequence 3′ to the intron junction site. A possible interpretation of this high conservation is that it reflects contraints on the intron junction, and it is therefore reasonable to speculate from these data that the genomic clones corresponding to the 10 cDNAs all contain an intron at this point.

How many genes are needed for each plant FAS component? It is now clear that several plants have multiple genes for ACP. Based on quantitative Southern analysis, De Silva *et al.* have estimated there are approximately 35 seed expressed ACP genes in *Brassica napus* (De Silva *et al.*, 1990). Even in *Arabidopsis thaliana,* which is generally considered to have smaller gene families, there is evidence for at least 3 or 4 ACP genes. Why is there this level of complexity and does it extend to other components of FAS? One possibility is that the different isoforms of ACP perform different physiological roles. ACP-I and ACP-II of spinach leaf differ significantly in their reactivity with the oleoyl-ACP thioesterase and acyl-ACP:glycerol-3-phosphate acyltransferase. Because these two reactions constitute a branch point between the prokaryotic and eukaryotic pathways of leaf lipid metabolism, it was suggested that the existence of ACP isoforms may play a role in controlling export or retention of acyl chains produced in the chloroplast (Guerra, Ohlrogge and Frentzen, 1986). However, this interpretation is complicated by the observation that ACP-II, which predominates in seeds where presumably most fatty acids are exported from plastids via the thioesterase reaction, was found to be the preferred isoform for the acyltransferase reaction (Guerra, Ohlrogge and Frentzen, 1986).

Multiple genes for FAS components may also be related to the diverse functions of fatty acids in the plant. Fatty acid synthesis is required both for housekeeping synthesis of membranes in all cells and for tissue-specific functions such as oil storage in developing seeds and cuticular wax production in epidermal cells. One mechanism available for control of FAS for these diverse functions might be through the use of tissue-specific genes. Results with ACP in spinach and *Arabidopsis thaliana* indicate that plants have genes for ACP which might preliminarily be characterized as "constitutive" based on their expression in all tissues analyzed (leaf, root, seed). We speculate that these genes code for a baseline production of ACP/FAS present in all cell types. In addition, there appears to be superimposed on this baseline, the expression of additional ACP genes in specific tissues. For example, ACP-I is found only in spinach leaves and indeed becomes the more abundant isoform in this organ (Ohlrogge and Kuo, 1985). The expression of additional ACPs in leaf occurs in all species examined and may reflect a requirement for controlled assembly of the lipid-rich photosynthetic membranes.

Can foreign FAS componencts function in vivo *in a heterologous system?* The availability of clones for ACP has made it possible to overexpress ACP *in vivo* by transforming plants with ACP sequences under the control of a strong promoter. This was recently accomplished by linking a spinach ACP-I gene to the transit peptide and promoter of a tobacco ribulose-1,5-bisphosphate carboxylase small subunit gene (*rbcS*) (Post-Beittenmiller, Schmid and Ohlrogge, 1989). Plants transformed with this construct exhibited a three- to four-fold increase in leaf holoACP levels. In addition, spinach apoACP, a form of ACP lacking the phosphopantetheine prosthetic group and previously not detected in plants, was observed. Both the holoACP and apoACP were localized in the chloroplasts, suggesting that attachment of the prosthetic group is not required for the uptake of ACP into chloroplasts. Overexpression of the spinach ACP did not markedly alter the level of endogenous tobacco ACPs, suggesting that tobacco ACP expression is not subject to direct regulation by ACP levels.

The spinach ACP-I was biologically active in tobacco as evidenced by the presence of spinach acyl-ACP-I intermediates of fatty acid synthesis. Thus, the foreign ACP was able to integrate with the endogenous tobacco FAS components. Although the component enzymes of plant FAS are readily separable *in vitro,* their molecular organization *in vivo* is not known. There is speculation that the individual polypeptides of plant FAS may exist *in vivo* as a supramolecular complex structure similar to that proposed for other pathways such as the Calvin or citric acid cycle. If such a FAS complex exists, the biological activity of spinach ACP in tobacco suggests that the complex is able to accommodate the overexpression of a foreign ACP.

Plants which expressed the spinach holo- and apoACP-I appeared normal and no changes in fatty acid or lipid metabolism could be detected. This is in contrast to several *in vitro* studies where increases in ACP concentration led to production of shorter chain fatty acids. These results point up the uncertainties in extrapolating *in vitro* assays and may also reflect the many *in vivo* controls which serve to regulate the quantity and quality of fatty acid products in leaves.

Biochemical Regulation of FAS

In some cases it may be useful to alter not only the fatty acid composition of an oilseed but also the quantity of edible oil produced. For example, if a crop such as soybean could be engineered to produce a specific optimum fatty acid profile, then increasing the percent of oil might lead to a more valuable crop. Alternatively, for some food products, the oil may interfere with yield, extraction, or stability of a more valuable component such as protein. In such cases reduction of oil quantity may be desirable.

At present the factors which determine the quantity of oil produced in a seed are not understood. In order to design rational strategies for modification of oil quantity, it would be helpful to identify regulatory steps in the oil biosynthetic pathway. While several approaches can be used to identify regulatory steps, in each case, *in vivo* evidence is needed to substantiate any proposal. *In vivo* information on the activity of enzymes in a pathway can be obtained from examination of the steady state pools of intermediates in the pathway. For the fatty acid biosynthetic pathway acyl-ACPs represent the metabolic intermediates beyond acetyl- and malonyl-CoA, and therefore information on their *in vivo* pool sizes can be used to infer the relative activity of enzymes in the pathway. We have recently developed methods which permit the extraction and analysis of acyl-ACPs from leaf and seed tissue and these methods have led to the data in Table 2. (Post-Beittenmiller *et al.*, in press).

Table 2. Relative ACP levels

	Leaf	Seed
Malonyl-ACP-II	4.2%	9.7%
Free ACP-II	64%	56%
Acetyl-ACP-II	5.6%	9.3%
4:0-ACP-II	3.4%	3.1%
6:0-ACP-II	4.6%	3.6%
≥8:0-ACP-II	18%	19%

(Reproduced with permission.)

Because regulation of metabolic pathways frequently occurs at the beginning of the pathway, it can be expected that fatty acid biosynthesis in plants might be regulated at the reactions supplying the acetate or malonate moieties or at the initial condensation reaction (see Figure 5). A minimum requirement for any regulatory enzyme is that the reaction it catalyzes must be far from equilibrium under steady state conditions (Rolleston, 1972). This is because if an enzyme has enough activity to allow its substrates to reach equilibrium, it will catalyze the forward and reverse reactions at the same rate, and consequently no net change in flux will be accomplished by altering the activity of the enzyme. In animals, the formation of malonyl-CoA is the first committed step in this pathway, and acetyl-CoA carboxylase is considered to be the principle site of regulation (Kim *et al.*, 1989).

It is improbable that either the acetyl or malonyl transacylases could be regulatory. Both transacylases catalyze fully reversible reactions and the thermodynamic equilibrium constants for both transacylation reactions are approximately two. In addition, malonyl transacylase is one of the most active FAS enzymes *in vitro* and thus is not likely to be rate-limiting. Although acetyl transacylase is slower *in vitro*, both the substrate and product of the acetyl transacylase-catalyzed reaction can enter directly into FAS via keto-acyl-ACP synthase reactions, and consequently acetyl transacylase is not a likely candidate to play a regulatory role. Based on analysis of acyl-ACP pools in spinach leaf and seed as shown in Table 2, both acetyl-CoA carboxylase and keto-acyl-ACP synthase are probably far from equilibrium. The steady state levels of acetyl- and malonyl-ACP probably reflect the levels of acetyl- and malony-CoA, respectively, because of the reversible nature of the transacylases which catalyze their interchange. Therefore, since the *in vivo* level of acetyl-ACP was comparable to or higher than the *in vivo* level of malonyl-ACP and, furthermore, since the acetyl-CoA carboxylase-catalyzed reaction strongly favors formation of malonyl-CoA, it follows that this step is far from equilibrium, and potentially regulatory. Similarly, the keto-acyl-ACP synthase reactions are far from equilibrium, since overall product formation by FAS is strongly favored, and yet the keto-acyl-ACP synthase substrates, acetyl-CoA and 2:0-16:0-ACPs, are observed in significant quantities.

In leaf tissue the rate of fatty acid synthesis can be altered by light, which offers a more definitive method of identifying a regulatory step. When a metabolic pathway undergoes a change in overall rate, the concentration of the substrate for the regulatory enzyme will change in the opposite direction to the change in the rate (Rolleston, 1972). Thus, when the rate of a pathway slows, the substrate concentration for the regulatory enzyme will increase. The rate of fatty acid biosynthesis is six-fold less in the dark compared to the light (Browse *et al.*, 1981). Analysis of the acyl-ACP pools under these two conditions revealed major changes in steady state levels of acetyl-ACP which can be correlated to a potential regulatory step in the pathway. In leaf sampled during the light cycle, there were very low levels of malonyl-ACP (<2–4 percent), and the level of acetyl-ACP, although slightly higher than malonyl-ACP levels, was not significantly higher than levels of 4:0-ACP and 6:0-ACP. However, in the dark, the level of acetyl-ACP increased 4- to 5-fold, and was accompanied by a comparable decrease in the level of free ACP leading to a 13- to 15-fold change in the ratio of acetyl-ACP/free ACP. In contrast, the level of malonyl-ACP did not change significantly. The reversible activity of the acetyl transacylase suggests that the accumulation of acetyl-ACP would also reflect a significant increase in acetyl-CoA level. Consideration of the reactions shown in Figure 5 indicate that an increase in acetyl-ACP/acetyl-CoA pools, but not malonyl-ACP/malonyl-CoA pools, would most likely result from a reduced rate of acetyl-CoA carboxylase. A lower activity of malonyl transacylase or keto-acyl-ACP synthase would not have produced a similar result. Therefore, the most reasonable explanation for the major increase in acetyl-ACP is that the activity of acetyl-CoA carboxylase decreased in the dark and these data provide the first *in vivo* evidence that regulation of acetyl-CoA carboxylase activity is responsible, at least in part, for the light/dark control of fatty acid biosynthesis in spinach leaf.

Conclusions

The fatty acid composition of plant oils appears to be particularly amenable to modification both by classical breeding practices and through molecular genetic approaches. Information from studies of acyl carrier proteins has indicated that plants have multiple genes for fatty acid synthesis components and some of these are expressed in a tissue-specific fashion. Therefore, it may be possible to use the promoters of these genes together with other plant, microbial, or animal genes to provide alterations in lipid metabolism that are expressed only in seeds and with the appropriate developmental timing. Changes in the type and amount of unsaturation of dietary oils may provide several benefits toward reducing health problems. Therefore, the recent progress in isolation of desaturase genes may now provide new opportunities to alter the nutritional quality of dietary plant oils.

Acetyl-CoA carboxylase has been identified as a regulatory enzyme which provides light/dark control of leaf fatty acid synthesis. Further studies of acyl-ACP and other metabolite pools in developing seeds will provide information on whether acetyl-CoA carboxylase or other enzymes also regulate the flux of carbon into oilseed triacylglycerols. This information should allow the design of strategies useful for genetic engineering of the quantity of oil produced by oilseeds.

References

Battey, J.F. and Ohlrogge, J.B. 1990. Evolutionary and tissue specific control of multiple acyl carrier protein isoforms in plants and bacteria. Planta 180: 352-360.

Browse, J., Roughan, P.G. and Slack, C. 1981. Light control of fatty acid synthesis and diurnal fluctuations of fatty acid composition in leaves. Biochem. J. 196: 347-354.

Chuman, L. and Brody, S. 1989. Acyl carrier protein is present in the mitochondria of plants and eucaryotic microorganisms. Eur. J. Biochem. 184: 643-649.

De Silva, J., Loader, N.M., Jarman, C., Windust, J.H.C., Hughes, S.G. and Safford, R. 1990. Plant Mol. Biol. 14: 573-581.

Eastwell, K.C. and Stumpf, P.K. 1983. Regulation of plant acetyl-CoA carboxylase by adenylate nucleotides. Plant Physiol. 72: 50-55.

Frentzen, M. 1986. Biosynthesis and desaturation of the different diacylglyerol moieties in higher plants. J. Plant Physiol. 124: 193-209.

Guerra, D.J., Ohlrogge, J.B. and Frentzen, M. 1986. Activity of acyl carrier protein isoforms in reactions of plant fatty acid metabolism. Plant Physiol. 82: 448-453.

Gurr, M.L., Blades, J., Appleby, R.S., Smith, C.G., Robinson, M.P. and Nichols B.W. 1974. Studies on seedoil triglycerides. Eur. J. Biochem. 43: 281-290.

Hannapel, D.J. and Ohlrogge, J.B. 1988. Regulation of acyl carrier protein messenger RNA levels during seed and leaf development. Plant Physiol. 86: 1174-1178.

Hansen, L. 1987. Three cDNA clones for barley leaf acyl carrier proteins I and II. Carlsberg Res. Commun. 52: 381-392.

Harwood, J.L. 1988. Fatty acid metabolism. Ann. Rev. Plant. Physiol. 39: 101-138.

Hloušek-Radojčić, A., Post-Beittenmiller, M.A. and Ohlrogge, J.B. 1989. Characterization of acyl carrier protein genes from *Arabidopsis*. Plant Physiol. 89: S67.

Jaworski, J.G. 1987. Lipids: Structure and Function, Vol. 9. In "The Biochemistry of Plants," Stumpf, P.K. and Conn, E.E., (Eds), pp. 159-173, Academic Press, NY.

Jaworski, J.G., Clough, R.C. and Barnum, S.R. 1989. A cerulenin insensitive short chain 3-ketoacyl-acyl carrier protein synthetase in *Spinacia oleracea* leaves. Plant Physiol. 90: 41-44.

Kim, K.-H., Lopez-Casillas, F., Bai, D.H., Luo, X. and Pape, M.E. 1989. FASEB J. 3: 2250-2256.

Lands, W.E.M.L. 1986. "Fish and Human Health." Academic Press, Orlando.

McKeon, T.A. and Stumpf, P.K. 1982. Purification and characterization of the stearoly-acyl carrier protein desaturase and the acyl-acyl carrier protein thioesterase from maturing seeds of safflower. J. Biol. Chem. 257: 12141-12147.

Ohlrogge, J.B., Browse, J. and Somerville, C.R. The genetics of plant lipids. Biochim. Biophys. Acta. In press.

Nikolau, B.J. and Hawke, J.C. 1984. Purification and characterization of maize leaf acetyl-CoA carboxylase. Arch. Biochem. Biophys. 228: 86-96.

Ohlrogge, J.B., and Kuo, T.M. 1984a. Control of lipid synthesis in developing soybean seeds: enzymatic and immunochemical assay of acyl carrier protein. Plant Physiol. 74: 622-625.

Ohlrogge, J.B. and Kuo, T.M. 1984b. Spinach acyl carrier protein: Primary structure, mRNA translation and immunoelectrophoretic analysis. In "Structure, Function, and Metabolism of Plant Lipids", pp. 63-67. Elsevier, Amsterdam.

Ohlrogge, J.B., Kuhn, D.K. and Stumpf, P.K. 1979. Subcellular localization of acyl carrier protein in leaf protoplasts of *Spinacia oleracea*. Proc. Natl. Acad. Sci. 76: 1194-1198.

Ohlrogge, J.B. and Kuo, T.M. 1985. Plants have isoforms of acyl carrier proteins that are expressed differently in different tissues. J. Biol. Chem. 260: 8032-8037.

Ohlrogge, J.B. 1987. The biochemistry of plant acyl carrier proteins. In "The Biochemistry of Plants," Vol. 9, p. 137, Academic Press, New York.

Post-Beittenmiller, M.A., Schmid, K.M. and Ohlrogge, J.B. 1989. Expression of holo and apo forms of spinach acyl carrier protein in leaves of transgenic tobacco plants. The Plant Cell 1: 889-899.

Post-Beittenmiller, D., Jaworski, J.G. and Ohlrogge, J.B. *In vivo* pools of free and acylated acyl carrier proteins in spinach: evidence for sites of regulation of fatty acid biosynthesis. J. Biol. Chem. In press.

Post-Beittenmiller, M.A., Hloušek-Radojčić, A. and Ohlrogge, J.B. 1989. DNA sequence of a genomic clone encoding an *Arabidopsis* acyl carrier protein. Nucleic Acids Res. 17: 1777.

Rizek, R.L., Welsh, S.O., Marston, R.M., and Jackson, E.M. 1983. Levels and sources of fat in the U.S. food supply and in diets of individuals. In "Dietary Fats and Health." pp. 13–43. Perkins, E.G. and Visek, W.J. Eds. American Oil Chemists Society.

Rolleston, F.S. 1972. A theoretical background to the use of measured concentrations of intermediates in study of the control of intermediary metabolism. Curr. Top. Cell. Regul. 5: 47-75.

Roughan, P.G. and Slack, C.R. 1982. Cellular organization of glycerolipid metabolism. Ann. Rev. Plant Physiol. 33: 97-132.
Safford, R., Windust, J.H.C., Lucas, C., De Silva, J., James, C.M., Hellyer, A., Smith, C.G., Slabas, A.R. and Hughes, S.G. 1988. Plastid-localized seed acyl-carrier protein of *Brassica napus* is encoded by a distinct, nuclear multigene family. Eur. J. Biochem. 174: 287-295.
Scherer, D.E. and Knauf, V.C. 1987. Isolation of a cDNA clone for the acyl carrier protein-I of spinach. Plant Mol. Biol. 9: 127-134.
Schmid, K.M. and Ohlrogge, J.B. 1989. Isolation of a cDNA clone for spinach acyl carrier protein II. Plant Physiol. 89 Suppl. 68. (Abstr.)
Schmid, K.M. and Ohlrogge, J.B. 1990. A root acyl carrier protein-II from spinach is also expressed in leaves and seeds. Plant Mol. Biol. In press.
Shanklin, J. and Somerville, C.R. Purification of stearoyl-ACP desaturase from avocado and characterization of cDNA clones from Castor and cucumber. Submitted.
Simoni, R., Criddle, R. and Stumpf, P. 1967. Purification and properties of plant and bacterial acyl carrier proteins J. Biol. Chem. 212: 573-581.
Slabas, A.R., Harding, J., Hellyer, A., Roberts, P. and Bambridge, H.E. 1987. Induction, purification and characterization of acyl carrier protein from developing seeds of oil seed rape (*Brassica napus*). Biochim. Biophys. Acta 921: 50-59.
Slabas, A.R., Sidebottom, C.M., Hellyer, A., Kessell, R.M.J. and Tombs, M.P. 1986. Induction, purification and characterization of NADH-specific enoyl acly carrier protein reductase from developing seeds of oil seed rape (*Brassica napus*). Biochim. Biophys. Acta 877: 271-280.
Stymne, S. and Stobart, A.K. 1987 Lipids: Structure and Function, Vol. 9. In "The Biochemistry of Plants" Stumpf, P.K. and Conn, E.E., (Eds.) pp. 175-214, Academic Press, NY.
Turnham, E. and Northcote, D.H. 1983. Changes in the activity of acetyl-CoA carboxylase during rape-seed formation. Biochem. J. 212: 223-229.
Weising, K., Schell, J. and Kahl, G. 1988. Foreign genes in plants: transfer, structure, expression, and applications. Ann. Rev. Genet. 22: 421-471.

Biotechnological Alterations of Lipid Metabolism in Plants

David F. Hildebrand, Hong Zhuang, Thomas R. Hamilton-Kemp, Roger A. Andersen, W. Scott Grayburn and Glenn B. Collins
Departments of Agronomy & Horticulture
University of Kentucky
Lexington, KY 40546

There are a number of aspects of lipid metabolism for which desired changes are approachable using biotechnology. One is the alteration of fatty acid composition of seed oils, and another is the change in the formation of fatty acid-derived flavor and aroma compounds. We have identified polypeptides associated with control of linolenate content in plant tissues using mutants and a chemical modulator of lipid composition. Work is in progress involving use of an ω-9 desaturase gene to decrease saturated fatty acid content of plant lipids. We have elucidated the role of specific lipoxygenase isozymes in the formation of C_6 aldehydes in soybean seed preparations and have produced transgenic plants with altered lipoxygenases. Finally, we have developed improved somatic embryogenesis systems for soybeans, facilitating genetic transformation and the study of the control of lipid biosynthesis and peroxidative metabolism.

Introduction

Our laboratories are involved in three major areas related to biotechnological alterations of lipid metabolism in plants. These are the manipulation of unsaturated fatty acid biosynthesis; control of polyunsaturated fatty acid peroxidation; and development of gene transfer systems for a major oilseed, soybeans. In this report we will give a review and update on the first and third areas and present more detailed results in the second area.

I. Manipulation of Fatty Acid Biosynthesis

The prospects for improvement of quality of plant oils by use of biotechnology appear excellent. Major goals in this area include improving the nutritional quality and improving the oxidative stability of plant lipids. Alteration of components in plant oils such as carotenoids and tocopherols could facilitate accomplishing both of these goals. Likewise, alteration of triglyceride composition will impact both nutritional quality and oxidative stability. We will focus on this latter area.

Triglyceride composition is mostly controlled by the acyl transferases and the available pool of fatty acids. The fatty acid pool can be controlled by the fatty acid biosynthetic enzymes. Key enzymes in unsaturated fatty acid biosynthesis are the desaturases, which is the area this report will focus on. All higher plants contain ω-9, -6 and -3 desaturases and some plants contain ω-12 desaturases (which are common in animals). McKeon and Stumpf reported on the partial purification and characterization of an ω-9 desaturase in 1982. Recently many groups have cloned ω-9 desaturase genes from plants, animals and yeast (Bossie and Martin, 1989; Shanklin and Somerville, 1990; Thiede *et al.*, 1986; Thompson, 1990; Yadav and Hitz, 1990). Work is in progress in our lab to overexpress the rat liver ω-9 desaturase (Thiede *et al.*, 1986) in the cytoplasm of soybean cells. An ω-12 desaturase from animals has been characterized biochemically (Fujiwara *et al.*, 1983), but little biochemical information is available concerning this enzyme from plants. Likewise, little biochemical data has been generated on ω-6 and -3 desaturases. Murata *et al.* (1990) reported on the cloning of a cyanobacterium ω-6 desaturase. No ω-3 desaturase has been cloned to date.

In order to facilite cloning of ω-3 desaturase genes we have examined polypeptide differences between single gene low linolenate mutants and wildtype plants in soybeans (C1640; Wilcox *et al.*, 1984; Wilcox and Cavins, 1985) and *Arabidopsis* (fad-D; Browse *et al.*, 1984). The soybean mutant, C1640, resulted in a 50 percent reduction in the linolenate (18:3) content of seed [but not leaf] lipid (Wilcox *et al.*, 1984) with the mutation affecting the linoleate (18:2) to 18:3 step (Wang *et al.*, 1989). The *Arabidopsis* fad-D mutant, on the other hand, results in a large reduction of 18:3 and hexadecatrienoate (16:3) in leaf, but not seed lipids and again affects the 18:2 to 18:3 [and 16:2 to 16:3] steps (Browse *et al.*, 1984; Browse *et al.*, 1986a and b). We were unable to see any clear differences in polypeptides between mutant and wildtype soybeans or *Arabidopsis* on 1-D SDS or IEF gels visualized by Coomassie blue or silver staining or autoradiography or fluorography of *in vivo* labeled proteins. We were likewise unable to see any clear or reproducible differences between the mutants or wildtypes on silver or Coomassie stained 2-D gels. However, we were able to find a polypeptide that was greatly reduced in the mutant versus the wildtype in soybeans and *Arabidopsis* on 2-D gels followed by autoradiography or fluorography of *in vivo* labeled proteins. Densitometric analyses of the 2-D gel autoradiograms indicate that this polypeptide is about 0.15 percent of the labeled polypeptides from developing embryos of wildtype soybeans and (gt) 0.0005 percent in C1640. This cosegregates with 18:3 levels in F2 progeny of crosses between mutant and wildtype plants of both *Arabidopsis* and soybeans (Brockman and Hildebrand, 1990; Wang *et al.*, 1987b). In *Arabidopsis* this polypeptide alteration maps to within 10 map units of the fad-D locus (Brockman and Hildebrand, 1990). The pyridazinone derivative, San 9785, which reduces 18:3 to levels similar to those in the mutants, also results in a large reduction in this polypeptide in both *Arabidopsis* and soybean (Brockman *et al.*, 1990; unpublished data). Studies indicate that the reduction of levels of this polypeptide in low 18:3 *Arabidopsis* plants is not a result of the low 18:3 levels (Brockman and Hildebrand, 1990; Brockman *et al.*, 1990). It is likely that this polypeptide which has been identified in soybeans and *Arabidopsis* is an ω-3 desaturase [or part of one], but it is also possible that it is part of an ancillary protein involved in control of ω-3 desaturations.

Antibodies have been generated to the 18:3-associated polypeptide from soybeans, and studies are in progress to ascertain whether this polypeptide is part of the ω-3 desaturase or involved in the control of 18:3 in another way. The effects of these antibodies on 18:2 desaturation is being examined as well as the copurification of this polypeptide with 18:2 desaturase activity. The 18:3 associated polypeptides of soybeans and *Arabidopsis* have similar molecular weights and pIs, but the antibodies made to the soybean polypeptide apparently show little cross reaction with the *Arabidopsis* polypeptide. Preliminary evidence suggests that the polypeptides identified in soybeans and *Arabidopsis* represent different ω-3 desaturase isozymes that probably show different developmental or tissue-specific regulation. However, more work is needed to clarify this situation. If the current studies provide further evidence that these polypeptides are likely to be involved in control of 18:3 levels in plant tissues, partial amino acid sequencing will be done and corresponding oligonucleotides synthesized. These together with the antibodies will be used to identify cDNA clones which are candidates for ω-3 desaturases.

II. Polyunsaturated Fatty Acid Peroxidation

The peroxidation of polyunsaturated fatty acids in plant tissues can be controlled to a large extent by levels of particular lipoxygenase isozymes. We have been investigating the diversity and expression of lipoxygenase isozymes in plant development using soybeans as a model system. We have also been evaluating the effects of particular lipoxygenase isozymes on C_6 aldehyde production and, finally, we have transferred a soybean lipoxygenase gene into tobacco and are examining the effects of altered lipoxygenase expression.

Developmental Expression of Lipoxygenase Isozymes

Introduction. Lipoxygenase (linoleate: oxygen oxidoreductase, E.C. 1.13.11.12) is widely distributed in plant tissues and is particularly abundant in the seeds of leguminous plants. In soybean seed tissues, three lipoxygenase isozymes have been isolated and characterized. The first lipoxygenase isolated and characterized from soybean seeds was designated lipoxygenase 1 (L1) (Theorell et al., 1947). Since then, two additional soybean lipoxygenase isozymes have been purified, characterized and designated lipoxygenase 2 (L2) and lipoxygenase 3 (L3) (Christopher et al., 1970 and 1972). The three seed lipoxygenases are monomeric proteins similar in size ranging from 94 to 97 kD with distinct isoelectric points ranging from about 5.7 to 6.4 (Hildebrand et al., 1988; Shibata et al., 1987 & 1988; Yenofsky et al., 1988). Charge differences have been the principal distinguishing property by which the lipoxygenases have been separated and characterized. Earlier reports suggest that, in mature soybean seeds, lipoxygenase 3 is the most abundant of the isozymes on a protein basis followed by lipoxygenase 1, with lipoxygenase 2 being the least abundant (Axelrod et al, 1981; Kitamura, 1984).

Many researchers have classified plant lipoxygenases as one of two types. Type I characterizes lipoxygenases with a high pH optimum, e.g. pH 9 or above (such as lipoxygenase 1), while type II characterizes lipoxygenases with pH optima of about pH 7 (such as lipoxygenase 2 and lipoxygenase 3).

While soybean seed lipoxygenase has been extensively studied with three different isozymes described, the diversity and characteristics of lipoxygenase(s) in soybean leaves has not been examined. In a survey of lipoxygenase activity in leaves of 28 plant species, Sekiya *et al.* (1983) found soybean leaves to be among the highest in terms of activity. Soybean roots and hypocotyls have recently been found to possess lipoxygenase isozymes distinct from those found in seeds (Hildebrand et al., 1989a; Park and Polacco, 1989). These studies were undertaken to quantify changes in expression of lipoxygenase isozymes during soybean embryo development, to examine lipoxygenase(s) in soybean leaves, and to compare them to the previously characterized seed and hypocotyl isozymes.

Materials and Methods

Lipoxygenase activities were determined by spectrophotometric measurement of the formation of conjugated dienes at 235 nm (Hildebrand and Hymowitz, 1981). The combined activity of the lipoxygenase 2 plus lipoxygenase 3 isozymes (Type II) was determined using 18:2 or 18:3 as the substrate at pH 6.8 while lipoxygenase 1 activity was determined at pH 9.0. All substrate solutions were adjusted to an optical density of 0.45 A235 against the appropriate buffer blank, using partially oxidized substrate. This created a solution which was approximately 18 μm hydroperoxylinoleic [or hydroperoxylinolenic] acid. This was done to standardize hydroperoxylinoleic acid levels, which would vary otherwise due to autooxidation. This also served to activate inactive forms of lipoxygenases (Cheesbrough and Axelrod, 1983).

Both substrate solutions were prepared immediately prior to assaying and kept in a light-proof container on ice during use. Vials containing stock solutions of linoleic acid diluted 1:10 in ethanol were opened only in an atmosphere of argon and stored in the dark under argon at -17 C. The molar extinction coefficient value of 23,000 (Gibian and Vandenberg, 1987) was used in calculations of lipoxygenase activity.

The protein content of the tissue extracts was determined using the Modified Lowry procedure (Bensadoun and Weinstein, 1976) with bovine serum albumin as a standard.

Native isoelectric focusing (IEF) was done in a 6 percent gel using an LKB flat bed system as described by Funk *et al.* (1985). Western blots were performed as described by Wang and Hildebrand (1987) using a mixture of antibodies prepared to soybean leaf LOXs used with the Western blot shown in Figure 4. The seed extract loaded on the gels was made from mature soybean cv. "Century" seeds homogenized with 20 vol/wt water in a mortar and pestle. This was centrifuged at 12,000 g for 30 min and the supernatant used for loading.

For studies of lipoxygenase mutants, near-isogenic lines in the "Century" background were used (Davies and Nielsen, 1987). Leaf analyses were performed with leaves 25-35 mm wide from plants with half-filled pods and extracted with 1/2 volume water/weight. Hypocotyls were from seeds grown for 4 days in germination towels in darkness. After homogenization in an equal volume of water, hypocotyl extracts were centrifuged as described above. Mature seeds were ground in 40 volumes of water and centrifuged as described above.

Results and Discussion

A large increase in the level of type I (ph 9.0) and type II (ph 6.8) lipoxygenase activity was observed in axes and cotyledons when the seeds reached 6–7 mm in length and continued until 9–10 mm (Figure 1). A 2–3-fold increase in water-soluble proteins occurs at the same time as the increase in lipoxygenase activity. The overall increase in soluble proteins did not account for the increase in lipoxygenase activity alone. When activity is viewed on the basis of activity per mg soluble protein, type I activity for both axes and cotyledons shows a 13-fold increase between the 5 mm embryo tissues and the 10 mm embryo tissues, while type II shows a 10-fold increase for the same intervals. Type I lipoxygenase showed a continual increase from the same intervals. Type I lipoxygenase showed a continual increase from 6–7 mm stage similar to that of total soluble protein accumulation. Type II lipoxygenase also showed further increases beyond the 6–7 mm stage, but the increase was not continuous and it was less than total soluble protein increase at the later developmental stages (11-14 mm). Since soybean seeds reach full length well before maximum dry matter accumulation occurs, the results reported here are biased toward the earlier stages of development. Additional studies will be needed to clearly delineate quantitative changes in lipoxygenase levels during the later stages of soybean embryo development.

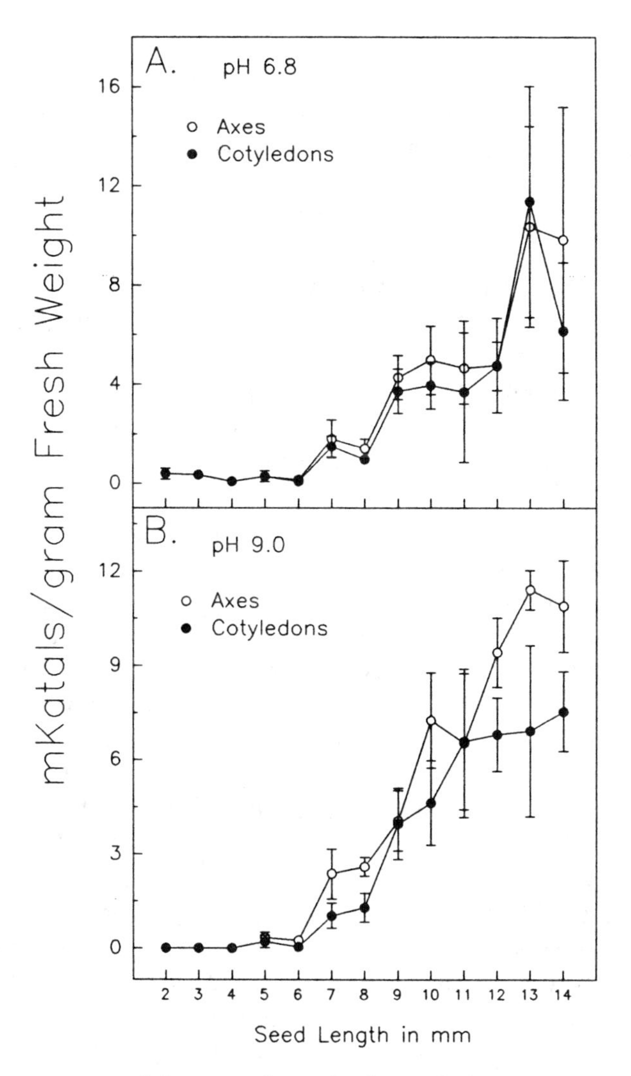

Figure 1. Lipoxygenase activity per g tissue fresh weight based on conjugated diene formation during soybean embryo development. The abcissa labels refer to the size of the developing seed in millimeters. Axes and cotyledon tissues were not analyzed separately until the seed reached 5 mm in length. Graph A shows the activity at pH 6.8 (known as type II activity) which is due to lipoxygenases 3, 2 and 1. Graph B shows the activity at pH 9.0 (type I activity) which represents lipoxygenase 1 activity (with only negligible contributions from lipoxygenase 2 and 3).

The individual lipoxygenase isozymes increased in parallel during the development of soybean cotyledons and axes. Neither lipoxygenase 1, 2 or 3 were detectable from samples below 6 mm in length, even when higher protein concentrations were loaded on the gels

(Hildebrand et al., 1989b). Using S1 nuclease protection analysis, lipoxygenase 1, 2 and 3 transcripts were undetectable in 3–5 mm developing seeds (unpublished data). Low but detectable type II lipoxygenase activity was seen at the 3–5 mm seed size which was prior to the appearance of lipoxygenases 1–3 (Figure 1). This activity could be due to isozymes different from lipoxygenases 1–3 being expressed in the embryo tissues at early stages of development. Seed coats were carefully removed in all experiments so that maternal tissues were not mixed in with embryonic tissues. The seed coats which were removed were also analyzed for lipoxygenase activity and pI. Seed coats at early stages of development were found to have much higher activity than the embryo tissues and this activity was due to isozymes which had more acidic pIs than lipoxygenases 1–3 (data not shown), somewhat analogous to the more acidic lipoxygenases seen in soybean seedling tissues (Hildebrand et al., 1989a; Park and Polacco, 1989), but this relationship requires further study. The activity seen at the stages of development prior to the expression of lipoxygenases 1–3 is not likely to be due to the presence of seed coat fragments as seed coats are very easily and clearly removed from embryo tissues using the techniques described by Lazzeri *et al.* (1985). It is possible that there was some residual endosperm tissue on the embryos used, but nothing is known concerning the endosperm lipoxygenases at this time. Sekiya *et al.* (1986) found relatively higher levels of lipoxygenase activity at the early stages of soybean seed development than was found in this study, but this is probably due to the fact that seed coats, which have high lipoxygenase activity at these stages, as mentioned above were included in their analyses. Kermasha and Metche (1987) found little change in lipoxygenase activity during French bean development, but few early developmental stages were examined, and again seed coats were included in the analyses. Domoney *et al.* (1990) reported that the two major lipoxygenase polypeptides of developing pea seeds were expressed during pea seed development in a manner similar to that for lipoxygenases 1–3 shown here. Domoney *et al.* (1990) also described a minor lipoxygenase polypeptide that is the predominant form synthesized early in seed development before the major forms appear. This early pea seed lipoxygenase was also detected in all other tissues and was considered to be a "housekeeping" lipoxygenase. It is not clear whether the expression of this early lipoxygenase was in the developing pea embryos, seed coats, or both. Preliminary studies indicate that lipoxygenases 1–3 are not expressed to any appreciable degree in any other soybean tissue (Hildebrand *et al.*, 1989b, unpublished data).

The relationship between leaf size and total lipoxygenase activity was investigated with two lipoxygenase substrates (Figure 2). Leaves of approximately 30-40 mm had the greatest lipoxygenase activity with either linoleic or linolenic acid as substrate. Considerable variability was seen in this study. Similar results with less variability were seen in an earlier study using leaves from individual soybean plants, with linoleic acid as the substrate (Hildebrand *et al.*, 1988).

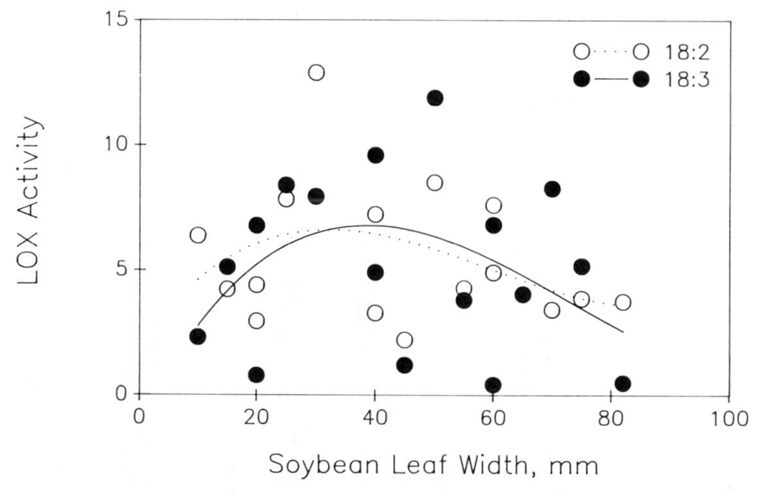

Figure 2. Lipoxygenase activity of soybean leaves. Activities were determined using linoleic (18:2) or linolenic (18:3) acids as substrates at pH 6.8. Activity is expressed as mKatals/mg soluble protein.

The soybean leaf lipoxygenase isozymes appear to be different from those found in soybean seeds although the molecular weights are very similar (Hildebrand *et al.*, 1989b). Direct evidence that leaf lipoxygenase isozymes differ from those found in seeds is shown in Figures 3 and 4. Mutations in the seed lipoxygenases affect lipoxygenase activity differently in seeds and leaves (Figure 3). Also, mutants lacking 1 or 2 seed lipoxygenase isozymes still show proteins that react with leaf lipoxygenase antibodies (Figure 4). In addition, leaf and seed lipoxygenases have different pIs in wild type and mutant backgrounds. Northern analysis suggests that mRNA for the three LOX seed isozymes is abundant only in embryos (Altschuler *et al.*, 1989).

We have found at least three isozymes of lipoxygenase in soybean leaves which appear to be different from any of the three previously reported soybean seed isozymes. At least one of these leaf isozymes appears to be different from the previously reported hypocotyl/radical isozymes (Figure 4). However, soybean leaves and hypocotyls appear to possess some isozymes in common. Several of the most acidic bands are seen in both tissues, while the most basic bands are detected only in leaf extracts. Our results support the conclusion of Park and Polacco (1989) that soybean seeds and hypocotyls possess distinct lipoxygenase isozymes.

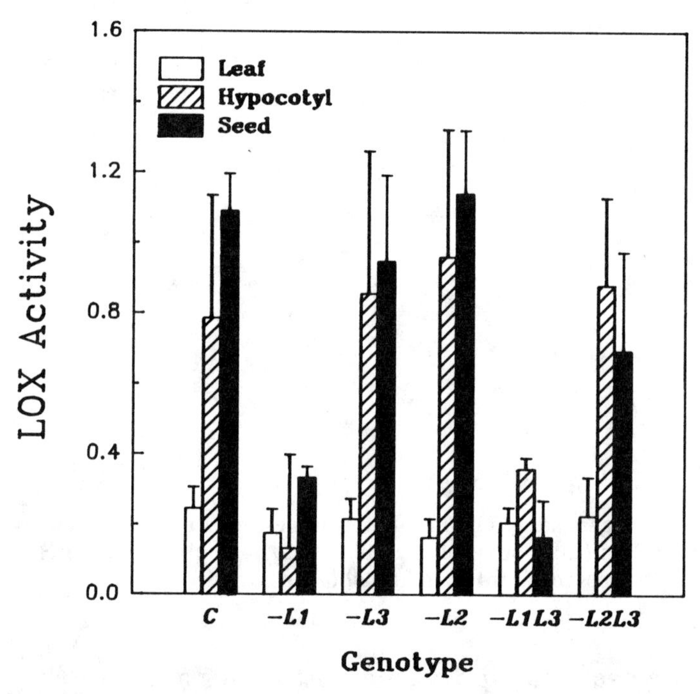

Figure 3. **Lipoxygenase activity of soybean leaves, hypocotyls, and seeds. *G. max* 'Century' (C) and derivatives lacking one or more seed lipoxygenase are compared. Activities were determined using linoleic acid as substrate at pH 6.8. Activity is expressed as mKatals/mg soluble protein.**

Effects of Lipoxygenase Isozymes on C_6 Aldehyde Production

Lipoxygenases are postulated to be involved in plant senescence, germination, the generation of growth regulatory compounds, pest resistance and the formation of flavors and aromas (which may be desirable or undesirable) of many plant products (Hatanaka et al, 1987; Hildebrand *et al.*, 1988, Gardner, 1989), which have been of interest to food scientists. Lipoxygenase activity can lead to formation of intracellular free radicals and fatty acid hydroperoxides, which

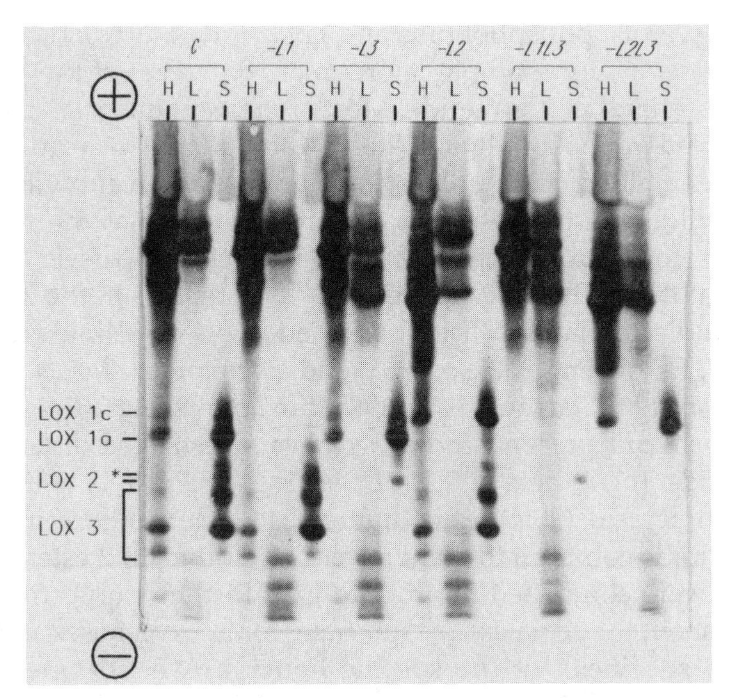

Figure 4. Western immunoblot of IEF-PAGE gel immuno-decorated with soybean leaf lipoxygenase antibodies. Undiluted extracts from leaves (L) and hypocotyls (H) were loaded while a 6:100 dilution of seeds (S) was used. The positions of seed lipoxygenase 1, 2, and 3 are indicated. C is the "Century" cultivar, which contains lipoxygenase 1, 2 and 3. Mutants lacking one or two seed lipoxygenases are indicated.

might be related to plant senescence (Leshem *et al.*, 1981; Lynch and Thompson, 1984) and syntheses of some growth regulatory compounds (Bousquet and Thimann, 1984; Legge and Thompson, 1983; Kacperska and Kubacka-Zebalska, 1985, 1989; Vick and Zimmerman, 1983; Zimmerman and Coudron, 1979). The products from cleavage of fatty acid hydroperoxides by hydroperoxide lyase, C_6-aldehydes, might be in part responsible for the pest-resistance of some plants (Lyr and Banasiak, 1983; Gardner *et al.*, 1990) and the flavor of many food products (Axelrod, 1974, Eskin *et al.*, 1977; Gardner, 1989).

Even though the three abundant lipoxygenase isozymes found in mature soybean seeds have been well characterized biochemically, and the lipoxygenase pathway has been largely elucidated (Hatanaka *et al.*, 1987; Vick and Zimmerman, 1987; Hildebrand, 1989a), the actual molecular structures of *in vivo* substrates for lipoxygenases in plants are unknown. A majority of cellular fatty acids are esterified to triacylglycerols, phospholipids and glycolipid (Harwood, 1980). In soybean seeds, for example, more than 90 percent of the 18:2 and 18:3 is esterified in triglyceride. Most of the remaining 18:2 and 18:3 is esterified in phospholipids. Fresh, good quality, mature soybean seeds generally contain only a few tenths of a percent of free fatty acids (Hildebrand, 1989). It is widely thought that the principal substrate for lipoxygenases are free fatty acids. However, it is not clear that this is the case. There are several reports in the literature which indicate that fatty acids esterified in phospholipids (Brash *et al.*, 1987; Eskola and Laakso, 1983) and glycolipids (Koch *et al.*,1958; Guss *et al.*, 1968; Yamauchi et al 1985) can act as substrates for lipoxygenases. In our present studies we have investigated different free fatty acids [oleic acid (18:1), *trans*-18:2, 18:2, 18:3, γ-18:3, and arachidonic acid (20:4)] and different 18:2 derivatives, including glycerolipids common to plant tissues and other 18:2 esters. These studies were conducted in order to gain further supportive information about the actual *in vivo* substrates for lipoxygenases and lyase, and the specificity of the specific lipoxygenase isozymes to the substrates.

Materials and Methods

Soybean lipoxygenase mutant backcross lines designated -L2L3, -L1L3, -L2, -L3 and -L1, were kindly provided by Dr. Niels Nielsen, Purdue Univ., and Century is a commercial cultivar (Davies and Nielsen, 1987). One-millimolar substrate solutions were prepared by mixing the lipids in water with a Tissuemizer (Tekmar SDT-1810) for about 1 min to disperse the substrates in water.

The C_6-aldehyde production analyses were performed in 1.8 ml screwtop vials. To each vial was added 5 mg seed meal, 50 microliters 0.1 M sodium phosphate, pH 6.8, and 150 microliters of 1 mM substrates in water. Samples were stirred for 30 sec with a magnetic stir bar and then incubated at 300C for 10 min after which 250 μl headspace vapor samples were analyzed by gas chromatography using direct injection onto a 30 m x 0.54 mm DB-5 (methyl silicone) fused silica column. The column oven was held isothermal at 50°C for 3 min and then the temperature was programmed at 5°C per min to 200°C.

Results and Discussion

The level of C_6-aldehydes was highest in the homogenates of -L2L3, -L1L3, and -L3 mutants, with 18:2, followed by 18:3, and then arachidonic acid 20:4 (Figure 5). The addition of trans-18:2 and oleic acid, however, reduced the C_6-compound formation in all homogenates. Figure 5b shows the results obtained by using the different esterified 18:2 derivatives as substrates for hexanal production, using the same method mentioned above. With the mutant line without L1L3, C_6-aldehyde was formed mostly by 18:2, followed by me18:2 and mono18:2, CoA18:2, MGDG and propyl18:2. With the addition of tri18:2 and di18:2, no n-hexanal generation was observed (Figure 5b).

Lipoxygenase 2 gave the largest formation of C_6-aldehydes except for CoA 18:2 and arachidonic acid (Figure 5a and b). Lipoxygenase 1 promoted C_6-aldehydes formation with all substrates where lipoxygenase 2 was also effective. In the presence of CoA18:2, the lines with wildtype levels of lipoxygenase 1, in general, produced relatively more C_6-compounds. With 20:4, lipoxygenase 1 and lipoxygenase 2 generated the same amount of C_6-aldehydes. Reduction of C_6-aldehydes by lipoxygenase 3 was very clearly observed with all substrates (Figure 5a and b).

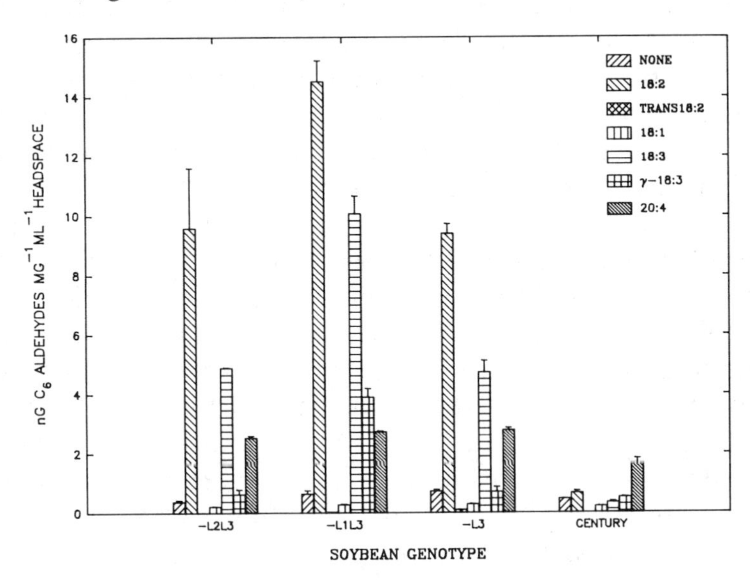

Figure 5a: The effect of free fatty acids on C_6-aldehyde formation of soybean homogenates.

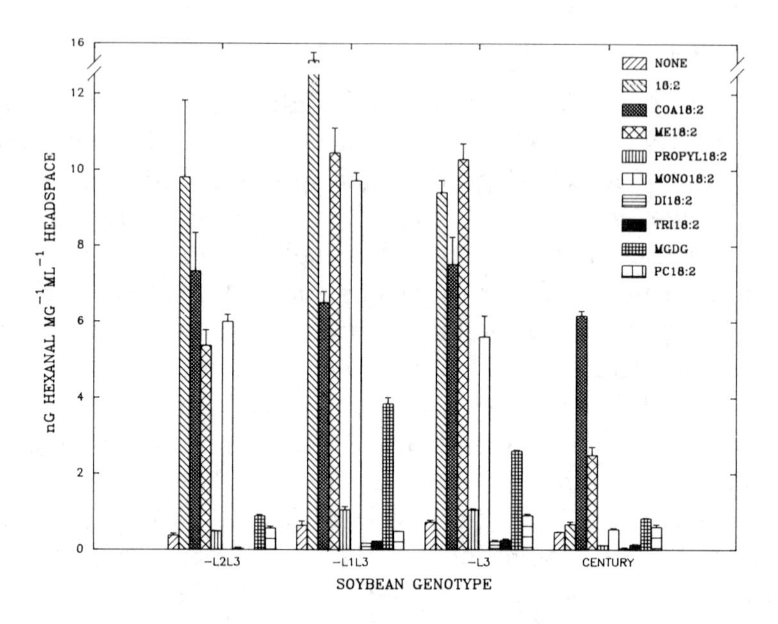

Figure 5b: The effect of 18:2 derivatives on C_6-aldehyde formation of soybean homogenates.

Free 18:2 is the best substrate for C6-aldehyde production, and diacylglycerol, triacylglycerol, phospholipid, *trans*- 18:2 and 18:1 do not promote formation of C_6-compounds with the system used here. In our observations at pH 6.8, the highest formation of C_6-aldehydes among free fatty acids and esterified 18:2 derivatives was with free fatty acid 18:2 (Figure 5a and b).

Lipoxygenase 2 was largely responsible for the generation of C_6-aldehydes except for both CoA18:2 and 20:4. Lipoxygenase 3 reduced these aldehydes. Matoba *et al.* (1985) reported that with 18:2 as the substrate, lipoxygenase 2 resulted in the highest yield of n-hexanal. In our present experiments we found similar results for other polyunsaturated free fatty acids and esterified 18:2 derivatives except for CoA18:2, where lipoxygenase 1 was most effective, and 20:4, where lipoxygenase 1 had the same activity as lipoxygenase 2. Lipoxygenase 3 was found to reduce C_6-aldehyde formation with all substrates tested, consistent with the report of Hildebrand *et al.* (1990) where lipoxygenase 3 was found to reduce hexanal production with free 18:2 and endogenous substrates.

Except for both tri18:2 and di18:2, lipoxygenase activity [lipid dependent oxygen consumption] is qualitatively similar to C_6-aldehyde production (data not shown), but the relationship between these chemical processes is complex. As mentioned above, most of the lipid compounds used in our experiments caused fatty acid-dependent oxygen uptake and formation of C_6-aldehydes with the homogenates from mature soybean seeds. This indicates that polyunsaturated long chain free fatty acids with *cis,cis*-1,4-pentadiene structures and some of their esterified derivatives can be substrates for lipoxygenases and that the hydroperoxides formed are substrates for lyase. Nevertheless, there exists a clear difference, especially in relative quantity with 18:2, 20:4, di18:2 and tri18:2. Our results show that free 18:2 gave the highest relative level of C_6-aldehydes; however, its fatty acid-dependent O_2 uptake was lower than many other compounds. With 20:4 and particularly tri18:2 and di18:2, the highest oxygen uptake was seen, but much lower or no hexanal formation was observed. In addition, the presence of lipoxygenase 2 resulted in the largest yield of C_6-aldehydes, and presence of lipoxygenase 3 resulted in the lowest yield of these compounds. In contrast, seeds with lipoxygenase 3 increased the amount of fatty acid-dependent oxygen consumption.

These results indicate that a number of unsaturated fatty acids and their derivatives might be potential *in vivo* substrates for lipoxygenases and involved in lipid-dependent O2 consumption and C_6-aldehyde formation in the plant tissues, but the relationship between lipoxygenase and lyase in C_6-aldehyde formation is obscure or complex, and further studies are needed to elucidate the roles of these two enzymes in C_6-aldehyde formation with various lipid substrates.

C_6-aldehyde production from leaves of the soybean mutant backcross lines was also measured and the following results for total C_6-aldehyde production [area units per 10 mm leaf disk (Hildebrand *et al.*, 1990)] were obtained: -L2-L3, 255 82; -L1-L3, 350 49; -L2, 100 12; -L3, 226 31; -L1, 202 55; Century, 332 41. Although large differences were seen, they did not correspond to the presence or absence of any of the seed lipoxygenase isozymes. These results are consistent with those presented in the previous section, indicating that the lipoxygenases present in soybean leaves are different than those in seeds and are under different genetic control.

Transformation with Lipoxygenases

Work is in progress on examining the effects of altered lipoxygenase 2 and 3 expression in transgenic tobacco and soybean plants. The work with lipoxygenase 2 has only recently been started. A full length lipoxygenase 2 cDNA (Shibata *et al.*, 1988; kindly supplied by B. Axelrod) has been inserted into a modified KYLX (Schardl *et al.*, 1987) vector and plant transformation has been initiated. A full length lipoxygenase 3 cDNA (Yenofsky *et al.*, 1988; kindly supplied by R. Yenofsky) has also been cloned 3′ to the 35S promoter in the plus and minus sense orientation in a modified KYLX vector. These were transferred into *Agrobacterium* using standard triparental mating procedures. About 80 transgenic tobacco plants with lipoxygenase 3 in the plus sense and about 40 with lipoxygenase 3 in the minus sense were produced using a modification of the leaf disc method of Horsch *et al.* (1985). Only a small number of these transgenic plants showed good expression of the introduced lipoxygenase 3 transcript. One of these showed increased lipoxygenase expression in progeny seed. Seed from untransformed tobacco lacked detectable lipoxygenase activity. Preliminary analyses of transgenic tobacco leaves were complicated by highly variable expression of the endogenous lipoxygenase isozymes and apparent instability of endogenous and introduced lipoxygenases. C_6-aldehyde production of headspace vapors of transgenic versus control leaves was also examined, but again these analyses were made difficult due to highly variable C_6-aldehyde production [due to variable lipoxygenases and perhaps also hydroperoxide lyases]. Measurement of C_6-aldehydes from headspace samples was further complicated by the fact that variable amounts of C_6-aldehydes were released from the leaf samples and that the first 18:3 product, *cis*-3-hexenal, was rapidly converted into several secondary products with unequal volatilities.

It is now apparent that multiple measurements of different tissues at different developmental stages will need to be done of transgenic plants with high expression of the introduced transcripts versus multiple controls to clearly evaluate the effects of the introduced lipoxygenases. It will also be useful to use additional vectors that should give higher expression of the introduced lipoxygenases, and such work is in progress.

Examination of C_6-aldehyde production from tobacco leaves with admixture of soybean flour with and without lipoxygenase 3 indicated as much as a 67 percent reduction in C_6-aldehyde production due to lipoxygenase 3 (Table 1).

III. Progress in Soybean Transformation

The leaf-disc transformation system developed by Horsch *et al.* (1985) has facilitated the *Agrobacterium*-mediated genetic transformation of those dicotyledonous species able to regenerate in culture from leaf tissue. This development has led to the transformation of

Table 1.

C_6-aldehyde	Tobacco leaf	Leaf + -L3	Leaf + Century
hexenal	3.0 ± 1.0	5.4 ± 2.8	1.8 ± 0.6
cis-3-hexenal	0.8 ± 0.6	0.4 ± 0.2	0.0 ± 0.0
trans-2-hexenal	7.4 ± 1.5	36.6 ± 15.7	16.5 ± 3.4
total	11.6 ± 2.8	42.4 ± 18.7	18.4 ± 4.0

It would therefore be predicted that lipoxygenase 3 should reduce and lipoxygenase 2 should increase C_6-aldehyde production in various tissues of transgenic plants.

various crop species, with soybean being, until recently, a major exception among dicotyledonous plants.

Successful gene transfer has recently been achieved for soybeans by three different techniques. One approach involved the delivery of DNA coated particles (eg. gold or tungsten) into young meristems followed by regeneration of plants from treated meristems (McCabe *et al.*, 1988; Christou *et al.*, 1989). A second approach involved *Agrobacterium* treatment of cotyledons from germinating seedlings, followed by regeneration of shoots and plants from proliferating cotyledon tissues, a protocol similar to the leaf-disc method (Hinchee *et al.*, 1988). A modification of this approach was utilized by Chee *et al.* (1989) which involved inoculation of the plumule, cotyledonary node and adjacent cotyledon tissues of germinating seedlings using a syringe fitted with a 30 1/2 gauge needle. They were able to obtain transformed lines at a frequency of 0.07 percent using this simple approach. The third approach, taken in our laboratory, involved *Agrobacterium* treatment of macerated immature cotyledons followed by regeneration through the somatic embryogenesis pathway (Parrott *et al.*, 1988). All of these protocols suffer from low efficiency. The particle acceleration method requires elaborate, expensive equipment to produce about 2 percent of transformed shoots of which many are chimeric. The *Agrobacterium*/organogenesis method gives an even lower frequency of transformed shoots and is so far highly genotype specific. The *Agrobacterium*/embryogenesis technique has produced three primary transformants out of 18 regenerants, but all were apparently chimeric. Chimerism is believed to result from multicellular initiation, which is typical of the embryos first formed on cultured immature cotyledons (Hartweck *et al.*, 1988). Later-formed embryos are more likely to be derived from proliferating embryogenic tissues arising from single cells. Regenerated transgenic plants can also be obtained via the organogenic pathway following particle acceleration (Christou *et al.*, 1989).

We expect to reduce chimerism by selecting later-forming embryos, particularly those initiated after re-culture of cotyledon tissues when the first flush of embryos have been removed (Lazzeri *et al.*, 1988). In addition, coupling *Agrobacterium* treatment of early-formed embryos to a recurrent embryogenesis protocol of the type used by McGranahan (1988) to transform walnut will provide an ancillary inexpensive method for regenerating transformed plants. Both of these modified protocols will reduce chimerism by allowing for more cell division cycles between transformation events and regeneration of plants. Dupont is now marketing a particle delivery device originally developed by John Sanford (Sanford, 1988). We recently leased a Dupont particle accelerator device and are currently also using this approach for soybean transformation. Finer and McMullen (1990) recently reported the successful use of the Dupont device with their previously reported embryogenic suspension cultures (Finer and Nagasawa, 1988). They obtained transgenic plants and progeny at apparently higher efficiency than has previously been reported in soybeans. However, their system has so far only been successful with the soybean cultivar "Fayette."

We are currently using the following protocols for alteration of lipoxygenase or fatty acid biosynthetic genes. The cotyledonary somatic embryogenesis system developed in our laboratory will be coupled with *Agrobacterium* treatment. The original system (A) produces transformed somatic embryos obtained after cocultivation of macerated cotyledons from immature sexual embryos with *Agrobacterium*. An improved or alternative system (B) will utilize transformed secondary embryos obtained after cocultivation of primary somatic embryos with *Agrobacterium*.

A. Zygotic cotyledon system.

Cotyledons from immature sexual embryos (3–5 mm long) will be macerated through steel mesh, placed on N10 embryogenic medium (Lazzeri *et al.*, 1988; Parrott *et al.*, 1988; 1989a and b), and cocultivated 48 h with overnight suspensions of *Agrobacterium*. The cotyledons will then be washed and transferred to fresh N10 medium containing 500 μg/ml Mefoxin (sodium cefoxitin) to inhibit growth of free bacteria, and 100 μg/ml Kanamycin or 50 μg/ml G418 (John Finer, personal communication) for preferential selection of transformed cells. After 14 d, cotyledons will be transferred to hormone-free medium containing Mefoxin and Kanamycin or G418, for development of somatic embryos (Parrott *et al.*, 1989a). After 28 d from initial excision of cotyledons, visible somatic embryos will be removed and diverted to protocol B. These embryos will be assumed to be untransformed or chimeric. The parent cotyledon tissue, including the fertile crescent described by Hartweck *et al.* (1988), will be returned to culture on N10 for a further 14 d and then to hormone-free medium for 14 d. Selection with Mefoxin and Kanamycin or G418 will be maintained throughout this secondary culture procedure. After 28 d from removal of the first somatic embryos, the second flush of embryos will be removed for maturation. These and subsequent embryos will be assumed to be potentially transformed. If healthy embryogenic parent tissue remains, this will be recycled through tertiary culture to obtain a third flush of embryos, etc.

B. Somatic embryo system.

Primary somatic embryos will be excised either from first-cycle cultures of zygotic cotyledons described under A (above), or from similar but unmacerated cotyledons cultured 14 d on N10 and 14 d on hormone-free medium without *Agrobacterium* treatment. Somatic embryos will be treated at the detachment wound site (root pole) with overnight suspensions of *Agrobacterium* for 48 h on N10 medium. The embryos will then be recultured 14 d on N10 medium and 14 d on hormone-free medium. Selection against free bacteria and for transformed cells will be maintained with Mefoxin and Kanamycin or G418 during this period. After 28 d, secondary embryos will be returned for a further culture cycle to produce further secondary embryos. Secondary embryos will be assumed to be transformed. However, if untransformed "escapes" are found among the first putative transformants, subsequent transformation procedures will be carried out using a vector additionally containing the phosphinothricin acetyl transferase gene (White *et al.*, 1990). After initial selection on Kanamycin or G-418, additional selection in the proliferation of additional secondary (or tertiary) somatic embryos will be done on a bialaphos-containing medium. This will promote the selection of uniformly transformed somatic embryos which could be regenerated into non-chimeric transformed plants.

We have been able to obtain reasonably large numbers of embryos on selective medium after coculture of wounded somatic embryos with *Agrobacterium*. No control or untransformed cultures have been able to form embryos on this selective medium.

C. Alternative systems.

If we have problems with the above systems, we will utilize the particle accelerator-mediated transformation system. We have recently put in place the second generation particle acceleration (gene delivery) device of Dupont. This system is currently maximally efficient utilizing tungsten as the delivery particles. Expression vector plasmids prepared as described above containing the lipoxygenase and desaturase plus sense and antisense genes and the dual 35S promoter or the phaseolin promoter will be precipitated onto tungsten particles as described in the Dupont Biolistic Particle Delivery Systems manual. These will then be delivered into somatic embryos, shoot meristems and seedling meristems as described for Agrobacterium above. The somatic embryos will be cultured and secondary embryos proliferated as described above. Plants will be regenerated from shoot meristems after particle delivery by organogenesis on a cytokinin-enriched medium as described by McCabe *et al.* (1988).

Potentially transformed embryos from protocols A and B will be matured for 4–6 weeks on hormone-free medium, desiccated for 1 week and returned to hormone-free medium for germination to plantlets (Parrott *et al.*, 1989a). Plantlets will be transferred to potting mix and grown in a greenhouse. A minimum of fifty transgenic plants will be produced with each of the four constructs and screened for high levels of production of the introduced transcripts in mid-maturation embryo tissue because of the apparently random nature of integration events.

References

Altschuler, M., Grayburn, W.S., Collins, G.B. and Hildebrand, D.F. 1989. Developmental expression of lipoxygenase isozymes in soybeans. Plant Science 63: 151-158.

Axelrod, B. Lipoxygenases. In: Whitaker, J.R. (ed)., "Adv. Chem Ser. Food Related Enzymes," 1974, 136, 324-348.

Axelrod, B., Cheesbrough, T.M. and Laakso, S. 1981. Lipoxygenases in soybean. Methods Enzymol. 71: 441-451.

Bensadoun, A. and Weinstein, D. 1976. Assay of proteins in the presence of interfering materials. Analyt. Biochem. 70: 241-250

Bousquet, J.F. and Thimann, K.V. 1984. Lipid peroxidation forms ethylene from 1-aminocyclopropane-1-carboxylic acid and may operate in leaf senescence. Proc. Natl. Acad. Sci. U.S.A. 81: 1724-1727.

Bossie, M.A. and Martin, C.E. 1989. Nutritional regulation of yeast D-9 fatty acid desaturase activity. J. Bact. 171: 6409-6413.

Brash, A.R., Ingram, C.D. and Harris, T.M. 1987. Analysis of a specific oxygenation reaction of soybean lipoxygenase-1 with fatty acids esterified in phospholipids. Biochemistry 26: 5465-5471.

Brockman, J.A. and Hildebrand, D.F. 1990. A polypeptide alteration associated with a low linolenate mutant of *Arabidopsis thaliana*. Plant Physiol. Biochem. 28: 11-16.

Brockman, J.A., Norman, H.A. and Hildebrand, D.F. 1990. Effects of temperature, light and a chemical modulator on linolenate biosynthesis in mutant and wild type *Arabidopsis* calli. Phytochemistry 29: 1447-1453.

Browse, J., McCourt, P. and Somerville, C.R. 1984. Glycerolipid metabolism in leaves: new information from *Arabidopsis* mutants. In: "Structure, Function and Metabolism of Plant Lipids," Siegenthaler, P.A. and Eichenberger, W. ed., Elsvier Science Publishers, Amsterdam, pp. 167-170.

Browse, J., McCourt, P. and Somerville, C.R. 1986a. A mutant of *Arabidopsis* deficient in C18:3 and C16:3 leaf lipids. Plant Physiol. 81: 859-864.

Browse, J., Warwick, N., Somerville, C.R. and Slack, C.R. 1986b. Fluxes through the procaryotic and eucaryotic pathways of lipid synthesis in the "16:3" plant *Arabidopsis thaliana*. Biochem. J. 235: 25-31.

Chee, P.P., Fober, K.A. and Slightom, J.L. 1989. Transformation of soybean (*Glycine max*) by infecting germinating seeds with *Agrobacterium tumefaciens*. Plant Physiol. 91: 1212-1218.

Cheesbrough, T.M. and Axelrod, A. 1983. Determination of the spin state of iron in native and activated soybean lipoxygenase 1 by paramagnetic susceptibility. Biochemistry 22: 3837-3840.

Christopher, J.P., Pistorius, E.K. and Axelrod, B. 1970. Isolation of an isozyme of soybean lipoxygenase. Biochim. Biophys. Acta 198: 12-19.

Christopher, J.P., Pistorius, E.K. and Axelrod, B. 1972. Isolation of a third isozyme of soybean lipoxygenase. Biochim. Biophys. Acta 284: 54-62.

Christou, P., Swain, W.F., Yang, N.-S. and McCabe, D.E. 1989. Inheritance and expression of foreign genes in transgenic soybean plants. Proc. Natl. Acad. Sci. USA 86: 7500-7504.

Davies, C.S. and Nielsen, N.C. 1987. Registration of soybean germplasm that lacks lipoxygenase isozymes. Crop Sci. 27: 370-371.

Domoney, C., Firmin, J.L., Sidebottom, C., Ealing, P.M., Slabas, A. and Casey, R. 1990. Lipoxygenase heterogeneity in *Pisum sitivum*. Planta 181: 35-43.

Eskin, N.A.M., Grossman, S. and Pinksy, A. 1977. Biochemistry of lipoxygenase in relation to food quality. Crit. Rev. Food Sci. Nutr. 9: 1–41.

Eskola, J. and Laakso, S. 1983. Bile salt-dependent oxygenation of polyunsaturated phosphatidylcholines by soybean lipoxygenase-1. Biochem. Biophys. Acta 751: 305-311.

Finer, J.J. and Nagasawa, A. 1988. Development of an embryogenic suspension culture of soybean (*Glycine max* Merrill.). Plant Cell Tissue Organ Cult. 15: 125-136.

Finer, J.J. and McMullen, M.D. 1990. Transgenic soybean plants obtained via particle bombardment of embryogenic cultures. Presented at the 3rd Biennial Conference on Molecular and Cellular Biology of the Soybean. July 23 to 25, Ames, Iowa.

Fujiwara, Y., Okayasu, T., Ishibashi, T. and Imai, Y. 1983. Immunochemical evidence for the enzymatic difference of Δ^6-desaturase form Δ^9- and Δ^5-desaturase in rat liver microsomes. Biochem. Biophys. Res. Communic. 110: 36-41.

Funk M.O., Carroll, R.T., Thomson, J.F.and Dunham, W.R. 1986. The lipoxygenases in developing soybean seeds, their characterization and synthesis *in vitro*. Plant Physiol: 1139-1144.

Funk, M.O., Whitney, M.A., Hausknecht, E.C. and O'Brian, E.M. 1985. Resolution of the isozymes of soybean lipoxygenase using isoelectric focusing and chromatofocusing. Anal. Biochem. 146: 256-251.

Gardner, H.W. 1989. How the lipoxygenase pathway affects the organoleptic properties of fresh fruit and vegetables. In, "Flavor Chemistry of Lipid Foods," Min, D.B. and Smouse, T.H. (Eds), American Oil Chemists Society, Champaign, IL, pp. 98-112.

Gardner, H.W. 1989b. Soybean lipoxygenase-1 enzymatically forms both (9S)- and (13S)-hydroperoxides from linoleic acid by a pH-dependent mechanism. Biochim. Biophys. Acta 1001: 274-281.

Gardner, H.W., Dornbos, Jr.,L., and Desjardins, A.E. 1990. Hexanal, trans-2-hexenal, and trans-2-nonenal inhibit soybean, glycine max, seed germination. J. Agric. Food Chem. 38: 1316-1320.

Gibian, M.J. and Vandenberg, P. 1987. Product yield in oxygenation of linoleate by soybean lipoxygenase: the value of the molar extinction coefficient in the spectrophotometric assay. Analyt. Biochem. 163: 343-349.

Guss, P.L., Richardson, T. and Stahmann, M.A. 1968. Oxidation of various lipid substrates with unfractionated soybean and wheat lipoxidase. J. Am. Oil Chem. Soc. 45: 272-276.

Hartweck, L.M., Lazzeri, P.A., Cui, D., Collins, G.B. and Williams, E.G. 1988. Auxin-orientation effects on somatic embryogenesis from immature soybean cotyledons. In Vitro Cell. Devel. Biol. 24: 821-828.

Harwood, W.M. 1980. Plant acyl lipids: Structure, distribution, and analysis. In, "The Biochemistry of Plants: A Comprehensive Treatise Vol, 4 Lipids: Structure and Function," Stampf, P.K. and Conn, E.E. (Eds.), pp. 1–55, Academic Press, New York.

Hatanaka, A., Kajimara, T. and Sekiya, J. 1987. Biosynthetic pathway for C_6-aldehyde formation from linolenic acid in green leaves. Chem. Phys. Lipids 44: 341-361.

Hildebrand, D.F. and Hymowitz, T. 1981. Two soybean genotypes lacking lipoxygenase-1. J. Am. Oil Chem. Soc. 58: 583-586.

Hildebrand, D.F. and Hymowitz, T. 1983. Lipoxygenase activities in developing and germinating soybean seeds with and without lipoxygenase-1. Bot. Gaz. 144: 212-216.

Hildebrand, D.F., Hamilton-Kemp, T.R., Legg, C.S. and Bookjans, G. 1988. Plant lipoxygenases: occurrence, properties and possible functions. Current Topics in Plant Biochem. Physiol. 7: 201-219.

Hildebrand, D.F. 1989. Lipoxygenase. Physiol. Plant. 76: 249-253.

Hildebrand, D.F., Snyder, K.M., Hamilton-Kemp, T.R., Bookjans, G., Legg, C.S. and Andersen, R.A. 1989a. Expression of lipoxygenase isozymes in soybean tissues. In: "Biological Role of Plant Lipids," Biacs, P.A., Gruiz, K and T. Kremmer (Eds.), pp. 51–56. Akademiai Kiado, Budapest and Plenum Publishing Co., New York and London.

Hildebrand, D.F., Grayburn, W.S., Versluys, R.T., Hamilton-Kemp, T.R., Andersen, R.A. and Collins, G.B. 1989b. Genetic control and modification of lipoxygenase activity in plants. In, "Proceedings of the International Symposium on New Aspects of Dietary Lipids," International Union of Food Sci. and Technology, Goteburg.

Hildebrand, D.F., Hamilton-Kemp, T.R., Loughrin, J.H., Ali, K. and Anderson, R.A. 1990. Lipoxygenase 3 reduces hexanal production from soybean seed homogenates. J. Agric. Food Chem. 38: 1934-1936.

Hinchee, M.A.W., Connor-Ward, D.V., Newell, C.A., McDonnell, R.E., Sato, S.J., Gasser, C.S., Fischoff, D.A., Re, D.B., Fraley, R.T. and Horsch, R.B. 1988. Production of transgenic soybean plants using Agrobacterium-mediated DNA transfer. Biotechnology 6: 915-922.

Horsch, R.B., Fry, J.E., Hoffman, N.L., Eichholtz, D., Rogers, S.G. and Fraley, R.T. 1985. A simple and general method for transferring genes into plants. Science 227: 1229-1231.

Kacperska, A. and Kubacda-Zebalska, M. 1985. Is lipoxygenase involved in the formation of ethylene from ACC? Physiol. Plant 64: 333-338.

Kacperska, A. and Kbacka-Zebalska, M. 1989. Formation of stress ethylene depends both on ACC synthesis and on the activity of free radical-generating system. Physiol. Plant. 77: 231-237.

Kermasha, S. and Metche, M. 1987. Changes in lipoxygenase and hydroperoxide isomerase activities during the development and storage of French bean seed. J. Sci. Food Agric. 40: 1–10.

Kitamura, K. 1984. Biochemical characterization of lipoxygenase-lacking mutants, L-1-less, L-2-less and L-3-less soybeans. J. Agric. Biol. Chem. 48: 2339-2343.

Koch, R.B., Stern, B. and Ferrari, C.G. 1958. Linoleic acid and trilinolein as substrates for soybean lipoxygenase. (s). Arch. Biochem. Biophys. 78: 165-179.

Lazzeri, P.A., Hildebrand, D.F. and Collins, G.B. 1985. A procedure for plant regeneration from immature cotyledon tissue of soybean. Pl. Molec. Biol. Rep. 3: 160-167.

Lazzeri, P.A., Hildebrand, D.F., Sunega, J., Williams, E.G. and Collins, G.B. 1988. Soybean somatic embryogenesis: interactions between sucrose and auxin. Plant Cell Rep. 7: 517-520.

Legge, R.L. and Thompson, J.E. 1983. Involvement of hydroperoxides and an ACC-derived free radical in the formation of ethylene. Phytochemistry 22: 2161-21-66.

Leshem, Y.Y. and Wurzburger, J., Grossman, S. and Frimer, A.A. 1981. Cytokinin interaction with free radial metabolism and senescence: effects on endogenous lipoxygenase and purine oxidation. Physiol. Plant 53: 9–12.

Lynch, D.V. and Thompson, J.E. 1984. Lipoxygenase mediated production of superoxide anion in senescing plant tissue. FEBS Lett. 173: 251-254.

Lyr, H. and Banasiak, L. 1983. Alkenals, volatile defense substances in plants, their properties and activities. Acta Phytopathol. Hungaricae, 18: 3–12.

Matoba, T., Hidaka, H., Narita, H., Kitamura, K., Kaizuma, N. and Kito, M. 1985. Lipoxygenase-2 isozyme is responsible for generation of n-hexanal in soybean homogenate. J. Agric. Food Chem. 33: 852-855.

McCabe, D.E., Swain, W.F., Martinell, B.J. and Christou, P. 1988. Stable transformation of soybean (*Glycine max*) by particle acceleration. Biotechnology 6: 923-926.

McGranahan, G.H., Leslie, C.A., Uratsu, S.L., Martin, L.A. and Dandekar, A. M. 1988. *Agrobacterium*-mediated transformation of walnut somatic embryos and regeneration of transgenic plants. Biotechnology 6: 800-804.

McKeon, T.A. and Stumpf, P.K. 1982. Purification and characterization of the stearoyl-acyl carrier protein desaturase and the acyl-acyl carrier protein thioesterase from maturing seeds of safflower. J. Biol. Chem. 25: 12141-12147.

Murata, N., Nishida, I., Wada, H., Gombos, Z., Ohta, H., Sakamoto, T. and Ishizaki, O. 1990. Transformation of plants and algae with genes for transfer and desaturation of fatty acids. Presented at the 9th International Symposium on Plant Lipid Biochemistry: Structure and Utilization, July 8 to 13, at Wye, England.

Park, T.K. and Polacco, J.C. 1989. Distinct lipoxygenase species appear in the hypocotyl/radical of germinating soybean. Plant Physiol. 90: 285-290.

Parrott, W.A., Dryden, G., Vogt, S., Hildebrand, D.F., Collins, G.B. and Williams, E.G. 1988. Optimization of somatic embryogenesis and embryo germination in soybean. In Vitro Cellular & Devel. Biol. 8: 817-820.

Parrott, W.A., Hoffman, L.M., Hildebrand, D.F., Williams, E.G. and Collins, G.B. 1989a. Recovery of primary transformants of soybean. Plant Cell Rep. 7: 615-617.

Parrott, W.A., Williams, E.G., Hildebrand, D.F. and Collins, G.B. 1989b. Effect of genotype on somatic embroygenesis from immature cotyledons of soybean. Plant Cell Tissue Organ Cult. 16: 15–21.

Sanford, J.C. 1988. The biolistic process. TIBTECH. 6: 299-302.

Schardl, C.L., Byrd, A.D., Benzion, G., Altschuler, M.A., Hildebrand, D.F. and Hunt, A.G. 1987. Design and construction of a versatile system for the expression of foreign genes in plants. Gene 61: 1–11.

Sekiya, J., Kajiwara, T., Munechika, K. and Hatanaka, A. 1983 Distribution of lipoxygenase and hydroperoxide lyase in the leaves of various plant species. Phytochemistry 22: 1867-1869

Sekiya, J., Monma, T., Kajiwara, T. and Hatanaka, A. 1986. Changes in activities of lipoxygenase and hydroperoxide lyase during seed development of soybean. Agric. Biol. Chem. 50: 521-522.

Shanklin, J. and Somerville, C. 1990. Characterization of stearoyl-ACP desaturase. Presented at the 9th International Symposium on Plant Lipid Biochemistry: Structure and Utilization, July 8 to 13, at Wye, England.

Shibata, D., Steczko, J., Dixon, J.E., Hermodson, M., Yazdanparast, R. and Axelrod, B. 1987. Primary structure of soybean lipoxygenase L-1. J. Biol. Chem. 262: 10080-10085.

Shibata, D., Steczko, J., Dixon, J.E., Andrews, P.C., Hermodson, M. and Axelrod, B. 1988. Primary structure of soybean lipoxygenase L-2. J. Biol. Chem. 263: 6816-6821.

Theorell, H., Holtman, R.T. and Akeson, A. 1947. Crystalline lipoxidase. Acta Chem. Scand. 1: 571-576.

Thiede, M.A., Ozols, J.and Strittmatter, P. 1986. Construction and sequences of cDNA for rat liver stearyl coenzyme A desaturase. J. Biol. Chem. 261: 13230-13235.

Thompson, G.A. 1990. Cloning, expression in *E. coli* and characterization of a stearoyl-ACP desaturase from safflower seeds. Presented at the 9th International Symposium on Plant Lipid Biochemistry: Structure and Utilization, July 8 to 13, at Wye, England.

Vick, B.A. and Zimmerman, D.C. 1987. Oxidative systems for modification of fatty acids: the lipoxygenase pathway. In, "The Biochemistry of Plants: a Compehensive Treatise, Vol. 9" Stumpf, P.K. (ed.), Academic Press, Orlando, FL, pp. 53-90.

Vick, B.A. and Zimmerman. D.C. 1983. The biosynthesis of jasmonic acid: a physiological role for plant lipoxygenase. Biochem. Biophys. Res. Commun. 111: 470-477.

Wang, X.M. and Hildebrand, D.F. 1987. Effect of a substituted pyridazinone on the decrease of lipoxygenase activity in soybean cotyledons. Plant Science 51: 29-36.

Wang, X.M., Hildebrand, D.F., Norman, H.A. Dahmer, M.L., St.John, J.B. and Collins, G.B. 1987a. Reduction of linolenate content in soybean cotyledons by a substituted pyridazinone. Phytochemistry 26: 955-960.

Wang, X.M., Hildebrand, D.F. and Collins, G.B. 1987a. Identification of proteins associated with changes in the linolenate content of soybean cotyledons. In, "The Metabolism, Structure and Function of Plant Lipids," Stumpf, P.K., Mudd, J.B. and Nes, W.D. (Ed.), Plenum Press, New York, pp. 533-535.

Wang, X.M., Norman, H.A., St. John, J.B., Yin, T. and Hildebrand, D.F. 1989. Comparison of fatty acid composition in tissues of low linolenate mutants of soybean. Phytochemistry 28: 411-414.

White, J.S., Chang,-Y.P., Bibb, M.J. and Bibb, M. 1990. A cassette containing the bar gene of *Streptomyces hygroscopicus*: a selectable marker for plant transformation. Nucl. Acids Res. 18: 1062.

Wilcox, J.R., Cavins, J.F. and Nielsen, N.C. 1984. Genetic alteration of soybean oil composition by a chemical mutagen. J. Am. Oil Chem. Soc. 61: 97-100.

Wilcox, J.R. and Cavins, J.F. 1985. Inheritance of low linolenic acid content of the seed oil of a mutant in *Glycine max*. Theor. Appl. Genet. 71: 74-78.

Yadav, N.S. and Hitz, W. 1990. Cloning and bacterial expression of soybean cDNA for stearoyl-ACP desaturase. Presented at the 9th International Symposium on Plant Lipid Biochemistry: Structure and Utilization, July 8 to 13, at Wye, England.

Yamauchi, R., Kojima, M., Kato, K. and Ueno, Y. 1985. Lipoxygenase-catalyzed oxygenation of monogalactosyldilinolenoyl-glycerol in dipalmitoylphosphatidylcholine liposomes. Agric. Biol. Chem. 49: 2475-2477.

Yenofsky, R.L., Fine, M. and Liu, C. 1988. Isolation and characterization of a soybean (*Glycine max*) lipoxygenase-3 gene. Molec. Gen. Genetics 211: 215-222.

Zimmerman, D.C. and Coudron, C.A. 1979. Identification of traumatin, a wound hormone, as 12-oxo-*trans*-10-dodecenoic acid. Plant Physiol. 63: 536-541.

Role of Diacylglycerol Acyltransferase in Regulating Oil Content and Composition in Soybean

---

Prachuab Kwanyuen and Richard F. Wilson
United States Department of Agriculture
Agricultural Research Service and Crop Science Department
4114 Williams Hall, P.O. Box 7620
North Carolina State University
Raleigh, NC 27695-7620

Accessions of the USDA Soybean (*Glycine max* L. Merr.) Germplasm Collection exhibit genetic diversity for oil concentration ranging from 12 to 27 percent of dry weight. Although oil concentration is a highly heritable quantitative genetic trait, the genetic and biochemical basis for genotypic differences in the oil content of soybean seed is unknown. Recent knowledge on biological regulation of this trait has emerged from research on diacylglycerol acyltransferase (EC 2.3.1.20), the enzyme that catalyses triacylglycerol synthesis. Diacylglycerol acyltransferase purified from the cv. Dare has a native mass of about 1.5 MDa. Structural analysis suggests the protein consists of 10 monomers having three nonidentical subunits in a 1:2:2 molar ratio. Kinetics of the enzyme purified from soybeans exhibiting high or low oil content suggest that genotypic differences in oil content may be governed by gene dosage effects. However, subtle conformational changes in protein structure may influence oil composition, as evidenced by apparent substrate specificities observed in germplasm containing low-palmitic acid. Therefore, diacylglycerol acyltransferase may play a unique role in determining the content and composition of triacylglycerol in soybean.

Introduction

Protein and oil account for about 62.5 percent of soybean seed dry weight. These major constituents of soybean are quantitatively inherited traits, determined respectively by the action of many different genes. Although the exact number of genes regulating synthesis of these products is unknown, there is increasing need to understand how these gene systems interact. The problem may be stated simply: genetic selection to effect positive change in oil content results in a negative response in protein. Depending on market conditions, this association works to counteract gain in product value. The biochemical basis for this negative genetic correlation is unknown. Indeed, the biochemical basis for genetic regulation of protein and oil content in soybean is unknown. Thus, utilization of germplasm resources to enhance constituent value is limited by inability to target gene systems that regulate content, composition and constituent quality.

Although current genetic technology has exploited natural genetic variation in soybean, development of improved germplasm often has been achieved without knowledge of the biochemical basis for trait modification; but, the success of breeding programs depends on initial selection of parental lines. As breeding objectives become more complex, the choice of parental materials also becomes more difficult. Therefore, definitive information on the biochemical basis for genetic regulation of a given trait should directly enhance the capability, efficiency and effectiveness of producing soybean germplasm with value-added traits. Higher oil content and genetically altered fatty acid composition are traits that have significant potential to increase economic worth, nutritional value and product quality of soybean. Selection criteria based on knowledge of the biochemical basis for genotypic differences in glycerolipid metabolism, would elevate breeding technology beyond the scope of current soybean germplasm evaluation documentation. Many enzymatic reactions contribute to or mediate expression of glycerolipid synthesis. This discussion focuses on the catalysis of triacylglycerol synthesis, the major component of soybean oil.

The enzyme that catalyses the synthesis of triacylglycerol, acyl-CoA: 1,2-sn- diacylglycerol O-acyltransferase (EC 2.3.1.20), was first purified to homogeneity by Kwanyuen and Wilson (1986) from soybean (*Glycine max* L. Merr.) cotyledons. As in mammalian fat cells (Coleman and Bell, 1983), diacylglycerol acyltransferase from soybean seed was intimately associated with membranes in microsomal preparations. Although detergent solubilization diminished enzyme activity, the achieved purification of diacylglycerol acyltransferase was estimated to be 3,000-fold. Enzyme purity was confirmed by ultracentrifugation and high-performance liquid chromatography using gel filtration columns. This fraction was devoid of lipase (EC

3.1.1.3), glycerolphosphate acyltransferase (EC 2.3.1.15), lysolecithin acyltransferase (EC 2.3.1.23), acylglycerol acyltransferase (EC 2.3.1.22), and 1-acylglycerolphosphate acyltransferase (EC 2.3.1.51) activities. Therefore within reasonable certainty, the method yielded a homogeneous preparation of diacylglycerol acyltransferase.

Numerous attempts have been made to characterize diacylglycerol acyltransferase kinetics in plant (Ichihara *et al.*, 1988) and mammalian (Polokoff and Bell, 1980; Coleman and Bell, 1983) tissue with impure enzyme. However, purified diacylglycerol acyltransferase is required to predict the apparent mass of the native enzyme, number and type of subunits that comprise the enzyme, amino acid composition of those components, and contribution of enzyme activities with various substrates to product composition. Ability to obtain such information has established the fundamental basis for evaluating the role of diacylglycerol acyltransferase in the expression of genotypic differences and genetic regulation of oil content and composition of triacylglycerol in soybean.

Experimental Procedures

Purification of Diacylglycerol Acyltransferase. Soybean (*Glycine max* L. Merr.) seed of germplasm exhibiting genotypic variation in oil content or composition were imbibed in the dark at 25 C for 18 h. Cotyledons were excised and chilled to 4 C before use. Procedures for preparation of microsomes, i.e., solubilization, isolation, assay and confirmation of diacylglycerol acyltransferase homogeneity were as previously described (Kwanyuen and Wilson, 1990).

Chromatography. Chromatography of purified diacylglycerol acyltransferase in native form was conducted with a Hewlett-Packard Model 1084B HPLC equipped with a TSK G4000 SW (7.5 x 300 mm) gel filtration column. The elution solvent contained: 0.1 M phosphate buffer (pH 7.0), 10 mM B-mercaptoethanol, and 20 percent (w/v) ethylene glycol. Retention times for standard proteins (Blue dextran, 2000 kDa; thyroglobulin, 670 kDa; ferritin, 470 kDa; bovine serum albumin, 66.2 kDa) and native diacylglycerol acyltransferase were determined at 280 nm. Subunits dissociated from the native enzyme were chromatographed in a similar manner using a TSK G3000 SW (7.5 x 300 mm) gel filtration column. The elution solvent for this system contained: 0.1 M phosphate buffer (pH 7.0), 0.1 percent (w/v) sodium dodecyl sulfate, and 0.1 percent (w/v) B-mercaptoethanol.

Estimation of molecular weight. Linear regression analysis of the molecular weight of standard proteins vs the natural log (ln) of respective retention times during gel filtration chromatography was used to estimate the mass of native diacylglycerol acyltransferase prior to delipidation. Phospholipid content was determined by direct assay of esterified phosphate liberated during digestion of the native protein with concentrated perchloric acid (Lanzetta *et al.*, 1979). The putative mass of diacylglycerol acyltransferase was deduced from the difference between the mass of native protein prior to delipidation and the weight of phospholipid relative to phosphatidylcholine (PC, MW 763.9).

Dissociation of subunits. An acetone powder of purified diacylglycerol acyl-transferase was prepared at 4 C prior to dissociation. Prechilled acetone was added dropwise with stirring to a final concentration of 65 percent (v/v). After centrifugation at 10,000 xg for 10 min, the acetone powdered protein was washed once with 0.1 M phosphate buffer (pH 7.0) containing 65 percent (v/v) acetone, and resuspended in 0.1 M phosphate buffer (pH 7.0) containing 2 percent (w/v) sodium dodecyl sulfate and 2 percent (w/v) B-mercaptoethanol. The resuspended sample was then incubated with stirring at 37 C for 72 h to achieve dissociation and dialyzed at 25 C for 12 h in 0.1 M phosphate buffer (ph 7.0) containing 0.1 percent (w/v) sodium dodecyl sulfate and 0.1 percent (w/v) B-mercaptoethanol.

Electrophoresis and electroelution of subunits. Sodium dodecyl sulfate poly-acrylamide gel electrophoresis (SDS-PAGE) of dissociated proteins was performed according to the method of Chua (1980) with a linear gradient from 12.5 percent to 20.0 percent (w/w) gel. Slab gel electrophoresis was carried out at 6 mA/gel for 16 h. The gels were stained with 0.25 percent (w/v) Coomassie Brilliant Blue R-250 and destained in methanol-acetic acid-water, 4:1:5 v/v/v (Weber and Osborn, 1969). Linear regression analysis of the molecular weight of the calibration standards (phosphorylase B, 92.5 kDa; bovine serum albumin, 66.2 kDa; ovalbumin, 45 kDa; carbonic anhydrase, 31 kDa; trypsin inhibitor, 21.5 kDa; lysozyme, 14.4 kDa) vs ln relative mobility after electrophoresis was used to estimate the apparent mass of each subunit. Isolated subunits were extracted from destained gels by electroelution at 25 C in a Bio-Rad Model 422 apparatus equipped with 3.5 kDa cut-off membrane caps. This treatment was carried out in 50 mM ammonium bicarbonate buffer containing 0.1 percent (w/v) sodium dodecyl sulfate.

Amino acid analysis. Total amino acid composition of native diacylglycerol acyltransferase and each of the dissociated subunits were determined by the method of Spackman *et al.* (1958) in a LKB Alpha-Plus amino acid analyzer after hydrolysis *in vacuo* with 6 N HCl at 110 C for 24 h. Amino acid composition of each sample was verified by reverse-phase HPLC chromatography of derivatized hydrolysates (Heinrikson and Meredith, 1984) in a Waters Pico-Tag amino acid analysis system. Amino acid residues/mole protein were calculated and verified by two methods: 1) total protein mass/weighted mean of amino acid mass; and 2) sum (x*y); where, x = the molar ratio of individual amino acids relative to histidine; y = protein mass divided by the summation of (x * amino acid mass).

Lipid composition and analysis of metabolic activity. Glycerolipids were isolated and characterized by the method of Carver *et al.* (1984). Triacylglycerol molecular species composition was determined as described by Carver and Wilson (1984). Glycerolipid synthetic capacity of developing soybean cotyledons was determined via *in vivo* saturation kinetics derived from incorporation of [2–14C]acetate (55 Ci/mol) plus 0 to 1 mmol potassium acetate with seed at 30 days after flowering. Reactions were conducted for two hr at 25 C in 10 ml 0.2 N Mes buffer, pH 5.5. Saturation kinetics for diacylglycerol acyltransferase activity with different substrate combinations were determined using *sn*-diolein, *sn*-1,2 palmitoylolein, *sn*-dilinolein, *sn*-1,2 oleoylpalmitin or *sn*-1,2 palmitoyl linolein with various concentrations of [14C]16:0-CoA, [14C]18:1-CoA or [14C]18:2-CoA (50 Ci/mol). All treatments were replicated at least three times.

Results and Discussion

According to data compiled by the USDA Foreign Agricultural Service (USDA-FAS, 1990) on global oilseed products, projections for 1989/90 show that the U.S. may produce only 34.5 percent of world's supply of soybean oil, down significantly from the 1981/82 average of 40.3 percent (Table 1). This trend may be attributed to expansion of foreign soybean production area and supply, while U.S. soybean production area has declined and supply has remained unchanged compared to 1981/82 levels. However, very low growth in the supply of U.S. produced soybean oil has not diminished domestic consumption, which has increased steadily from 4.3 to 5.2 MMT (ca. 86.7 percent of supply). In the near future, action by the

Table 1. Trends in U.S. production and utilization of soybean oil

Statistic	Annual FAS Estimate 1981/82	1989/90	Share of World Total 1981/82	1989/90
	Amount		Percent	
Area (MHa)	26.8	24.0	53.5	41.8
Supply (MMT)	5.8	6.0	40.3	34.5
Export (MMT)	0.9	0.7	25.7	17.5
Consumption (MMT)	4.3	5.2	32.8	32.5
Surplus (MMT)	0.5	0.1	35.7	7.1

USDA-FAS, 1990

U.S. Congress will stimulate further domestic consumption of U.S. soybean oil by encouraging development of new technologies to expand soybean oil utilization. The impact of this action may create greater global demand for soybean oil; but, with U.S. stocks near all-time lows and the long-term trend toward the loses of the global market share, the U.S. is not in a favorable position to capitalize on expanded world demand for soybean oil. Indeed in the near future, the U.S. may be forced to import soybean oil just to meet domestic demand. Thus, to compete effectively in global oilseed product markets and avoid larger trade deficits, the U.S. will not only need to increase soybean production, but also produce soybeans with enhanced economic or nutritional value. The latter objective may be accomplished, in part, by development of soybeans with higher-oil content or altered fatty acid composition.

State of Current Technology for Genetic Regulation of Oil Content in Soybean

Soybean oil content is a highly heritable trait, and may be manipulated higher or lower through application of plant breeding technology (Wilson, 1987). The wide range of genetic variability for oil concentration among accessions of the USDA Soybean Germplasm Collection provides ample resource for accomplishment of such objectives within all Maturity Groups (Table 2). Most commercial soybean cultivars contain about 20 percent (w/w) oil, which is very close to the mid-range of high-oil and low-oil accessions. Few plant introductions exceed 27 percent oil or have less than 12 per-

Table 2. Examples of genetic variability for oil content in soybean

Maturity Group	High-Oil Genotypes		Low-Oil Genotypes	
	Accession	%Oil	Accession	%Oil
0	PI 297513	24.9	PI 181571	14.6
II	PI 79885	23.5	PI 81767	13.5
IV	PI 88349	23.5	PI 181550	15.5
VI	PI 374221	26.7	PI 212605	13.9
VIII	PI 221716	26.6	PI 323579	13.6

Wilson, 1987

cent oil. It is not known whether soybean germplasm may be developed with oil levels that exceed those apparent natural limits; but, selection of agronomic germplasm with oil content that is significantly different from the norm is within the range of current technology. The primary difficulty however, is initial selection of parental material to establish breeding populations.

Because of rather limited description of genotypic differences among germplasm resources, selection of parental material for altering oil content in soybean is basically a function of the experience or intuition of the geneticist. There is little information on the genetic regulation of oil content or concentration in soybean beyond the knowledge that the trait is maternal or determined by the genotype of the maternal plant (Table 3; Singh and Hadley, 1968). Lack of evidence for cytoplasmic influence on oil content or concentration indicates that genes governing the trait reside in the nuclear genome. Still, the number of genes is unknown, and may not be determined accurately with current technology. Therefore, one may only speculate on the genetic basis of genotypic differences in oil content in soybean. In that regard, it is often necessary to make breeding decisions based on generalizations. But, assumptions formed from statistical correlations with other traits can be misleading. As shown in Table 3, it would appear that seed size in soybean was positively correlated with oil content, at least from this data set. However, if the two traits are regressed over all accessions of the germplasm collection, no correlation would be found. Indeed, within any Maturity Group, one may find a wide range of seed size for any given oil and protein content (Table 4; Hartwig and Edwards, 1975).

Table 3. Inheritance of oil concentration in soybean

Cross	Generation	%Oil	Seed DWT	Oil Content
		w/w	mg/seed	mg/seed
Blackhawk (selfed)	P1	21.9	154.2	33.8
PI × P2	F1	21.1	134.5	28.4
P2 × P1	F1	12.8	51.8	6.6
PI 79648 (selfed)	P2	11.0	40.3	4.4
F1 (selfed)	F2	16.8	96.2	16.2
Mid-Parent		16.5	97.2	16.0

Singh and Hadley, 1968

In spite of these uncertainties, progress can be made to increase oil content through genetic selection (Table 5; Burton and Brim, 1981), albeit progress may be slow. In this example, oil concentration was increased by 1.1 percent (w/w) after three selection cycles which could take from 3 to 6 years depending on experimental facilities, selection approach, and operating resources. These data also demonstrate that a consequence of selection for higher oil content is the decline in protein concentration. Unfortunately, this correlation is real, and represents a serious problem if the breeding objective is to increase the total economic value of soybean. In almost every case known, where higher oil content was the selection goal, loss of protein was greater than gain in oil. Inasmuch as the constituent value of protein or meal in soybean may be greater than oil, the increased value from oil may not compensate for the decreased return from meal. Obviously, the optimum goal for increasing total economic value of soybean would be higher levels of both oil and protein. However, the biochemical basis for this negative genetic correlation between oil and protein is unknown. Breaking the apparent deleterious genetic linkages that affect the negative correlation between oil and protein will require information that may be beyond the grasp of current genetic technology. A comprehensive approach to gain such information should include determination of the biochemical basis for genetic variability in soybean oil content. As a working hypothesis, one may initially subscribe to the concept that genetic regulation of oil content could be a function of glycerolipid synthetic capacity, as determined by the degree of amplification of genes encoding enzymes in the glycerolipid pathway. However, there

Table 4. Impact of seed size on oil and protein concentration in soybean

	Seed Size	Oil	Protein
	mg/seed	% (w/w)	% (w/w)
	49.5	21.8	40.8
	123.7	21.2	42.2
	144.2	21.6	39.8
	195.5	21.6	41.1
	301.0	21.4	39.4
LSD 0.05	53.4	0.2	0.9

Data based on accessions from Maturity Group VI; Hartwig and Edwards, 1975 (Reproduced with permission.)

Table 5. Recurrent selection for increased oil concentration in soybean

Cycle	Oil	Protein	Total
		percent of seed dry weight	
C0	18.8	42.0	60.8
C1	19.1	41.7	60.8
C2	19.3	41.2	60.5
C3	19.9	40.5	60.4
LSD 0.05	0.3	0.5	0.3

Burton and Brim, 1981 (Reproduced with permission.)

is equal probability that genotypic differences in oil content could simply be the result of competition for common substrates with other metabolic processes, such as protein synthesis. To address these questions, four soybean genotypes, exhibiting high-oil or low-oil content, were selected to evaluate potential genotypic differences in lipid metabolism (Table 6). These germplasm represented extremes in the range of natural genetic variation in oil and protein content with a two-fold genotypic difference in protein/oil ratio between high-oil and low-oil genotypes.

Evaluation of Lipid Synthetic Capacity in Soybean Cotyledons

Analysis of glycerolipid metabolism in developing soybean cotyledons using *in vivo* acetate saturation kinetics revealed that apparent maximal velocities for phospholipid (TPL), diacylglycerol

Table 6. Characteristics of selected soybean genotypes

Genotype	Size	Oil	Protein	P/O
	mg/seed	% (w/w)		Ratio
High-Oil				
Dare	142.0	24.8	38.2	0.65
N85-3097	182.0	25.7	36.4	0.71
Low-Oil				
PI 175188	132.0	16.6	45.5	0.36
PI 123439	63.0	16.3	46.4	0.35
LSD 0.05	47.0	4.2	5.1	0.20

P/O, %protein/%oil

(DG) and triacylglycerol (TG), respectively, were significantly greater in high-oil compared to low-oil genotypes (Table 7). Although triacylglycerol accounted for greater than 85 percent of total glycerolipid in these genotypes, phospholipids exhibited the greatest metabolic activity as determined by acetate saturation kinetics. This should be expected considering the role of phospholipids as precursors in the triacylglycerol biosynthetic path way (Figure 1). In this system, with labeled acetate, fatty acids esterified to glycerolipids accounted for 95 percent of the incorporated radioactivity. The fatty acid esterified in triacylglycerol is then derived from *sn*-1,2 diacylglycerol and acyl-CoA liberated primarily from phosphatidylcholine by separate enzymatic reactions (Stymne and Stobart, 1984).

These data (Table 7) demonstrated a positive relation between oil content and capacity for total lipid (TL) synthesis in a given genotype. In addition, the entire glycerolipid synthetic pathway appeared to be effected; and, since only 13 to 19 μmole acetate (predicted from apparent Km) was required for saturation of total lipid synthesis in these genotypes, there was a high probability that genotypic differences in oil content were not a result of differential partitioning or competition for acetate among other metabolic processes. Given that N85-3097 exhibited greater glycerolipid synthetic capacity than PI 123439 at saturating levels of acetate, the biochemical basis for genotypic difference in triacylglycerol synthesis may be attributed primarily to genetic effects on fatty acid synthetase activity. However, other enzymes in the triacylglycerol synthetic pathway also may be effected in a similar manner. Unfortunately, defensible con-

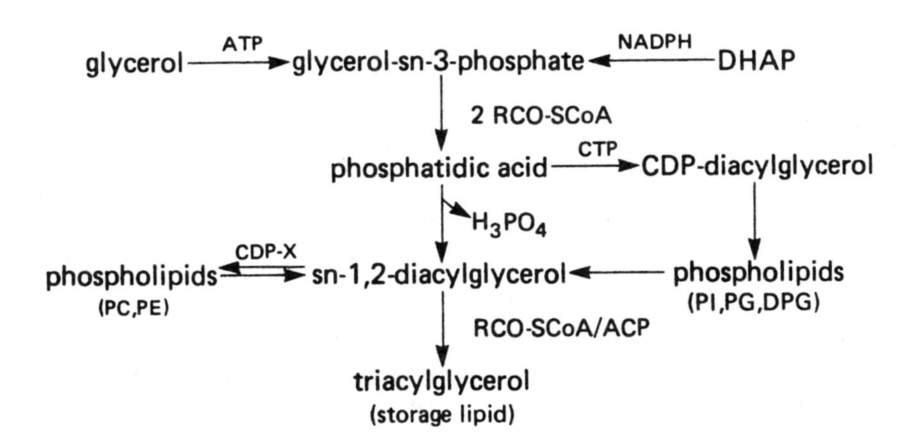

Figure 1. General pathway of complex glycerolipid metabolism and triaclglycerol synthesis in soybean and other oilseed species.

Table 7. Acetate saturation kinetic analysis of glycerolipid synthesis

Genotype		Glycerolipid			
		TBL	DG	TG	TL
N85-3097 (High-Oil)	Vmax	104.0	33.8	39.7	189.5
	Km	1.9	2.3	3.5	2.1
PI 123439 (Low-Oil)	Vmax	74.2	13.4	20.5	120.7
	Km	1.2	1.6	2.1	1.4

Vmax, nmole [14C]acetate/h/g DWT; Km, μmole acetate

clusions regarding potential genetic effects on specific glycerolipid synthetic enzymes may not be formed from this information. Kinetic analysis of individual reactions is required to evaluate the synthetic capacity of specific enzymes in the triacylglycerol metabolic pathway. The problem with that approach is that most of these enzymes have not yet been purified from soybean. The only glycerolipid synthetic protein known to be purified to homogeneity from soybean is diacylglycerol acyltransferase, the enzyme that catalyses the final step in triacylglycerol biosynthesis.

Genotypic Differences in Diacylglycerol Acyltransferase Activity

Using the method of Kwanyuen and Wilson (1986), diacylglycerol acyltransferase was purified from soybean seed exhibiting high-oil or low-oil content, and subjected to saturation kinetic analyses with

[14C]18:1-CoA and diolein (Table 8). Results revealed significantly greater maximal reaction velocity for the enzyme purified from the high-oil genotypes. However, the apparent Km for 18:1-CoA was not significantly different among all genotypes. These findings suggested that gene dosage effects mediated the tissue level of diacylglycerol acyltransferase in soybean. If this is true, and since diacylglycerol acyltransferase is at the end of the triacylglycerol synthetic pathway, other enzymes involved in complex glycerolipid metabolism, which ultimately provide the substrates for triacylglycerol synthesis, also may be under similar genetic regulation. Therefore, the biochemical basis for genetic variability in oil content of soybean could involve several tightly linked genes that encode glycerolipid synthetic enzymes in the triacylglycerol pathway, in addition to genes regulating fatty acid synthetase. Then it follows that a genetic probe for one of these proteins should target gene clusters in genomic DNA that encode the associated gene products.

Table 8. Saturation kinetic analysis of diacylglycerol acyltransferase with [14C]18:1-CoA plus diolein

Statistic	High-Oil Genotypes		Low-Oil Genotypes	
	Dare	N85-3097	PI123439	PI175188
Vmax	82.8	75.4	48.1	51.1
Km	12.5	11.9	11.9	11.3

Vmax, nmole/min/mg protein; Km, μmole 18:1-CoA

Ability to purify diacylglycerol acyltransferase has provided an opportunity to construct specific synthetic oligonucleotides from amino acid sequence analysis to probe genomic DNA for genes that encode a glycerolipid synthetic enzyme. The biophysical characterization of structural properties also provides a means to test the hypothesis that synthetic capacity is relative to the size of this enzyme, and determined by gene dosage effects in a given genotype. Much has been learned already about this unique protein. As determined by gel filtration chromatography, the apparent mass of active native diacylglycerol acyltransferase from the cv. Dare was ca. 1840 kDa. However after delipidation, the mass of the denatured protein was ca. 1540 kDa. Agglomeration of the resuspended protein in buffer indicated that phospholipids, which account for ca. 95 percent (w/w) of lipid associated with the protein, were required for functional structure.

Incubation of the delipidated protein in detergent at 37 C with stirring for at least 72 h was required to achieve dissociation of subunits. SDS-PAGE of the dissociation products revealed three nonidentical subunits with respective mass estimates at 40.8 kDa, 28.7 kDa and 24.5 kDa. The concentration of amino acids in subunits electroeluted from SDS-PAGE gels and in the purified native enzyme have been expressed on a mole percent basis in Table 9. The amino acid composition of each subunit was characteristic of membrane-bound proteins, with polar indices less than 20.3 percent, well below the 40 percent boundary claimed for membrane-bound and detergent solubilized proteins (Capaldi and Vanderkooi, 1972). Hydrophobic amino acids accounted for 48.5 percent, 56.1 percent, and 51.7 percent of the total concentration in the respective subunits. In addition, cysteine and methionine were present in very low amounts in the 28.7 and 24.5 kDa subunits. Based upon these data, a greater proportion of the surface area of the 40.8 kDa subunit could be exposed to the cytosol, whereas both of the smaller subunits might be more deeply embedded in the surrounding membrane. Thus, the 40.8 kDa peptide could contain the active site. Preliminary data on the amino acid sequence of the 40.8 kDa subunit suggest that lysine occupies the N-terminal amino acid residue.

Table 9. Composition of diacylglycerol acyltransferase from soybean (*Glycine max* L. Merr. cv. Dare) cotyledons

Amino Acid	Subunit (kDa)			Predicted Composition	
Composition	40.8	28.7	24.5	Monomer	Native
Nonpolar (%)	48.5	56.1	51.7	52.7	52.7
Neutral (%)	20.2	18.6	21.3	20.1	20.2
Acidic (%)	20.3	14.4	17.0	16.7	16.5
Basic (%)	11.0	10.9	10.0	10.5	10.6
Residues/Mole	365	468	420	1246	12525
Molar Ratio	1	2	2	10	1

Kwanyuen and Wilson, 1990 (Reproduced with permission.)

Total amino acid analysis also demonstrated that the three subunits occurred in a 1:2:2 molar ratio. Using that ratio, the weighted mean for the concentration of individual amino acids among subunits was not significantly different from the observed composition in the native enzyme. This finding obviated consideration of other polypeptides in the protein structure. The predicted number of amino acid residues in each polypeptide and the native enzyme was determined. These data suggested the protein contained ca. 1246 residues/mole monomer: 365 (40.8 kDa), 468 (28.7 kDa), 420 (24.5 kDa); and about 12525 residues/mole native diacylglycerol acyltransferase.

Thus, it was concluded that the monomeric form of diacylglycerol acyltransferase from soybean contained five peptides with a putative molecular weight of 153.1 kDa. This proposed association suggested that the complete structure of the native enzyme was composed of 10 identical monomers. In comparison, a 350 kDa triacylglycerol synthetase complex has been isolated from rat intestinal villus cells which contained monoacylglycerol acyltransferase (EC 2.3.1.22), acyl-CoA synthetase (EC 6.2.1.1) and diacylglycerol acyltransferase activities in equal proportions (Manganaro and Kuksis, 1985). The monoacylglycerol acyltransferase was reported to exist as at least a dimer having identical 37 kDa subunits. If acyl-CoA synthetases of rat liver have subunit mass of 27 and 76 kDa, then undissociated diacylglycerol acyltransferase in the complex may be similar in size (ca. 173 kDa) to a single monomer in soybean cotyledons. However, Ozasa et al. (1989) have estimated, from radiation inactivation curves, the functional monomeric size of diacylglycerol acyltransferase in rat liver at 72 kDa.

Although it is premature to form any definitive conclusions regarding structure of diacylglycerol acyltransferase in animal systems, information derived with the enzyme purified from soybean seed promulgates the concept that biophysical properties inherent to the protein may differ among genotypes. As shown, the enzyme from high-oil genotypes exhibits greater triacylglycerol synthetic capacity than low-oil genotypes without significant differences in apparent Km. Therefore, genotypic differences in enzyme activity may be partially determined by genetic regulation of the number of monomeric units that compose diacylglycerol acyltransferase. Under this premise, if the enzyme in Dare contains 10 monomeric units, a direct relation between size and activity of the enzyme may be evident in low-oil genotypes, assuming no alteration in subunit composition. Work is underway to test that assumption with diacylglycerol acyltransferase purified from soybean germplasm exhibiting genetic differences in oil content.

Influence on Diacylglycerol Acyltransferase on Oil Composition

In addition to biophysical properties of diacylglycerol acyltransferase that may mediate genotypic differences in soybean oil content, attention also should be given to potential effects that diacylglycerol acyltransferase may exert on triacylglycerol composition in soybean germplasm exhibiting genetically altered fatty acid content. Considerable progress has been made in the development of soybean germplasm with enhanced oil quality through genetic alteration of fatty acid composition (Wilson *et al.*, 1981). Another example of such accomplishments is the soybean germplasm line designated as N79-2077, which was the first known germplasm resource containing low-palmitic acid content (Table 10).

Palmitic acid is a saturated 16-carbon fatty acid found in virtually all vegetable oils. Soybean oil typically contains 11-14 percent (w/w) palmitic acid. Recent nutritional studies have linked high levels of dietary palmitic acid with elevated serum cholesterol, a condition that may enhance the incidence of arteriosclerosis (Rudel *et al.*, 1981). Because soybean oil may account for over 50 percent of dietary palmitic acid in the U.S., lower palmitic acid has become a value-added trait for health, if not economic reasons. Selection of N79-2077 demonstrated that palmitic acid content could be lowered genetically in soybean oil. This germplasm contained only ca. 47 percent of the palmitic acid content in Dare (Wilson *et al.*, 1988).

Recent research (Table 10) has shown that significantly reduced synthetic capacity was involved in the biochemical basis for genetic control of palmitic acid in N79-2077, which may be attributed to gene action on fatty acid synthetase. However, these saturation kinetics for glycerolipid metabolism also showed that metabolic transfer of palmitic acid from phospholipids through diacylglycerol to triacylglycerol was impeded in N79-2077 compared to Dare. Therefore, to a lesser but statistically significant extent, incorporation of palmitic acid into triacylglycerol could be influenced by genetic regulation of metabolic events during glycerolipid synthesis. Additional evidence for such effects was shown by analysis of triacylglycerol molecular species composition (Table 11).

Analysis of triacylglycerol molecular species in seed of the cultivar Dare and N79-2077 revealed that essentially all of the palmitic acid in triacylglycerol from these genotypes was accounted for in five different molecular species. Calculation of the amount of palmitic acid in each species showed that 95.3 percent of the

Table 10. Acetate saturation kinetic analysis of glycerolipid synthesis

Genotype	%16:0 in TL	Glycerolipid TPL	DG	TG	TL
	mole%	% of total radioactivity in 16:0			Vmax
Dare	12.4	56.6	8.9	34.5	696.6
N79-2077	5.9	62.4	10.4	27.2	404.6
LSD 0.05	2.3	3.6	0.9	4.5	181.0

TL, Total Lipid; Vmax, nmol [14C]acetate/h/g DWT in 16:0

Table 11. Palmitic acid distribution among triacylglycerol species

TG Species	Dare	N79-2077	Ratio
	mmol 16:0/kg DWT in TG		%
SSM	17.2	4.4	25.6
SMM	10.3	9.5	92.2
SSD	8.6	6.6	76.7
SMD	21.4	9.5	44.4
SDD	45.6	13.2	28.9
Total	103.1	43.2	41.9

S, 16:0; M, 18:1; D, 18:2; Ratio, N79-2077/Dare; Wilson, Kwanyuen & Burton, 1988 (Reproduced with permission.)

genotypic difference in palmitic acid content was attributed to only three molecular species: SSM, SMD and SDD (S,16:0; M,18:1; D,18:2). Although synthesis of triacylglycerol molecular species is a nonrandom metabolic process, one might still expect to find uniform reduction of palmitic content among all of these respective compounds. These data demonstrated that genotypic differences in palmitic acid content were not distributed proportionately among all five species. Hence, this observation suggested that very specific natural genetic variation in triacylglycerol composition could be attributed to inherent substrate specificities exhibited by diacylglycerol acyltransferase for 16:0-CoA and/or diacylglycerol species containing palmitic acid.

Further evidence for that concept was accomplished by determination of relative maximal reaction velocities for synthesis of a given triacylglycerol molecular species from specific substrate com-

Table 12. Capacity of diacylglycerol acyltransferase from soybean to utilize substrates containing palmitic acid

TG Species	Substrate		Genotype		LSD 0.05
	sn-1,2 DG	Acyl-CoA	Dare	N79-2077	
			relative Vmax		
SSM	16:0/18:1	16:0	43.7	27.3	11.0
	18:1/16:0	16:0	52.7	38.5	9.5
SMD	16:0/18:1	18:2	53.3	33.3	13.4
	18:1/16:0	18:2	58.4	51.4	4.7
SDD	16:0/18:2	18:2	130.2	122.6	5.1
	18:2/18:2	16:0	54.2	51.1	2.1

Vmax relative to *sn*-1,2 palmitoyl linolein plus 16:0-CoA = 100%
Wilson, Kwanyuen and Burton, 1988 (Reproduced with permission.)

binations. The capacity of diacylglycerol acyltransferase purified from Dare and N79-2077 to utilize substrates containing palmitic acid was determined with specific *sn*-1,2 diacylglycerols in combination with 16:0-CoA or 18:2-CoA (Table 12). These substrates were chosen to evaluate apparent substrate specificity for reactions that produced SSM, SMD or SDD-type triacylglycerol species. With the exception of reactions using *sn*-1,2 palmitoyl linolein plus 18:2-CoA, the stereospecific position of a given acyl moiety within diacylglycerol appeared to have minor influence on the relative maximal reaction velocity of diacylglycerol acyltransferase from a given genotype. However in each treatment, the capacity for product formation by N79-2077 was statistically lower than complementary reactions with enzyme from Dare.

Although the relative availability or supply of substrates through the glycerolipid metabolic pathway to diacylglycerol acyltransferase engenders predominate influence on triacylglycerol product formation *in vivo*, these data showed that apparent genotypic differences in substrate utilization by diacylglycerol acyltransferase also could contribute to differences in triacylglycerol composition. Similar results were found in other genotypes with genetically altered fatty acid composition by Kwanyuen *et al.* (1988). The nature of the influence could be due to subtle structural modification at the active site. The extent to which such natural alteration in diacylglycerol acyltransferase structure exists among soybean genotypes may then explain or help define the action of modifier genes that have been proposed in the overall genetic regulation of fatty acid composition in soybean (Wilson and Burton, 1986). If sufficient evidence is found to support this theory, the resultant technologies developed from that knowledge could facilitate extremely novel approaches in genetic "tailoring" of oil composition for industrial and food uses. As proposed by Bafor *et al.* (1990), it may become possible to produce highly valued fats such as cocoa-butter in soybean and other oilseed species. In that regard, diacyl glycerol acyltransferase could play a strategic role among the glycerol acylating enzymes in effecting exotic oil compositions, because genetic manipulation of that enzyme should not jeopardize normal physiological function of metabolic processes by concommitant alteration of membrane lipids. Nevertheless, future biotechnological advances in genetic regulation of lipid metabolism will expand the ability to target gene systems that determine seed constituent composition. Still, the tangible benefit of this knowledge will be manifested in practical application by geneticists through development of enhanced germplasm resources. With the proper exercise of our understandings in this area of research, a new era will emerge in which significant changes in the improvement of overall quality, nutritional value and economic worth of soybean will become reality.

Conclusion

Although understanding of the genetic regulation of triacylglycerol content and composition in soybean is still in remedial stages, recent advances in the biochemistry of lipid metabolism have demonstrated that these traits are determined by gene action on not only fatty acid synthetase, but on various enzymes in the triacylglycerol synthetic pathway. The role that diacylglycerol acyltransferase plays in this system has emerged as more than an indiscriminate entity. Before this enzyme had been purified sufficiently for detailed characterization of its biophysical and biochemical properties, it was proposed that diacylglycerol acyltransferase exhibited only broad specificity for various substrates and exerted little control over triacylglycerol composition. At least in soybean, it may now be stated that this enzyme does contribute directly to genotypic differences in oil content. The exact mechanism is still unknown, but current theory favors the concept that oil content in soybean may be a function of enzyme size, through genetic regulation of the number of monomeric units composing the protein in a given genotype.

In addition, it has been shown that diacylglycerol acyltransferase exhibits preferential utilization of certain combinations of diacylglycerol and acyl CoA. These specificities may differ among genotypes, but may effect subtle modification of the final triacylglycerol composition. Hence, there is a high probability that the key to future advances in gaining control over genetic regulation of oil content and composition in oilseed species will be the knowledge derived from detailed structural analysis of this and other glycerolipid synthetic enzymes. This approach appears to offer the best opportunity to develop genetic probes for genomic DNA that identify regulatory gene sequences, and to create novel strategies for molecular genetic manipulation of gene products that mediate lipid metabolic processes. These technologies could then provide the basis for significant breakthroughs in the natural barriers that impede development of value-added traits in oil content and composition through classical genetic methodology.

References

Bafor, M., Stobart, A.K. and Stymne, S. 1990. Properties of the glycerol acylating enzymes in microsomal preparations from the developing seeds of safflower (*Carthamus tinctorius*) and turnip rape (*Brassica campestris*) and their ability to assemble cocoa-butter type fats. J. Am. Oil Chem. Soc., 67: 217.

Burton, J.W. and Brim, C.A. 1981. Recurrent selection in soybeans. III. Selection for increased percent oil in seeds. Crop Sci. 21: 31.

Capaldi, R.A. and Vanderkooi, G. 1972. The low polarity of many membrane proteins. Proc. Natl. Acad. Sci. U.S.A. 69: 930.

Carver, B.F., Wilson, R.F. and Burton, J.W. 1984. Developmental changes in acyl-composition of soybean seed selected for high oleic acid concentration. Crop Sci. 24: 1016.

Carver, B.F. and Wilson, R.F. 1984. Triacylglycerol metabolism in soybean seed with genetically altered unsaturated fatty acid composition. Crop Sci. 24: 1020.

Coleman, R.A. and Bell, R.M. 1983. Topography of membrane-bound enzymes that metabolize complex lipids. In: "The Enzymes," Vol. 16, p. 87, Academic Press, New York.

Chua, N.-H. 1980. Electrophoretic analysis of chloroplast proteins. In: "Methods of Enzymology," p. 434, Academic Press, New York.

Hartwig, E.E. and Edwards, C.J., Jr. 1975. Evaluation of soybean germplasm: maturity group V to X. USDA, ARS, Stoneville, MS.

Heinrikson, R.L. and Meredith, S.C. 1984. Amino acid analysis by reverse-phase high-performance liquid chromatography: precolumn derivatization with phenylisothiocyanate. Anal. Biochem. 136: 65.

Ichihara, K., Takahashi, T. and Fujii, S. 1988. Diacylglycerol acyltransferase in maturing safflower seeds: its influences on the fatty acid composition of triacylglycerol and on the rate of triacylglycerol synthesis. Biochim. Biophys. Acta 958: 125.

Kwanyuen, P. and Wilson, R.F. 1986. Isolation and purification of diacylglycerol acyltransferase from germinating soybean cotyledons. Biochim. Biophys. Acta 877: 238.

Kwanyuen, P., Wilson, R.F. and Burton, J.W. 1988. Substrate specificity of diacylglycerol acyltransferase purified from soybean. In: "Proceedings World Conference on Biotechnology for the Fats and Oils Industry," p. 294, Am. Oil Chem. Soc., Champaign, IL.

Kwanyuen, P. and Wilson, R.F. 1990. Subunit and amino acid composition of diacylglycerol acyltransferase from germinating soybean cotyledons. Biochim. Biophys. Acta 1039: 67.

Lanzetta, P.A., Alvarez, L.J., Reinbach, P.S. and Candia, O.A. 1979. An improved assay for nanomole amounts of inorganic phosphate. Anal. Biochem. 100: 95.

Manganaro, F. and Kuksis, A. 1985. Purification and preliminary characterization of 2-monoacylglycerol acyltransferase from rat intestinal villus cells. Can. J. Biochem. Cell Biol. 63: 341.

Ozasa, S., Kempner, E.S. and Erickson, S.K. 1989. Functional size of acylcoenzyme A: diacylglycerol acyltransferase by radiation inactivation. J. Lipid Res. 30: 1759.

Polokoff, M.A. and Bell, R.M. 1980. Solubilization, partial purification and characterization of rat liver microsomal diacylglycerol acyltransferase. Biochim. Biophys. Acta 618: 129.

Rudel, L.L., Park, J.S. and Carroll, R.M. 1981. Effects of polyunsaturated versus saturated dietary fat on non-human primate HDL. In: "Dietary Fat and Health," p. 649, Am. Oil Chem. Soc., Champaign, IL.

Singh, B.B. and Hadley, H.H. 1968. Maternal control of oil synthesis in soybean (*Glycine max* L. Merr.). Crop Sci. 8: 622.

Spackman, D.H., Stein, W.H. and Moore, S. 1958. Automatic recording apparatus for use in the chromatography of amino acids. Anal. Biochem. 30: 1190.

Stymne, S. and Stobart, A.K. 1984. The biosynthesis of triacylglycerols in microsomal preparations of developing cotyledons of sunflower (*Helianthus annuus* L.). Biochem J. 220: 481.

United States Department of Agriculture, Foreign Agricultural Service. 1990. World oilseed situation and market highlights. Circular Series FOP 7–90, Washington, DC.

Weber, K. and Osborn, M. 1969. The reliability of molecular weight determinations by dodecyl sulfate polyacrylamide gel electrophoresis. J. Biol. Chem. 244: 4406.

Wilson, R.F., Burton, J.W. and Brim, C.A. 1981. Progress in the selection for altered fatty acid composition in soybeans. Crop Sci. 21: 788.

Wilson, R.F. and Burton, J.W. 1986. Regulation of linolenic acid in soybeans and gene transfer to high yielding, high protein germplasm. In: "Proceedings World Conference on Emerging Technologies in the Fats and Oils Industry," p. 386, Am. Oil Chem. Soc., Champaign, IL.

Wilson, R.F. 1987. Seed metabolism. In: "Soybeans: Improvement, production and uses," Second Edition, p. 643, Am. Soc. Agron., Madison, WI.

Wilson, R.F., Kwanyuen, P. and Burton, J.W. 1988. Biochemical characterization of a genetic trait for low palmitic acid content in soybean. In: "Proceedings World Conference on Biotechnology for the Fats and Oils Industry," p. 290, Am. Oil Chem. Soc., Champaign, IL.

Arabidopsis as a Model to Develop Strategies for the Modification of Seed Oils

John Browse
Martine Miquel
Institute of Biological Chemistry
Washington State University
Pullman, WA 99164-6340

Modifying the fatty acid composition of the triacylglyerols in oilseeds is an attractive goal in plant biotechnology since it will allow the production of improved vegetable oils for food and manufacturing industries. Mutants of *Arabidopsis* with defects in lipid metabolism have greatly increased our understanding of the pathways of lipid biosynthesis in leaves and seeds. Mutations have been identified that, in seed tissue, cause deficiencies in the elongation of 18:1 to 20:1, desaturation of 18:1 to 18:2, and desaturation of 18:2 to 18:3. In each of these cases, the wild type exhibited incomplete dominance over the mutant allele. These results, along with results from earlier studies, point to a major influence of gene dosage in determining the fatty acid composition of seed lipids. The advantages of *Arabidopsis* as a model for plant molecular genetics will allow the mutants to be used as a means to clone the lipid metabolism genes by chromosome walking or insertion mutagenesis. Because of the time and effort required by these approaches, it is important to identify the genes which hold the greatest promise for usefully altering plant lipid composition. In this paper, we discuss strategies for modifying the fatty acid composition of seed oils based on our present understanding of the biochemistry and genetics of seed lipid synthesis.

Supported by the U.S. Department of Agriculture (Grant 89-37262-4388) and the Washington State University Agricultural Research Center.

Abbreviations: ACP, acyl carrier protein; X:Y, fatty acyl group containing X carbons with Y *cis*-double bonds; DAG, diacylglycerol; PA, phosphatidic acid; PC, phosphatidylcholine; TAG, triacylglycerol.

Introduction

Vegetable oils constitute one of the world's most important plant commodities, with current annual production in excess of 50 million metric tons (Table 1). Production has increased steadily since 1970 at an average annual rate of 4 percent—about twice the rate of growth in world population (Pryde and Doty, 1981; Stymne and Stobart, 1987). The health benefits (such as lack of cholesterol) of vegetable oils ensure that the demand for them will continue to grow. The major use of plant oils is for human consumption, but a significant proportion finds use in manufacturing industries, particularly in the production of detergents, paints, and specialty lubricants. For both food and industrial applications, it is the fatty acid composition of the oil which determines its usefulness and, therefore, its commercial value. The realtionship between fatty acid composition and oil characteristics is complex, since it depends not only on the overall fatty acid composition, but also on the combinations of fatty acids in the different molecular species of triacylglycerol and on which of the three distinct positions of the glycerol is occupied by a particular fatty acid (Table 2). The significance of these factors has been considered in many reviews and monographs (e.g., Weiss, 1983; Kramer *et al.*, 1983; Small, 1986; Hamilton and Bhati, 1987).

Table 1. Production data (for 1987/88) and composition of the major world oil crops

Oilseed	World Prod. of Oil (in 10^6 Metric Tons)	Oil Content of Tissue %by wt.	Typical Fatty Acid Composition (wt. %)						
			<16:0	16:0	18:0	18:1	18:2	18:3	Other
Soybean	15.3	19	-	11	3	22	55	8	1
Palm	8.4	48	-	47	4	38	10	-	-
Rapeseed (Canola)	7.5	43	-	4	2	59	23	8	3
Sunflower	7.2	40	-	7	3	14	75	-	-
Cotton	3.3	19	1	25	3	17	53	-	-
Peanut	2.8	48	-	11	2	51	31	-	2
Coconut	2.7	66	80	8	4	5	3	-	-
Palm Kernel	1.1	48	74	9	2	15	1	-	-

Data from Americal Oil Chemists Society, 1988; Stymne and Stobart, 1987; Weiss, 1983.

Table 2. Positional analysis of oilseed triacylglycerols

Plant Species	*sn* Position on Glycerol	Fatty Acid Composition (mol %) 16:0	18:0	18:1	18:2	18:3	20:1	22:1
Corn	1	16	4	21	58	1		
	2	1	0	22	77	1		
	3	7	2	19	71	2		
Rapeseed	1	4	2	23	11	6	16	35
	2	1	0	37	36	20	2	4
	3	4	3	17	4	3	17	51
Safflower	1	9	3	8	81			
	2	0	0	8	92			
	3	4	1	7	88			
Soybean	1	14	6	23	48	9		
	2	1	0	22	70	7		
	3	13	6	28	45	8		

References: Styme and Stobart, 1987; Slack and Browse, 1984.

To a large extent, the increases in oil production over the last 15 to 20 years have been fueled by the release of improved varieties and efficiencies of cultivation for a relatively few species—soybean, oil palm, rapeseed, and sunflower. In the case of soybean, demand for vegetable protein has also been a major factor since protein is the economically important product from this crop. As a result, expansion of oil production has continued even though these major vegetable oils exhibit fatty acid compositions, which make them less than ideal for human nutrition and the requirments of the food industry. For example, both soybean and rapeseed oils presently contain levels of linolenic acid which threaten the shelf life of products made from them (Weiss, 1983), while palm, coconut and, to a lesser extent, other oils contain high levels of saturated fatty acids (Table 1) which are undesirable because they may contribute to the development of atherosclerosis.

In this article, we outline what is known of the pathway of triacylglycerol synthesis in seeds, including information recently obtained from a series of *Arabidopsis* mutants with defects in the synthesis of seed lipids. As well as providing biochemical information, the *Arabidopsis* mutants can provide the means to directly isolate genes encoding key enzymes that control the fatty acid composition of seed oils. As such cloned genes become available, it is important to use our knowledge of the biochemistry and genetics of triacylglycerol synthesis to develop strategies for modifying the fatty acid composition of seed oils.

Triacylglycerol Components

Seed oils are composed almost entirely of triacylglycerols in which fatty acids are esterified to each of the three hydroxyl groups of glycerol. The use of triacylglycerols as a seed reserve maximizes the quantity of stored energy within a limited volume, because the fatty acids are a highly reduced form of carbon (Slack and Browse, 1984). A large variety of different fatty acid structures are found in nature (Hilditch and Williams, 1964), but just five account for 90 percent of the commercial vegetable oil produced: palmitic (16:0), stearic (18:0), oleic (18:1), linoleic (18:2), and α-linolenic (18:3) acids. As mentioned above, the factors governing the physical properties of a particular oil are complex but, in broad terms, the important characterisitics of these major fatty acid constituents are as follows:

Palmitic, stearic, and other saturated fatty acids are solid at room temperature, in contrast to the unsaturated fatty acids which remain liquid. Because saturated fatty acids have no double bonds in the acyl chain, they remain stable to oxidation at elevated temperatures. They are important components in margarines and chocolate formulations but, for most food applications, reduced levels of saturated fatty acids are desired.

Oleic acid has a single double bond, but is still relatively stable at high temperatures, and oils with high levels of oleic acid are suitable for cooking and other processes where heating is required. Recently, increased consumption of high oleic oils has been recommended, because oleic acid appears to lower blood levels of low density lipoproteins without affecting levels of high denisty lipoproteins.

Linoleic acid is the major "polyunsaturated" fatty acid in foods and is an essential nutrient (vitamin F) for humans, since it is the precursor for arachidonic acid and prostaglandin synthesis. It is a desirable component for many food applications, but it has limited stability when heated.

α-Linolenic acid is also an important component of human diet. It is used to synthesize the ω-3 family of very long-chain fatty acids and the prostaglandins derived from these. However, the three double bonds are highly susceptible to oxidation so that oils containing more than a few percent 18:3 deteriorate rapidly upon exposure to air, especially at high temperatures. Partial hydrogenation of such oils is often necessary before than can be used in food products.

Many industrial uses of vegatable oils depend on particular, less common fatty acids. Medium-chain acids (particularly lauric acid, 12:0) are used in detergents, the hydroxylated ricinoleic acid (12:OH 18:1), and long-chain eicosenoic (20:1) and erucic (22:1) acids are components of specialty lubricants (Pryde and Princen, 1985). Because of the different uses of vegetables oils, there are many opportunities to alter the fatty acid composition of seed triacylglycerols to better suit particular applications.

The Biochemistry of Seed Lipid Synthesis

The eventual application of DNA technology to plant improvement will depend on the identification of individual genes of agronomic significance. At present, very few genes are known in which variation in kind or amount of the gene product is projected to have a significant impact on plant composition or economic yield. If we understood the factors which regulate the fatty acid chain length or degree of unsaturation of storage lipids, then it might be possible to modify the composition of vegetable oils by genetic engineering techniques and, thus, meet the increasing demand for new oil products in the food and manufacturing industries. Unfortunately, as in many other areas of plant biology, the lack of detailed information on the regulation of seed lipid metabolism is a barrier to developing strategies to bring about such modifications.

In leaf cells, two distinct pathways contribute to the synthesis of glycerolipids and the associated production of polyunsaturated fatty acids. The evidence for the "two pathway" model has been summarized by Roughan and Slack (1982). In brief, the model proposes that fatty acids synthesized *de novo* in the chloroplast may either be used directly for production of chloroplast lipids by a pathway in the chloroplast (the prokaryotic pathway), or may be exported to the cytoplasm as CoA esters where they are incorporated into lipids in the endoplasmic reticulum by an independent set of acyltransferases (the eukaryotic pathway). The genetic approach has contributed considerably to our understanding of leaf lipid biosynthesis (Kunst *et al.*, 1988; Browse *et al.*, 1989) and of the role of lipids in chloroplast structure and function (Browse *et al.*, 1985; Hugly *et al.*, 1989).

The biochemistry of storage lipid synthesis in developing oilseeds is less well understood. Some of the steps involved in triacylglycerol synthesis and many aspects of subcellular compartmentation

(Stumpf, 1980; Stymne and Stobart, 1987) are the same as for membrane lipid biosynthesis in leaves (Browse and Somerville, 1991). However, it is useful to consider seed lipid biosynthesis in terms of a quite different scheme, shown in Figure 1. This scheme was developed using results from many oilseed species (Slack and Browse, 1984; Stymne and Stobart, 1987), but is drawn to describe the metabolism of developing *Arabidopsis* seeds. *Arabidopsis* seed lipids contain substantial proportions of both unsaturated 18-carbon fatty acids (30 percent 18:2, 20 percent 18:3) and long-chain fatty acids (22 percent 20:1) derived from 18:1 (Table 3). This suggests that *Arabidopsis* is a good model for the biochemistry of both 18:2/18:3-rich oilseeds and those species containing longer fatty acids.

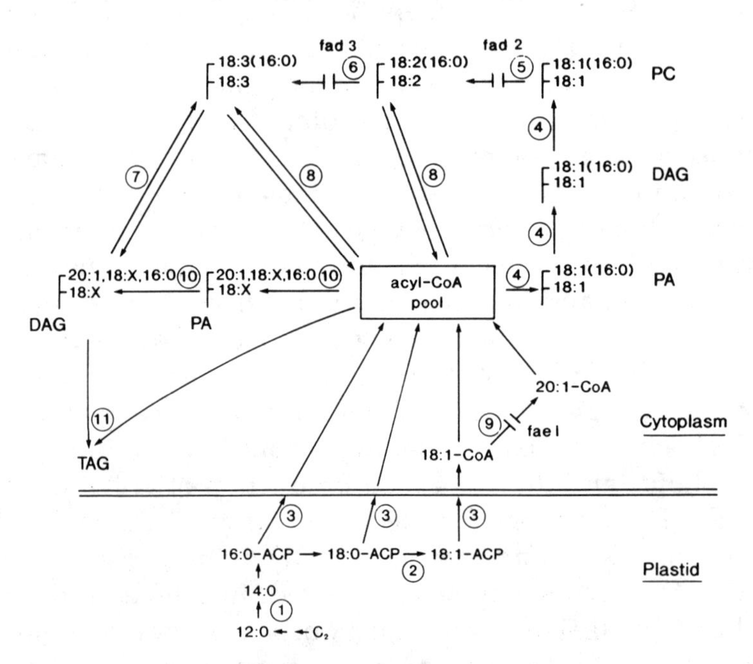

Figure 1. An abbreviated scheme showing our present understanding of triacylglycerol synthesis in oilseeds. The major reactions are explained in the text. Suggested lesions for some of the mutants listed in Table 3 are indicated.

As in leaf cells, 16:0- and 18:1-ACP thioesters are the major products of plastid fatty acid synthesis (1) and 18:0-ACP desaturase activity (2), and there is only a minor proportion of 18:0-ACP. The immediate steps in the metabolism of acyl-ACPs in oil seed plastids have not been studied. By analogy with leaf chloroplasts, they are

Table 3. Fatty acid composition of total seed lipids from some mutants of *Arabidopsis*

Mutant Line	Fatty Acid Composition (wt. %) 16:0	18:0	18:1	18:2	18:3	20:1	22:1
Wild Type	9	3	15	29	20	21	3
fad2	9	4	54	**3**	**6**	24	3
fad3	8	3	23	46	**2**	17	3
fae1	9	3	22	37	21	**0**	**0**
fab1	**19**	3	8	22	19	17	1
fab2	7	**8**	9	26	21	17	2
ela	12	5	13	24	**25**	18	2

References: Lemieux *et al.*, 1990; James and Dooner, 1990.

probably hydrolyzed by a specific thioesterase (3) (Ohlrogge *et al.*, 1978) to free fatty acids (Browse and Slack, 1985), which are then converted to acyl-CoA thioesters after transport through the plastid envelope (Roughan and Slack, 1977). Certainly 16:0- and 18:1-CoAs are the primary substrates for subsequent reactions in other cellular compartments (Stumpf, 1977). 18:1- and 16:0-CoA are used for the synthesis of phosphatidylcholine (PC) by reactions of the Kennedy pathway (4) (Stymne and Stobart, 1987), and PC is the substrate for 18:1 and, presumably, 18:2 desaturation (5,6) by microsomal enzymes (Stymne and Appelqvist, 1978; Browse and Slack, 1981). The synthesis of PC from diacylglycerol (DAG) by choline phosphotransferase is reversible (7) (Slack *et al.*, 1985), so that, in many oilseeds, PC is a direct precursor of the highly unsaturated species of DAG used for TAG synthesis (Slack *et al.*, 1978; 1983). However, the acyl CoA pool in the seed is a much more complex collection of fatty acyl groups than that in leaf mesophyll cells. Exchange of 18:1 from CoA with the fatty acids at position *sn*-2 of phosphatidylcholine (8) provides inputs of 18:2 and 18:3 (Stymne and Glad, 1981; Stymne *et al.*, 1983). Elongation (9) of 18:1-CoA to 20:1- and 22:1-CoA also occurs in *Arabidopsis* as it does in rapeseed and other cruciferous oilseeds (Stumpf and Pollard, 1983). Obviously, synthesis of DAG (10) via phosphatidic acid (PA) may also involve these other components of the acyl-CoA pool (Ichihara, 1984; Griffiths *et al.*, 1985), as does the final acylation (11) of DAG to form triacylglycerol (TAG) (Sun *et al.*,1988). Thus, the pool of DAG used for TAG synthesis is fed by phosphatidic acid phosphatase and choline phosphotransferase.

The relative contribution from these two routes varies among different species (Stobart and Stymne, 1985; Slack and Browse, 1984; Stymne and Stobart, 1987). The evidence for many of the steps in Figure 1 is incomplete; but, as described below, it is a good working model for considering biochemical and genetic evidence about oilseed lipid synthesis.

The desaturase enzymes which insert double bonds into the fatty acid chains of lipid intermediates are critically important in determining the composition of seed oils. However, the only plant desaturase that has been characterized in any detail is the soluble stearoly-ACP desaturase which inserts a double bond at the Δ9 position of 18:0 ACP. This enzyme has been purified to homogeneity by several groups and cDNAs encoding the desaturase have been cloned. The remaining desaturases, and many other enzymes of lipid metabolism, are membrane-bound proteins which have not been successfully solubilized.

A Genetic Approach

To circumvent the problems of studying lipid synthesis by traditional biochemical techniques, we have isolated a series of mutants of *Arabidopsis* with specific alterations in the fatty acid composition of their seed storage lipids. These mutants have provided an alternative approach toward understanding the mechanisms regulating synthesis and desaturation of triacylglycerols. More importantly, the advantages of *Arabidopsis* as a model for molecular biology (Meyerowitz, 1989) mean that the mutations can be used as markers to facilitate the cloning of the desaturase genes by chromosome walking or gene tagging.

Because there is no obvious way to select for mutants with altered lipid composition on the basis of gross phenotype, we screened for mutants by direct assay of fatty acid composition of seed samples by gas chromatography. A sample of M3 seed was harvested from mature M2 plants derived from ethylemethane sulfonate mutagenesis. From the first 5,000 lines sampled, a number of lines have been identified which exhibit one of six distinct mutant phenotypes (Table 3; James and Dooner, 1990; Lemieux *et al.*, 1990). These phenotypes include pronounced deficiencies in the desaturation of 18:1, the desaturation of 18:2, and the elongation of 18:1 to 20:1. Several lines with less pronounced alterations in seed fatty acid composition were also obtained for all three of these phenotypes. In addition, mutants were identified with increased levels of 16:0, 18:0, or 18:3 in their seed lipids (Table 3; James and Dooner, 1990).

Mutants Deficient in Polyunsaturated Fatty Acid Synthesis

Four mutant lines have been recovered that show reduced levels of 18:2 and 18:3 fatty acids in their seed lipids. These *fad2* mutants were all shown to be allelic by genetic complementation. The simplest (but not the only) explanation of the phenotype is that the plants have mutations in the structural gene encoding the 18:1 desaturase. Desaturation of 18:1-phosphatidylcholine by an enzyme in the endoplasmic reticulum is suggested to occur in developing oilseeds (Stymne and Appelqvist, 1978; Stymne and Stobart, 1987) and on the eukaryotic pathyway in leaf cells (Roughan and Slack, 1982). In order to determine whether the same gene controls desaturation in different tissues, we analyzed the overall fatty acid composition of seeds, leaves, and roots of plants from the *fad2* line and wild type *Arabidopsis.* We have previously characterized a mutant of *Arabidopsis* deficient in the chloroplast 18:1/16:1 desaturase (Browse *et al.*, 1989) which is the product of the *fadC* gene. All the chloroplast lipids appear to be accessible to this desaturase and we would, therefore, predict that chloroplast lipids in the *fad2* line would not be affected to a significant extent by a deficiency in the endoplasmic reticulum 18:-PC desaturase. Consistent with the prediction and the observation that 73 percent of all leaf glycerolipids are in the chloroplasts (Browse *et al.*, 1986), the *fad2* leaves show only a moderate increase in 18:1 compared with leaves of the wild type (Miquel and Browse, 1990). We anticipate that the phospholipids of the extrachloroplast membranes will be affected most in leaves of *fad2* plants, while the chloroplast lipids will show little or no change in fatty acid composition from wild type. In root tissue, the plastid membranes constitute a small proportion of the total cellular lipids, and the chloroplast 18:1/16:1 desaturase is not important in controlling the fatty acid composition (Browse *et al.*, 1989). In contrast, roots of the *fad2* plants show a very substantial decrease in the amounts of 18:2 + 18:3 compared with wild type, and a corresponding increase in the level of 18:1. All the above data are consistent with the hyposthesis that the mutation in *fad2* affects the activity of an endoplasmic reticulum 18:1 desaturase which is expressed throughout the plant. Despite the considerable changes in seed oil and membrane fatty acid composition, *fad2* plants are able to germinate, grow, and complete their life cycles normally under our standard growth conditions (22°C; 150μ/Einsteins/m^2/S).

The identification of a constitutively expressed desaturase gene in *Arabidopsis* contrasts with earlier results for other oilseeds which indicate that, in most cases, changes in triacylglycerol composition are not accompanied by changes in the fatty acid composition of leaf membrane lipids (Green and Marshall, 1984; Graef *et al.*, 1985; Martin and Rinne, 1986; Wilcox *et al.*, 1984). Constitutive expression of the *Arabidopsis fad2* gene will simplify gene cloning. For example, checking for complementation of the phenotype would be faster using root tissue. On the other hand, it may be difficult to obtain some changes in seed oil composition—such as very high levels of 18:0 (Graef *et al.*, 1985)— in *Arabidopsis* if these require mutations in constitutively expressed genes, and if such mutations disrupt membrane function in vegetable parts of the plant.

Searching for Mutants with Desaturase Genes Tagged by T-DNA

The isolation of genes involved in lipid metabolism is essential to understanding the regulation of the pathways shown in Figure 1 and to allow the modification of seed oil fatty acid composition by genetic engineering. Many of the soluble enzymes of plant lipid metabolism have been purified and this has allowed the isolation of corresponding cDNA or genomic clones (Browse and Somerville, 1991). However, most of the fatty acid desaturases and some other key enzymes of triacylglycerol synthesis are membrane-bound proteins, and attempts to solubilize and purify them from plant sources have not been successful. This precludes traditional methods of antibody production and cDNA library screening as a means of isolating the genes.

The advantages of *Arabidopsis* as a model organism for plant molecular biology allow the genetically marked mutant lines to provide the basis for cloning the desaturase genes by chromosome walking or gene tagging techniques. To facilitate the gene tagging approach, we have decided to determine whether there are any fatty acid mutants represented in the collection of T-DNA insertional mutants developed by Dr. Ken Feldmann (DuPont Experimental Station, Wilmington, DE). If this population does contain a fatty acid mutant which results from T-DNA insertion, then cloning the tagged gene will be relatively simple. Several genes have already been obtained from *Arabidopsis* by T-DNA tagging. These include *glabrous 1* (Marks and Feldmann, 1989) and *agamous* (Yanofsky *et al.*, 1990). Work on at least 20 other T-DNA tagged genes is currently underway.

The T-DNA population at present contains 1,800 lines, each of which is segregating for one or more (average 1.5) kanamycin resistance markers (this marker is part of the T-DNA). Since the *Arabidopsis* genome is ∾70,000 kb then, assuming random insertion of the T-DNA and an average gene size of 1.5 kb, there is a 5 percent probability of finding a mutant in any particular gene and, thus, 30–40 percent chance of finding one of the mutants (Table 1) after screening the 1,800 lines. As the T-DNA population is enlarged, additional lines (at least 6,000 by 1991) will be screened.

Because the T-DNA containing lines are segregating for the insert (and, thus, for any resultant mutation), it is necessary to sample several individuals from each line. In the case of seed mutants, this is easily achieved by taking a sample of 30 to 50 seeds of each line and pooling these for analysis. From the first 1,800 lines, we have identified three putative mutants. Work is now underway to check for cosegregation of the fatty acid defect with the kanamycin resistance marker (Feldmann *et al.*, 1989). This is important, even though the "seed" transformation protocol used to generate the mutants circumvents the problems of tissue culture-induced variation that have thwarted most other attempts at gene tagging with T-DNA. With an average of only 1.5 mutagenic T-DNA inserts in each line, the background mutation rate could be responsible for a proportion of the mutations observed.

Strategies for Modifying the Composition of Seed Oils

The prospect of being able to isolate and clone many genes controlling seed lipid biosynthesis, and the availability of genes from other organisms such as *E. coli* and mammals (Cronan and Rock, 1987; Bayley *et al.*, 1988), indicate that it is now important to predict which particular genes hold the most promise for usefully modifying the composition of plant lipids. Information from the *Arabidpsis* mutants is again useful in suggesting possible strategies. For example, all three of the mutants, *fad2, fad3,* and *fae1*, show a gene dosage effect with heterozygous seeds from crosses to wild type having an intermediate fatty acid composition (Lemieux *et al.*, 1990; James and Dooner, 1990). This result, which has been reported for mutations identified in other oilseed species (Green and Marshal, 1984), indicates that the desaturation and elongation reactions are limited by the extent to which these genes are expressed, even in wild type plants. Therefore, overexpression of cloned *fad2* and *fad3* genes in transgenic plants can be expected to result in more highly unsaturated oil. Similarly, increased levels of long chain fatty acids should be possible by overexpression of *fae1*.

The prospects for reducing the levels of saturated fatty acids (largely 16:0) in oilseeds may also be considered in the context of the biochemistry shown in Figure 1. 16:0-ACP is an intermediate in the synthesis of 18-carbon fatty acids, but hydrolysis by thioesterase and export of 16:0 from the plastid compete with the elongatin reaction. We have not isolated any mutants deficient in 16:0-ACP thioesterase activity and it is, therefore, possible that a single thioesterase is responsible for the hydrolysis of 16:0-, 18:0-, and 18:1-ACPs (Figure 1, [3]). If this is so, then increasing the rate of elongation of 16:0-ACP may provide an alternative means to reduce the level of 16:0 in oils.

References

American Oil Chemists Society. 1988. World Fats and Oils Report. J. Amer. Oil. Chem. Soc. 65: 1232.

Bayley, S.A., Moran, M.T., Hammond, E.W., James, C.M., Safford, R. and Hughes, S.G. 1988. Metabolic consequences of the medium chain hydrolase gene of the rate in mouse NIH 3T3 cells. Biotech. 6: 1219.

Browse, J.A. and Slack, C.R. 1981. Catalase stimulates linoleate desaturase activity in microsomes from developing linseed cotyledons. FEBS Lett. 131: 111.

Browse, J. and Slack, C.R. 1985. Fatty acid synthesis in plastids from maturing safflower and linseed cotyledons. Planta 166: 74.

Browse, J., McCourt, P. and Somerville, C.R. 1985. A mutant of *Arabidopsis* lacking a chloroplast specific liqid. Science 227: 763.

Browse, J. and Somerville, C.R. 1991. Genetic regulation of glycerolipid synthesis. Annu. Rev. Plant Physiol. Plant Mol. Biol. 42: (In press.)

Browse, J.A., Warwick, N., Somerville, C.R. and Slack, C.R. 1986. Fluxes through the prokaryotic and eukaryotic pathways of lipid synthesis in the 16:3 plant *Arabidopsis thaliana*. Biochem. J. 235: 25.

Browse, J., Kunst, L., Anderson, S., Hugly, S. and Somerville, C. 1989. A mutant of *Arabidopsis* deficient in the chloroplast 16:1/18:1 desaturase. Plant Physiol. 90: 522.

Cronan, J.E. and Rock, C.O. 1987. Biosynthesis of membrane lipids. In "*Escherichia coli* and *Salmonella typhimurium*: Cellular and Molecular Biology" (J.L. Ingraham, Ed.), Amer. Soc. Microbiol., p. 474, Washington, D.C.

Feldmann, K.A., Marks, M.D., Christianson, M.L. and Quatrano, R.S. 1989. A dwarf mutant of *Arabidopsis* generated by T-DNA insertion mutagenesis. Science 243: 1351.

Graef, G.L., Miller, L.A., Fehr, W.R. and Hammond, E.G. 1985. Fatty acid development in a soybean mutant with high stearic acid. J. Am. Oil Chem. Soc. 62: 773.

Green, A.G. and Marshall, D.R. 1984. Isolation of induced mutants in linseed (*Linum usitatissimum*) having reduced linolenic acid content. Euphytica 33: 321.

Griffiths, A.G., Stobart, A.K. and Stymne, S. 1985. The acylation of *sn*-glycerol-3-phosphate and the metabolism of phosphatidate in microsomal preparations from the developing cotyledons of safflower (*Carthamus tinctorius* L.) seed. Biochem. J. 230: 379.

Hamilton, R.J. and Bhati, A. (Eds.). 1987. Recent Advances in Chemistry and Technology of Fats and Oils. Elsevier, London, England.

Hilditch, T.P. and Williams, P.N. 1964. The Chemical Constituents of Natural Fats (4th Edition). Chapman and Hall, London, England.

Hugly, S., Kunst, L., Browse, J. and Somerville, C.R. 1989. Enhanced thermal tolerance and altered chloroplast ultrastructure in a mutant of *Arabidopsis* deficient in lipid desaturation. Plant Physiol. 90: 1134.

Ichihara, K. 1984. *sn*-Glycerol-3-phosphate acyltransferase in a particulate fraction from maturing safflower seeds. Arch. Biochem. Biophys. 232: 685.

James, D.W. and Dooner, H.K. 1990. Isolation of EMS-induced mutants in *Arabidopsis* altered in seed fatty acid composition. Theor. Appl. Genet. (In press).

Kramer, J.K.G., Sauer, F.D. and Pigden, W.J. (Eds.). 1983. High and Low Erucic Acid Rapeseed Oils. Academic Press, Toronto, Canada.

Kunst, L., Browse, J. and Somerville, C.R. 1988. Altered regulation of lipid biosynthesis in a mutant of *Arabidopsis* deficient in chloroplast glycerol-3-phosphate acyltransferase activity. Proc. Natl. Acad. Sci. 85: 4143.

Lemieux, B., Miquel, M., Somerville, C.R. and Browe, J. 1990. Mutants of *Arabidopsis* with alterations in seed lipid fatty acid composition. Theor. Appl. Genet. (In press). 80: 234-240.

Marks, M.D. and Feldmann, K.A. 1989. Trichome development in *Arabidopsis thaliana.* 1. T-DNA tagging of the glaborous 1 gene. The Plant Cell 1: 1043.

Martin, B.A. and Rinne, R.W. 1986. A comparison of oleic acid metabolism in the soybean (*Glycine max* L. Merr.) genotypes Williams and A5, a mutant with decreased linoleic acid in the seed. Plant Physiol. 81: 41.

Meyerowitz, E.M. 1989. *Arabidopsis*: A useful weed. Cell 56: 263.

Miquel, M. and Browse, J. 1990. Mutants of Arabidopsis deficient in 18:1-PC desaturation. In "Plant Lipid Biochemistry, Structure and Utilization," Biochemistry Society, London (In Press).

Ohlrogge, J.B., Shine, W.E. and Stumpf, P.K. 1978. Characterization of plant acyl-ACP and acyl-CoA hydrolases. Arch. Biochem. Biophys. 189: 382.

Pryde, E.H. and Doty, Jr., H.O. 1981. World fats and oils situation. In "New Sources of Fats and Oils" (E.H. Pryde, L.H. Princen and K.D. Mukherjee, Eds.), A.O.C.S., Champaign, Illinois.

Pryde, E.H. and Princen, L. 1985. Industrial Uses of Vegetables Oils. A.O.C.S., Champaign, Illinois.

Roughan, P.G. and Slack, C.R. 1977. Long-chain acyl-coenzyme-A synthetase activity of spinach chloroplasts is concentrated in the envelope. Biochem. J. 162: 457.

Roughan, P.G. and Slack, C.R. 1982. Cellular organization of glycerolipid metabolism. Ann. Rev. Plant Physiol. 33: 97.

Slack, C.R., Roughan, P.G. and Balasingham, N. 1978. Labelling of glycerolipids in the cotyledons of developing oilseeds by [1-^{14}C]acetate and [2-^{3}H]glycerol. Biochem. J. 170: 421.

Slack, C.R., Campbell, L.C., Browse, J.A. and Roughan, P.G. 1983. Some evidence for the reversibility of choline phosphotransferase-catalysed reaction in developing linseed cotyledons *in vivo.* Biochim. Biophys. Acta 754: 10.

Slack, C.R. and Browse, J.A. 1984. Lipid synthesis during seed development. In "Seed Physiology" (D. Murray, Ed.), Vol. 1, p. 209.

Slack, C.R., Roughan, P.G., Browse, J.A. and Gardiner, S.E. 1985. Some properties of cholinephosphotransferase from developing safflower cotyledons. Biochim. Biophys. Acta 833: 438.

Small, D.M. 1986. The Physical Chemistry of Lipids. Plenum Press, New York, New York.

Stobart, A.K. and Stymne, S. 1985. The regulation of the fatty-acid composition of the triacylglycerols in microsomal preparations from avocado mesocarp and the developing cotyledons of safflower. Planta 163: 119.

Stumpf, P.K. 1977. Lipid biosynthesis in developing seeds. In "Lipids and Lipid Polymers in Higher Plants" (M. Tevini and H.K. Lichtenthaler, Eds.), Springer-Verlag, Berlin, Germany.

Stumpf, P.K. 1980. Biosynthesis of saturated and usaturated fatty acids. In "The Biochemistry of Plants. A Comprehensive Treatise" (P.K. Stumpf and E.E. Conn, Eds.), Vol. 4, Academic Press, New York, New York.

Stumpf, P.K. and Pollard, M.R. 1983. Pathways of fatty acid biosynthesis in higher plants with particular reference to developing rapeseed. In "High and Low Erucic Acid Rapeseed Oils" (J.K.G. Kramer, F.D. Sauer and W.J. Pigden, Eds.), Academic Press, Toronto, Canada.

Stymne, S. and Appelqvist, A. 1978. The biosynthesis of lineoleate from oleoyl-CoA via oleoyl phosphatidylcholine in microsomes of developing safflower seeds. Eur. J. Biochem. 90: 223.

Stymne, S. and Glad, G. 1981. Acyl exchange between oleoyl-CoA and phosphatidylcholine in microsomes of developing soya bean cotyledons and its role in fatty acid desaturation. Lipids 16: 298.

Stymne, S., Stobart, A.K. and Glad, G. 1983. The role of the acyl-CoA pool in the synthesis of polyunsaturated 18-carbon fatty acids and triacylglycerol production in the microsomes of developing safflower seeds. Biochim. Biophys. Acta 752: 198.

Stymne, S. and Stobart, A.K. 1987. Triacylglycerol biosynthesis. In "The Comprehensive Treatise" (P.K. Stumpf and E.E. Conn, Eds.), Vol. 9, Academic Press, New York, New York.

Sun, C., Cao, Y.-Z. and Huang, A.H.C. 1988. Acyl coenzyme-A preference of the glycerol phosphate pathway in the microsomes from the maturing seeds of palm, maize and rapeseed. Plant Physiol. 88: 56.

Weiss, T.J. 1983. Food Oils and Their Uses. AVI Publishing, Westport, CT. Second edition. 310 p.

Wilcox, J.R., Cavins, J.F. and Nielsen, N.C. 1984. Genetic alteration of soybean oil composition by a chemical mutagen. J. Am. Oil Chem. Soc. 61: 97.

Yanofsky, M., Ma, H., Bowman, J., Drews, G., Feldmann, K.A. and Meyerowitz, E.M. 1990. The protein encoded by the *Arabidopsis* homeotic gene *agamous* resembles transcription factors. Nature 346: 35.

Designer Oils from Microalgae as Nutritional Supplements

David J. Kyle, Kimberly D.B. Boswell,
Raymond M. Gladue and Sue E. Reeb
Martek Corporation
6480 Dobbin Rd.
Columbia, MD 21045

The food industry is facing a challenge in the 1990's to produce more nutritional products for a consumer population which is becoming increasingly health conscious. Omega-3 fatty acids found in fish oils have been correlated with reduced incidence of coronary vascular disease, but in terms of food formulations, these oils exhibit problems in organoleptic characteristics and oxidative stability due to the presence of a complex array of polyunsaturated fatty acids. Many species of microalgae are prolific producers of edible oils with simple fatty acid profiles which may offer the industry a biotechnological solution to this dilemma. Microalgal production strategies for two unique oils, one containing eicosapentaenoic acid (EPASCO) and one containing docosahexenoic acid (DHASCO), are described. With the appearance of more of these "designer oils," food formulators will have at their disposal a much larger array of raw materials to prepare more nutritious consumer products in the future.

Introduction

Microalgae represent a major group of microorganisms that utilize light energy and produce biomass through photosynthesis. Consequently, they are the primary producers of nutritional components for animals in the sea. Microalgae have also been an important nutritional component in the food chain of man in many different areas of the world (Soeder, 1986). *Spirulina,* which is actually a photosynthetic prokaryote (blue-green alga), has long been recognized as an important source of protein in the diets of certain cultures (Durand-Chastel, 1980). *Chlorella* and *Scenedesmus* are also being cultivated in large ponds today for the production of high protein foodstuffs (Soong, 1980; Payer *et al.*, 1980). It has recently been recognized, however, that microalgae also offer an important source of dietary fatty acids (long chain polyunsaturated fatty acids or PUFAs) to marine life and possibly to man as well (Shifron, 1984; Kyle *et al.*, 1988; Volkman, 1989). Indeed, the active ingredients in fish oils which have been used in traditional folk medicine for hundreds of years, are now thought to be the PUFAs which originate from the microalgae in the food chain. Specifically, these PUFAs are of the omega-3 class, and are made up primarily of eicosapentaenoic acid (EPA; $C_{20:5}$) and docosahexenoic acid (DHA; $C_{22:6}$), but also include alpha linolenic acid (ALA; $C_{18:3}$) and octadecatetraenoic acid (OTA; $C_{18:4}$).

Although microalgae represent the primary producers of EPA and DHA in the biosphere, EPA has also been identified in certain fungi (Gellerman and Schlenk, 1979; Yamada *et al.*, 1988) and at least one bacterial species (Yazawa *et al.*, 1988). DHA on the other hand, has not yet been found in any microbial groups other than microalgae. Higher plants do not appear to contain these compounds, although small amounts of EPA have been isolated from certain bryophytes (Karunen, 1989). The role of EPA and DHA in the microalgae is virtually unknown, but their nearly exclusive association with membrane lipids suggests some role in membrane function.

Many strains of microalgae are also oleogenic, producing large amounts of vegetable oil-like triglyceride as food reserves when under stress (Shifron and Chisholm, 1980; Kyle *et al.*, 1988). In the cases of those EPA-producing microalgae, however, the triglyceride produced is generally deficient in the PUFAs while the polar lipid fractions remain enriched (Behrens *et al.*, 1989). This is also true in the EPA-containing fungi and suggests that a PUFA containing polar lipid may be a poor substrate for the transacylation reactions in these oleogenic species. There are some interesting exceptions to the rule, however, and those species may prove to be a valuable source of single cell oils (SCOs) rich in omega-3 PUFAs with important nutritional characteristics (Kyle, 1990).

Necessity for Omega-3 Dietary Supplementation

During the past ten years an extensive literature has been developed on the effects of fish oil dietary supplementation on coronary vascular disease as well as many other pathophysiologic conditions (Lands, 1986; Simopoulis *et al.*, 1986; Galli and Simopoulis, 1989; Lees and Karel, 1990). Most of the studies thus far have used a generic fish oil (MaxEPA or Menhaden oil) which is made up of a complex array of bioactive fatty acids. Although the data are controversial with regard to the effects on serum cholesterol, they are quite clear on the ability of fish oil to decrease both serum triglycerides and the aggregability of red blood cells (Illingworth and Ullmann, 1990). As a consequence, there is a decreased likelihood of forming a thrombus in a coronary artery which would lead to cardiac failure. This effect is now well understood in terms of eicosenoid metabolism (i.e., the formation of prostaglandin I_3 and thromboxane A_3 from EPA), and explains the epidemiological observations of reduced incidence of coronary disease in populations whose diets include large amounts of fish (Hennekens *et al.*, 1990). Thus, the principle clinical value of fish oil is that it contains a relatively large amount of EPA (10–15%). It is still unknown, however, how much EPA is required in our diet for a positive effect on coronary vascular status. Furthermore, large doses of fish oil (>20 g/d) have been reported to significantly lengthen bleeding time, and prolonged intake of large amounts of fish oil may lead to other pathologic conditions (eg., increased incidence of hemorrhagic stroke). Thus, like most other nutritional supplements, there is an optimal dose range, above and below which certain risk factors are elevated.

The role of DHA in coronary disease has been generally ignored in the literature since the role of DHA in eicosenoid metabolism appears to be minimal. On the other hand, the importance of DHA as a structural lipid in many different tissues, especially neural tissues, has been known for a long time (Table 1). With the recent confirmation that DHA can be retroconverted into EPA (Rosenthal *et al.*, 1990), an interest in the role of DHA in coronary vascular pathology has been rekindled. Since fish oil is rich in DHA as well as EPA, it is difficult to ascribe certain physiological effects to particular fatty acids in any clinical trial using fish oil. More appropriate trials for determining causal relationship should include oils which contain only a single bioactive fatty acid (i.e., DHA or EPA).

Mammals can convert dietary ALA into longer chain PUFAs such as EPA and DHA, but only at a very slow rate (Salem *et al.*, 1986). Because of the high levels of DHA in the brain and retinal tissues, there is a peak demand for dietary DHA at the time of most rapid brain development. In most mammals, including cows, the most

Table 1. Docosahexenoic acid (C22:6), eicosapentaenoic acid (C20:5), and arachidonic acaid (C20:4) levels as a percentage of the total lipids in various tissues.

Tissue	lipid class	% C22:6	% C20:5	% C20:4
Human brain (gray matter)[1]	PS	36.6	--	1.6
Human brain (gray matter)[1]	PE	24.3	--	13.8
Human brain (white matter)[1]	PS	5.6	--	2.0
Human brain (white matter)[1]	PE	3.4	--	6.4
Rat brain (synaptic vessicles)[2]	PS	37.0	--	2.8
Human retina[3]	PS	18.5	--	5.0
Human retina[3]	PE	22.2	--	13.4
Bovine retina (outer rod segments)[4]	total	37.6	--	8.3
Human sperm[5]	total	35.2	--	5.1
Ram sperm[5]	total	61.4	--	4.5
Human RBC[6]	total	6.6	2.1	9.9
Human platelets[7]	total	1.8	1.1	14.5
Human plasma[6]	PC	4.0	1.3	0.3
Human breast milk[6]	total	0.6	0.2	0.5

References: [1]O'Brian and Sampson (1965); [2]Breckenridge *et al.*, (1973); [3]Anderson (1970); [4]Nielson *et al.*, (1970); [5]Paulos *et al.*, (1973); [6]Sanders *et al.*, (1978); [7]Hirai *et al.*, (1987).

Abbreviations: PS, phosphatidylserine; PE, phosphatidylethanolamine; PC, phosphatidylcholine; PL, phospholipid.

rapid brain development occurs *in utero,* and the requisite DHA for the developing young comes from the mother via the vascular system. In humans, however, this period coincides with the first few weeks of postnatal life and, consequently, mother's milk represents a major source of DHA for the growing infant. Cow's milk-based infant formula is deficient in this important requirement, and formula-fed babies demonstrate a retarded rate of development of visual, and perhaps mental, acuity compared to breast-fed babies (Carleson *et al.*, 1990a; Uauy *et al.*, 1990). Recent experiments with fish oil supplemented formula (to provide the missing DHA) have shown that normal rate of visual development can be restored when the additional DHA is provided, but the inclusion of the other PUFAs in the fish oil (especially EPA) appears to result in a reduction in rate of weight gain in premature infants (Carlson *et al.*, 1990b). It is thought that the high levels of EPA in the fish oil inhibits the infant's endogenous arachidonic acid (ARA) biosynthesis which is essential for peripheral tissue development. It is clear, therefore, that the use of a generic fish oil to provide the DHA levels of mother's milk will be inadequate, since the ratio of EPA to DHA to ARA is quite different from breast milk (Table 1), and the infant's physiology appears to be very sensitive to dietary levels of these long chain PUFAs. As a consequence, the industry needs sources of specialty oils designed for particular applications (i.e., designer oils) which will be relatively inexpensive, readily attainable by conventional technologies, and easy to manufacture using acceptable GMP protocols.

Microbial EPA Production

Several species of fungi have been reported to produce EPA at high levels (Yamada *et al.*, 1988; Shimizu *et al.*, 1988). In deep tank fermentations, *Morteirella alpina* has been shown to produce 0.5 g EPA/liter (Shimizu *et al.*, 1989). These authors further showed that the pathway for EPA biosynthesis may involve ALA (18:3 omega-3) as a substrate, since supplementation of the growth medium with linseed oil (rich in ALA) significantly improved the EPA yields. Desaturase enzymes in this species must therefore be acting primarily on the carboxyl terminal of the fatty acid (i.e., $\Delta 8$, $\Delta 5$- and $\Delta 6$-desaturases). This is in marked contrast to another fungal species (*Saprolegnia parasitica*) which was shown to produce EPA by the omega-6 pathway which involves a $\Delta 17$-desaturation of ARA to EPA (Figure 1). Both of these fungal species can produce significant quantities of triglyceride, but as the oil content increases, the percentage of EPA in the extractable fatty acid decreases. Consequently, the overall EPA levels per unit biomass remains unchanged (i.e., the EPA remains in the polar membrane lipid fraction) and they do not appear to produce a triglyceride containing high levels of EPA in any appreciable quantities.

EPA production from the green microalgae *Chlorella minutissima* has also been documented (Seto, 1984). Levels as high as 35 percent EPA in the extractable fatty acid have been reported, but this marine *Chlorella* strain is generally slow growing and is not oleogenic. A more commercially acceptable strain has now been produced by protoplast fusion of the slow growing, EPA-containing marine *Chlorella* with a faster growing fresh water *Chlorella* (Seto, 1988). The feeding of this new strain to brine shrimp was shown to rapidly elevate the EPA level in the fatty acids of the brine shrimp and to provide an excellent nutritional supplement to the cultured shrimp (*Penaeus japonicus*) larvae. At the present time this new *Chlorella* strain

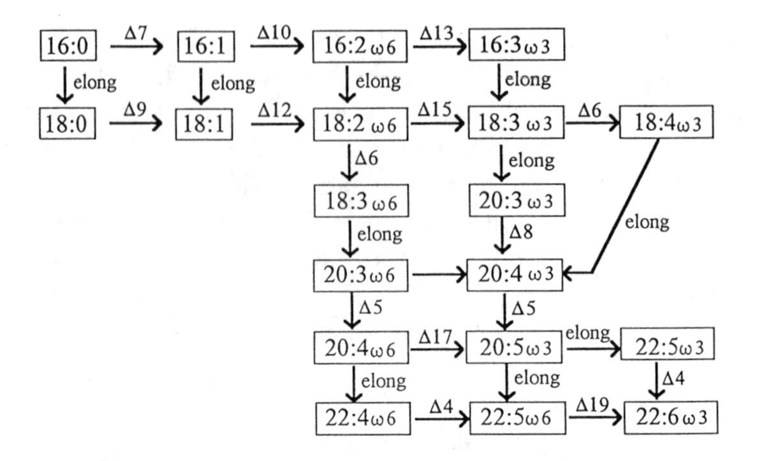

Figure 1. Omega-6 and Omega-3 pathways of PUFA biosynthesis and the possible branch points for interconversion of the pathways. Elongation steps (elong) and desaturation steps (Δn, where n indicates the position of the double bond) are indicated.

is being grown in commercial quantities and used for feed in shrimp aquiculture. Once again, however, this microbe appears to be unable to produce an oil containing high levels of EPA.

We have previously reported the cultivation of strains of oleagenous diatoms which do contain EPA in the oil (triglyceride) fraction (Kyle *et al.*, 1988; Behrens *et al.*, 1989). Of these, strain MK8620 appears to be the most prolific producer with an oil content of 20–30% of its biomass and an EPA content of 20–30% of the extractable fat (Hoeksema *et al.*, 1989). The fatty acid composition of the oil of this and many other diatoms is rather unusual compared to a typical vegetable oil in that these oils contain a relativley high amount of palmitoleic acid ($C_{16:1}$) in addition to the EPA (Table 2). Palmitoleic acid is also relatively abundant in fish oils and it undoubtedly originates from diatoms which make up a large portion of the phytoplankton-based food chain. Palmitoleic acid may also offer some interesting nutritional advantages since Yamori *et al.* (1986) have reported a preventive effect by palmitoleic acid on stroke in stroke-prone spontaneously hypertensive rats. The major problem with the development of MK8620 as a commerical source of a nutritional oil is that the organism is an obligate phototroph (i.e., requires light for growth). Thus, any highly controlled fermentation-like approach for production will require specially designed photobioreactors, resulting in high process costs. An open pond approach could be envisioned as a less costly alternative, but a system would have to be developed which would be acceptable for food grade material, and the pond (or organism) would need to be designed to carefully exclude contaminating microalgal weed species. Such contamination problems are major concerns in most open pond approaches to large scale algal cultivation.

Although they are generically defined as photosynthesizers, not all microalgae are obligate phototrophs. In fact, many microalgae will grow at an acceptable rate under completely heterotrophic conditions if given the appropriate carbon sources (Lewin and Lewin, 1960; Gladue *et al.*, 1988). We have screened for heterotrophic microalgae which are also oleogenic and can produce EPA. One such organism (MK8908) which grows very well in a conventional fermentation geometry was identified (Glaude *et al.*, 1990). Biomass densities as high as 40–50 g/l have been attained in 72 hours (Figure 2) and, with the appropriate medium limitations, this cell line will produce up to 50% of its weight in an easily extracted single cell oil (i.e., 20–25 g oil/liter), which is referred to as EPASCO. This level of oil production is about double the oil productivity of the oleagenous yeast *Candida curvata*, which has been well characterized as a possible commercial source of single cell oil (Glatz *et al.*, 1984). In the low fat condition (<10% oil), the EPA content of MK8908 is quite high (10–15% of the extractable fatty acid), but when the oil biosynthesis is up-regulated, the EPA content drops. Thus, similar to most other algae and fungi, the EPA in MK8908 is primarily found in the polar lipid membranes, and the EPA content of the extracted EPASCO varies from 1–4%. Nevertheless, the very high productivity of MK8908 using standard fermentation equipment and procedures makes the production of EPASCO from this organism a commercializable process.

Microbial DHA Production

Other than microalgae, there are no known microbial sources of DHA. Furthermore, the ability to synthesize DHA is unique to only a few groups of microalgae (primarily diatoms and dinoflagellates). As in the case of EPA, there are no known functions for DHA in the membranes of these microalgae. In mammalian tissues, DHA is preferentially associated with neuronal tissues and makes up 61% of the total structural lipid in the gray matter of the brain (O'Brien and Sampson, 1965). DHA is primarily associated

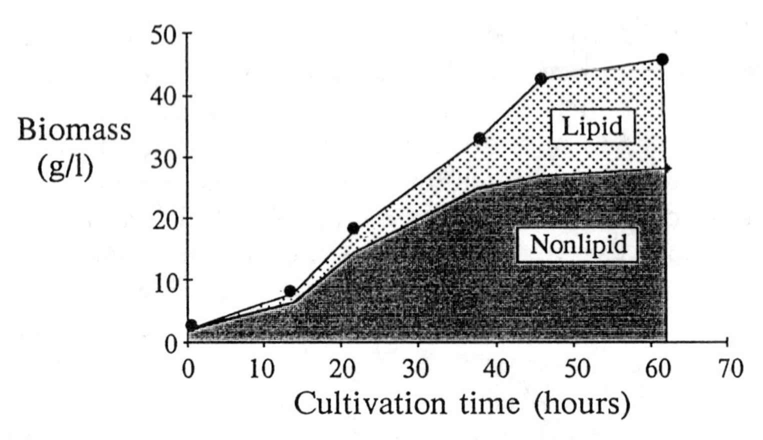

Figure 2. Growth of the EPASCO-producing microalgae (MK8908) by conventional fermentation. Data indicates the relative proportion of lipid (oil) and nonlipid components of the biomass during the course of the fermentation.

with phosphatidyl serine (PS) and phosphatidyl ethanolamine (PE) of synaptosomal membranes and vesicles (Baker, 1979), but it is also found in high abundance in the outer rod segments of the retina (Anderson, 1970), in spermatozoa (Paulos *et al.*, 1973), and in lower levels associated with EPA and ARA in peripheral tissues of the body (Table 1). A role for DHA in membrane excitation has been suggested (Chacko *et al.*, 1977) wherein a voltage-induced thinning of the membranes occurs at positions where DHA molecules are juxtaposed. This may result in the formation of a cation-conducting channel and membrane depolarization (Gudbjarnason *et al.*, 1978). How the suggested role of this unusual fatty acid in neuron conduction relates to its high concentration in certain microalgae still remains an enigma.

As a result of our screening activities, we have isolated a strain of microalgae (MK8805) which will grow heterotrophically and produce an oil (DHASCO) which contains over 30% DHA (Figure 3 and Table 2). The DHA levels in the polar lipid fractions from MK8805 can be as high as 60–70% and, although the transacylase(s) of this organism may not regard DHA as a preferred substrate, a considerable amount of DHA-containing polar lipid is nevertheless transacylated. The DHASCO from MK8805 also contains no other PUFAs in any appreciable quantities (Table 1), indicating that the fatty acid biosynthetic pathway in this organism may be quite unusual. Because of the lack of other PUFAs (especially EPA), this type of designer oil may be a more preferable starting material than fish oil for particular applications such as infant formula supplementation.

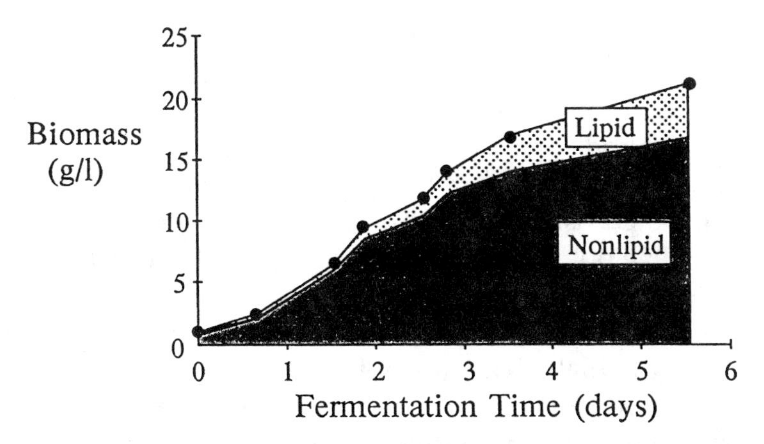

Figure 3. Growth of the DHASCO-producing microalgae (MK8805) by conventional fermentation. Data indicates the relative proportion of lipid (oil) and nonlipid components of the biomass during the course of the fermentation.

Table 2. Fatty acid profiles of various microorganisms, fish, and human breast milk.

Organism	Fatty Acid (% of total)									
	14:0	16:0	16:1	18:0	18:1	18:2	18:3	20:4	20:5	22:6
Fungi[1]										
Mortierella alpina (20–17)	--	16	--	8	21	14	--	28	<1	--
M. alpina+linseed oil	--	6	--	3	9	9	28	34	5	--
Yeast[2]										
Candida curvatum	--	28	1	14	44	10	2	--	--	--
C. curvatum mutant R22.102	--	26	--	41	16	8	4	--	--	--
Microalgae										
Cyclotella cryptica	12	19	42	3	2	--	--	--	6	--
Phaeodactylum tricornatum	6	22	31	--	16	3	3	--	13	--
MK8620 oil	5	17	39	4	7	2	--	--	22	--
MK8908 oil	22	29	1	4	26	3	1	1	4	--
MK8805 oil	18	27	--	--	11	--	--	--	--	36
Fish oil										
Menhaden	7	14	13	3	12	2	2	2	17	9
Human Breast Milk[3]	8	28	4	11	35	7	0.8	0.5	0.2	0.6

References: [1]Shimizu *et al.* (1989); [2]Ykema *et al.*(1989) [3]Sanders *et al.* (1978).

Fish Oils vs Microbial Oils as Omega-3 Nutritional Supplements

There have been several recent recommendations by international groups that the level of omega-3 PUFAs (especially EPA and DHA) in the American diet should be increased (Simopoulis, 1989; Health and Welfare Canada, 1990). Consequently, the food industry is looking to various sources of raw materials containing these compounds. Fish and fish oils are the most obvious choices, and both hydrogenated and partially hydrogenated Menhaden oils have been approved by the FDA for food use. However, only the unhydrogenated fish oils contain the high levels of EPA and DHA, and these have not yet been approved for food use in this country. Nevertheless, unhydrogenated fish oil admixed with margarines and salad oils are presently available to consumers in Europe (Young, 1990). The principal advantage of fish oil as a source of EPA is that it is inexpensive, and most fish oils are generally regarded as safe. However, there are many serious problems to the manufacturer who is using fish oils to supplement foods. These problems are outlined in Table 3 and include: 1) the strong odor and taste associated with fish oil; 2) the potential for heavy metal and chemical pollutant contamination of the fish and fish oil; 3) the large number of bioactive fatty acids in the fish oil, any of which may have antagonistic effects; 4) the high degree of polyunsaturation resulting in a hyersensitivity to oxidation; and 5) the limited raw material supply and dependence on the "luck-of-the-catch."

Table 3. Advantages and disadvantages of fish oils and single cell designer oils as raw materials for food/nutritional supplements.

Fish Oils	Designer Oils
Advantages • low cost • presently available • generally regarded as safe	Advantages • no odor/taste to the oils • controlled production (GMP) • single PUFA enrichment • proprietary protection of process • potential for biotechnological improvements
Disadvantages • strong odor/taste to oils • possibility of offshore pollutants • multiple PUFAs • oxidation sensitive • supply dependent on fishing industry • nonrenewable resource	Disadvantages • high cost • process development required

Many of the disadvantages associated with a fish oil source of omega-3 PUFAs can be overcome by use of single cell designer oils. By judicous selection and/or bioengineering of an appropriate microorganism, an oil with a lower level of overall polyunsaturation (although still high in the selected fatty PUFA), can be produced. Such oils are much less prone to oxidation, off-color and off-flavor development, and may even by more enriched in the particular PUFA of interest. DHASCO, for example, has double the DHA content of fish oil and yet a much lower total unsaturation index. Microbial sources of oil represent a renewable resource, and the potential diversity of the product is much larger than that available with fish oils. Because of the use of standard fermentation technologies, manufacturers can retain control of the raw material supply and be confident that the product is produced under good manufacturing practices (GMP). The above advantages do not come without a cost, however, and the most optimistic economic evaluations of the production of single cell oil put the cost much higher than unrefined, bulk fish oil. Thus, the value to the manufacturer of the various advantages listed in Table 3 will be what ultimately determines the utility of designer oils as nutritional additives.

Biotechnological Approaches to Production of Designer Oils

Microbial sources of oil may or may not be commercially feasible today, but various biotechnological approaches can be used to increase their value, or decrease their production costs. Unlike fish, the genetic manipulation of these microbes is relatively simple and new clones can be selected which might rapidly change the economics and ensure a future for a particular SCO as a food product. Oleagenous yeasts (eg. *Candida curvatum*) have been studied for many years as a potential source of single cell oil but even with the appropriate co-product values and minimal fermentation substrate costs (i.e., using whey as the carbon source), the product oil is still more costly than a competitive vegetable oil, such as palm, soy or corn oil. However, Ykema *et al.*, (1989) have recently selected clones of *C. curvatum* (=*Apotrichium curvatum)* with depressed levels of Δ9-desaturase activity and the resulting oil is enriched in stearate and may have a much higher value as a cocoa butter substitute (Table 2). With high cost cocoa butter as the competitive product, the SCO approach now becomes much more commercially attractive. Thus, the ability to genetically alter the oil composition in these microbes much more readily than the competitive agronomic product is the single most important factor ensuring the future of this technology.

Although microalgae are not as well genetically characterized as yeasts, a mutagenesis followed and clonal selection strategy can be similarly employed. We have used this approach to isolate mutant clones of the EPASCO producer which have improved EPA yields. With over 350 clones screened, we can define a population range with respect to EPA content (Figure 4), and we are in the process of improving yields by simply selecting clone lines in the upper 95 percentile for EPA production. Such a process is a slow route to EPA improvement, but in the absence of any knowledge of the function of EPA in the microbe, or the genetic determinants of EPA biosynthesis, it is difficult to devise a positive selection strategy for overproduction.

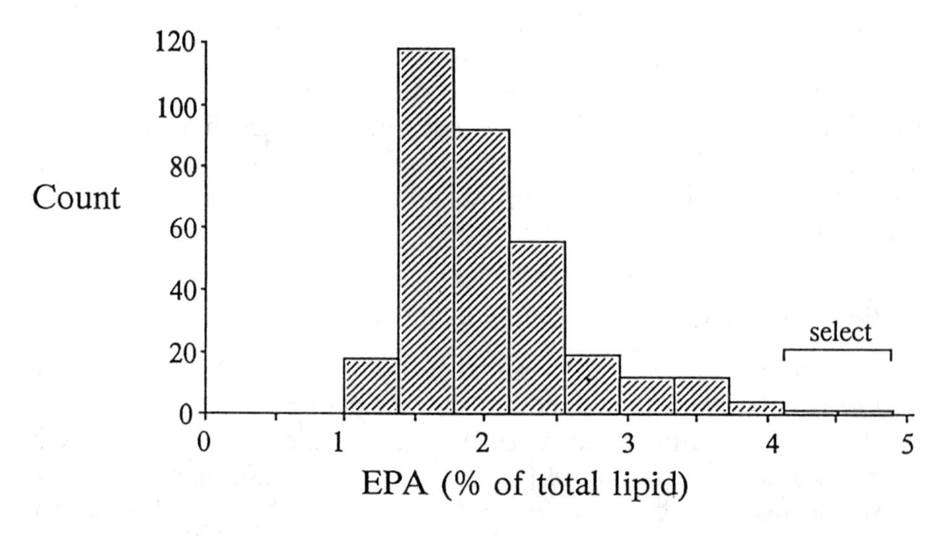

Figure 4. Frequency distribution of EPA levels in 350 mutants of MK8908 and the use of this protocol to select for clones with enhanced abilities to produce EPA.

In addition to the role of the PUFAs in the microorganism, one needs to know the routes of metabolism (both catabolism and synthesis) of the PUFAs before any major improvements in productivity will likely be seen. Unfortunately, little is known about the metabolic pathways of PUFAs in these microorganisms, and what is known appears to be species specific. The possible biosynthetic pathways for EPA and the requisite enzymes are shown in Figure 1. The pathway for the further elongation and desaturation of EPA to DHA in these organisms is totally conjecture. Using ^{13}C-labeled substrate fatty acids and subsequent analysis of microalgal fatty acids by GC/MS of their methyl ester derivatives, we showed that the initial steps in EPA biosynthesis in the EPASCO-producing microalgae (MK8908), are routed through palmitoleic acid (i.e., $C_{16:0} \rightarrow C_{16:1(7)} \rightarrow C_{18:1\,(9)}$) and then follow the predictable conversion to linoleic ($C_{18:2(9,12)}$) and linolenic ($C_{18:3(9,12,15)}$) acids (Chen *et al.*, 1990). Although the final steps of EPA biosynthesis in the EPASCO-producing microalgae have not yet been determined, we have also measured elongation and desaturation of $C_{18:2}$–CoA in a cell-free system for this organism. Furthermore, we have solubilized a PUFA elongase activity from a microsomal membrane fraction using octyl glucoside (data not shown). This research represents the initial steps toward the isolation of the genetic determinants responsible for PUFA biosynthesis in these microorganisms, and is the groundwork necessary for future genetic engineering approaches to enhancing the specific fatty acid levels in a particular designer oil.

Future Directions of SCO Research

The commerical potential for single cell designer oils will be the driving force behind future research in the development of these products. Presently, they are noncompetitive with the low cost agronomic production of commodity oils (both plant and marine oils) which has been practiced for centuries. Even with further optimizaton of the production processes, there is an inherent cost associated with the power requirements for the fermentors and the cost of the nutrient raw materials. In comparison to the more exotic oils such as olive oil, cocoa butter, primrose oil, etc., a fermentation approach would be quite competitive and, consequently, there is a driving force to improve and scale-up these processes. One should not be limited, however, to attempts to produce higher value oil substitues (or equivalents) using microorganisms. Rather, the potential for major modifications in the fatty acid compositions of these oils through either mutagenesis and selection, or more directed rDNA techniques, offers the possiblity of designing the most appropriate oil for a particular food or clinical application (eg., a DHA containing oil for infant formula). By genetically engineering these characteristics into an efficient fermentative oil producer, such as oleagenous yeast or heterotrophic oleagenous microalgae (i.e., MK8908; Figure 2), the economics for the production of these oils may be industrially attractive. Furthermore, if the genetic determinants for the requisite elongases and desaturases can be isolated, it may be possible to create transgenic agronomic plants which can produce the new designer oils at even lower costs.

Acknowledgements

Much of the work described herein was supported by NIH grants (R43-HL38547 and R43-DK41963) to DJK.

References

Anderson, R.E. 1970. Lipids of ocular tissues. IV. A comparison of the phospholipids of the retina of six mammalian species. Exptl. Eye Res. 10: 339.

Baker, R.R. 1979. The fatty acid composition of phosphoglycerides of nerve cell bodies isolated in bulk from rabbit cerebral cortex: changes during development. Can. J. Biochem. 57: 378.

Behrens, P.W., Hoeksema, S.D., Arnett, K.L., Cole, M.S., Heubner, T.A., Rutten, J.M. and Kyle, D.J. 1989. Eicosapentaenoic acid from microalgae. In "Novel Microbial Products for Medicine and Agriculture," p. 253. Elsevier Science Publ., Amsterdam.

Breckenridge, W.C., Morgan, I.G., Zanetta, J.P. and Vincedon, G. 1973. Adult rat brain synaptic vesicles. II. Lipid composition. Biochim. Biophys. Acta 320: 681.

Carleson, S.E., Cooke, R.J., Werkman, S.H., Peeples, J.M., Tolley, E. and Wilson, W.M. 1990a. Long-term docosahexaenoic acid (DHA) and eicosapentaenoic acid (EPA) supplementation of pre-term infants: effects on biochemistry, visual acuity, information processing and growth in infancy. INFORM 1: 306 abs. R3.

Carleson, S.E., Peeples, J.M., Werkman, S.H., Cooke, R.J. and Wilson, W.M. 1990b. Arachidonic acid (AA) in plasma and red blood cell (RBC) phospholipids (PL) during followup of pre-term infants: occurrence, dietary determinants and functional relationships. Presented at the 2nd International Conference on the Health Effects of Omega-3 Polyunsaturated Fatty Acids in Seafoods, March 20–23, Washington, D.C.

Chacko, G.K., Barnola F.V. and Villegas, R. 1977. Phospholipid and fatty acid compositions of axon and periaxonal plasma membranes of lobster leg nerves. J. Neurochem. 28: 445.

Chen, H., Bingham, S.E., Chantler, V., Pritchard, B. and Kyle, D.J. 1990. 13C-labeled fatty acids from microalgae. Dev. Industr. Microbiol. 31 (In press.)

Durand-Chastel. H. 1980. Production and use of *Spirulina* in Mexico. In "Algae Biomass," p. 51. Elsevier/North Holland Biomedical Press, Amsterdam.

Galli, C. and Simopoulis, A.P. 1989. "Dietary w-3 and w-6 Fatty Acids. Biological Effects and Nutritional Essentiality." Plenum Press, New York.

Gellerman, J.L. and Schlenk, H. 1979. Methyl directed desaturation of arachidonic acid to eicosapentaenoic acid in the Fungus *Saprolegnia parasitica*. Biochim. Biophys. Acta 573: 23.

Gladue, R., Hoeksema, S., Chen, H., Chantler, V., Lieberman, D., Pai, N., Pritchard, B., Rutten, J. and Kyle, D.J. 1988. Production of eicosapentaenoic acid (EPA) by photosynthetic and heterotrophic microalgae. Presented at the ISF-JOCS World Congress, September 26–30, Tokyo. Proceedings Vol. 2, p. 1007.

Gladue, R, Sicotte, V., Reeb, S. and Kyle, D.J. 1990. Screening and optimization of an EPA-containing oil from microalgae. INFORM 1: 356 abs. VVII.

Glatz, B.A., Hammond, E.G., Hsu, K.H., Baehman, L., Bati, N., Bednarski, W., Brown, D. and Floetenmeyer, M. 1984. Production and modification of fats and oils by yeast fermentation. In "Biotechnology for the Oils and Fats Industry," p. 163. Amer. Oil Chemists' Society Press, Champaign, IL.

Gudbjarnason, S., Doell, B. and Oskarsdottir, G. 1978. Docosahexenoic acid in cardiac metabolism and function. Acta Bikol. Med. Germ. 37: 777.

Health and Welfare Canada. 1990. "Nutrition Recommendations for Canadians." Canadian Government Publishing Center, Ottawa, Canada.

Hennekens, C.H., Buring, J.E. and Mayrent, S.L. 1990. Clinical and epidemiological data on the effects of fish oil on cardiovascular disease. In "Omega-3 Fatty Acids in Health and Disease," p. 71. Marcel Dekker, Inc., New York.

Hirai, A., Terano, T., Saitop, H., Tamura, Y. and Yoshida. 1987. Clinical and epidemiological studies of eicosapentaenoic acid in Japan. In "Proceedings of the AOCS Short Course on polyunsaturated Fatty Acids and Eicosanoids," p. 9. American Oil Chemists' Society Press, Champaign, IL.

Hoeksema, S.D., Behrens, P.W., Gladue, R., Arnett, K.L., Cole, M.S., Rutten, J.M. and Kyle, D.J. 1989. An EPA-containing oil from microalgae in culture. In "Health Effects of Fish and Fish Oils," p. 337. ARTS Biomedical Publ., St. John's, Canada.

Illingworth, D.R. and Ullmann, D. 1990. Effects of omega-3 fatty acids on risk factors for cardiovascular disease. In "Omega-3 Fatty Acids in Health and Disease," p. 39. Marcel Dekker, Inc., New York.

Karunen, P. 1989. Bryophytes as a source of polyunsaturated fatty acids. Presented at Symposium on New Aspects of Dietary Lipids on September 17–20, 1989, Göteborg Sweden.

Kyle, D.J., Behrens, P., Bingham, S., Arnett, K. and Lieberman, D. 1988. Microalgae as a source of EPA-containing oils. In "Biotechnology for the Fats and oils Industry," p. 117. American Oil Chemists' Society Press, Champaign, IL.

Kyle, D.J. 1990. Microbial omega-3-containing fats and oils for food use. Advan. Appl. Biotechnol. (In press).

Lands, W.E.M. 1986. "Fish and Human Health." Academic Press, Inc., Orlando, FL.

Lees, R.S. and Karel, M. 1990. "Omega-3 Fatty Acids in Health and Disease." Marcel Dekker, Inc., New York.

Lewin, J.C. and Lewin, R.A. 1960. Auxotrophy and heterotrophy in marine littoral diatoms. Can. J. Microbiol. 6: 127.

Nielson, N.C., Fleischer, S. and McConnell, D.G. 1970. Lipid composition of bovine retinal outer rod segments. Biochim. Biophys. Acta 211: 10.

O'Brien, J.S. and Sampson, E.L. 1965. Fatty acid and aldehyde composition of the major brain lipids in normal gray matter, white matter and myelin. J. Lipid Res. 6: 545.

Paulos, A., Darin-Bennett, A. and White, I.G. 1973. The phospholipid-bound fatty acids and aldehydes of mammalian spermatozoa. Comp. Biochem. Physiol. 46B: 541.

Payer, H.D., Pabst, W. and Runkel, K.H. 1980. Review of the nutritional and toxicological properties of the green alga *Scenedesmus obliquus* as a single cell protein. In "Algae Biomass," p. 787. Elsevier/North Holland Biomedical Press, Amsterdam.

Rosenthal, M.D., Garcia, M.C. and Sprecher, H. 1990. Retroconversion vs. desaturation: alternative metabolic fates for C22 omega-6 and omega-3 fatty acids in human cells. Presented at the 2nd International Conference on the Health Effects of Omega-3 Polyunsaturated Fatty Acids in Seafoods, March 20–23, Washington, D.C.

Salem, N., Kim, H.-Y. and Yergey, J.A. 1986. Docosahexaenoic acid: membrane function and metabolism. In "Health Effects of Polyunsaturated Fatty Acids in Seafoods," p. 263. Academic Press, London.

Sanders, T.A.B., Ellis, F.R., Path, F.R.C. and Dickerson, J.W.T. 1978. Studies of vegans: the fatty acid composition of plasma choline phosphoglycerides, erythrocytes, adipose tissue, and breast milk, and some indicators of susceptibility to ischemic heart disease in vegans and omnivore controls. Am. J. Clin. Nutr. 31: 805.

Seto, A. 1984. Culture conditions affect eicosapentaenoic acid content of *Chlorella minutissima*. J. Amer. Oil Chem. Soc. 61: 892.

Seto, A. 1988. Recent progress in the utilization of marine *Chlorella* for aquaculture. Presented at the ISF-JOCS World Congress, September 26–30, Tokyo. Proceedings Vol. 2, p. 1283.

Shifron, N.S. and Chisholm, S.W. 1980. Photoplankton lipids: environmental influences on production and possible commercial applications. In "Algae Biomass,", p. 627. Elsevier/North Holland Biomedical Press, Amsterdam.

Shifron, N.S. 1984. Oils from microalgae. In "Biotechnology for the Oils and Fats Industry," p. 145. American Oil Chemists' Society Press, Champaign, IL.

Shimizu, S., Kawashima, H., Shinmen, Y., Akimoto, K. and Yamada, H. 1988. Production of eicosapentaenoic acid by *Mortierella* fungi. J. Amer. Oil Chem. Soc. 65: 1455.

Shimizu, S., Kawashima, H., Akimoto, K., Shinmen, Y. and Yamada, H. 1989. Conversion of linseed oil to eicosapentaenoic acid-containing oil by *Mortierella alpina* 1S-4 at low temperatures. Appl. Microbiol. Biotechnol. 32: 1.

Simopoulis, A.P., Kifer, R.R. and Martin, R.E. 1986. "Health Effects of Polyunsaturated Fatty Acids in Seafoods." Academic Press, Inc., Orlando, FL.

Simopoulis, A.P. 1989. Executive summary. In "Dietary w-3 and w-6 Fatty Acids. Biological Effects and Nutritional Essentiality," p. 391. Plenum Press, New York.

Soeder, C.J. 1986. An historical outline of applied algology. In "Handbook of Microalgal Mass Culture," p. 25. CRC Press, Inc., Boca Raton, FL.

Soong, P. 1980. Production and development of *Chlorella* and *Spirulina* in Taiwan. In "Algae Biomass," p. 51. Elsevier/North Holland Biomedical Press, Amsterdam.

Uauy, R., Birch, D., Birch, E., Tyson, J. and Hoffman, D. 1990. Are omega-3 fatty acids essential for eye and brain development in humans? INFORM 1: 306 abs. R4.

Volkman, J.K. 1989. Fatty acids of microalgae used as feedstocks in aquaculture. In "Fats for the Future," p. 263. Ellis Horwood Ltd., Chihester, UK.

Yamada, H., Shimizu, S., Shinmen, Y., Kawashima, H. and Akimoto, K. 1988. Production of arachidonic acid and eicosapentaenoic acid by microorganisms. In "Biotechnology for the Fats and Oils Industry," p. 173. American Oil Chemists' Society Press, Chapaign, IL.

Yamori, Y., Nara, Y., Tsubouchi, T., Sogawa, Y., Ikeda, K. and Horie R. 1986. Dietary prevention of stroke and its mechanisms in stroke-prone spontaneously hypertensive rats—preventive effect of dietary fiber and palmitoleic acid. J. Hypertension 4: S449.

Yazawa, K., Araki, K., Okazaki, N., Watanabe, K., Ishikawa, C., Inoue, A., Numao, N. and Kondo, K. 1988. Production of eicosapentaenoic acid by marine bacteria. J. Biochem. 103: 5.

Ykema, A., Verbee, E.C., Nijkamp, J.J. and Smit, H. 1989. Isolation and characterization of fatty acid auxotrophs from the oleagious yeast *Apiotrichum curvatum*. Appl. Microbiol. Biotechnol. 32: 76.

Young, F.V.K. 1990. Using unhydrogenated fish oil in margarine. INFORM 1: 731.